# Applied Enhanced Oil Recovery

Aurel Carcoana

Prentice Hall
Englewood Cliffs, New Jersey 07632

**Library of Congress Cataloging-in-Publication Data**

Cârcoană, Aurel.
Applied enhanced oil recovery / Aurel Carcoana.
p. cm.
Includes bibliographical references and index.
ISBN 0-13-044272-0
1. Secondary recovery of oil. I. Title.
TN871.37.C37 1991
622'.33827–dc20 91-26308
CIP

Editorial production supervision
and interior design: *Laura A. Huber*
Production assistant: *Jane Bonnell*
Acquisitions editor: *Michael Hays*
Editorial assistant: *Dana L. Mercure*
Prepress buyer: *Mary Elizabeth McCartney*
Manufacturing buyer: *Susan Brunke*
Copy editor: *Sally Ann Bailey*
Cover artist: *Karen Stephens*
Marketing manager: *Alicia Aurichio*



**The publisher offers discounts on this book when ordered in bulk quantities. For more information, write: Special Sales/Professional Marketing, Prentice Hall, Professional & Technical Reference Division, Englewood Cliffs, NJ 07632.**



Printed in the United States of America
10 9 8 7 6 5 4 3 2 1

ISBN 0-13-044272-0

ISBN 0-13-044272

900

9 780130 442727

Prentice-Hall International (UK) Limited, *London*
Prentice-Hall of Australia Pty. Limited, *Sydney*
Prentice-Hall Canada Inc., *Toronto*
Prentice-Hall Hispanoamericana, S.A., *Mexico*
Prentice-Hall of India Private Limited, *New Delhi*
Prentice-Hall of Japan, Inc., *Tokyo*
Simon & Schuster Asia Pte. Ltd., *Singapore*
Editora Prentice-Hall do Brasil, Ltda., *Rio de Janeiro*

Dedicated to *Verona*
and *Laurence* and *Claudia* and *Olivia*.

# Contents

# Preface

Enhanced oil recovery is an area of great interest to those involved in and concerned with how to obtain more oil from existing oil reservoirs. Much research and development activity is being carried out, numerous technical papers are appearing, and excellent texts have been written in connection with the subject.

In teaching enhanced oil recovery (EOR), I felt the need for a textbook specifically addressed to petroleum engineering college students and to many other professionals involved in petroleum activity. This book contains thorough coverage of EOR methods describing its complex aspects in a simple and practical manner. As a textbook, the material could be taught in one four-credit-hour semester sequence with flexibility in detailing Chapters 3 through 5 (thermal methods) and 8 through 10 (miscible methods).

This book is the result of 30 years of experience in reservoir engineering, EOR studies, and projects in more than 100 oil reservoirs with a wide variety of characteristics and behaviors. The subjects are approached in a logical progression and include new advances that have occurred. The illustrative examples and case histories are in large part drawn from the technical literature, and the additional personal experience and comments are hoped to enhance the reader's understanding of the subject. The material contains practical mathematical calculations, but a conscious attempt was made to avoid the use of advanced mathematics, leaving an open door for the pursuit of deeper and more detailed knowledge through the existing texts concentrated on one particular type of EOR.

I am grateful to U.S. Department of Energy-Bartlesville Project Office, to the Society of Petroleum Engineers, and to other editors for allowing me free use of their literature.

I wish to acknowledge the support of Dr. J. Stanislao, dean of the College of Engineering and Architecture, North Dakota State University, Fargo, North Dakota, who encouraged these writings. In addition, thanks go to Mrs. Carol Saastad for typing and retyping the manuscript, to Mrs. Marilyn Goos who helped by correcting the grammar, to my son Laurence who prepared most of the line drawings, and to my brother Radu who encouraged me. I am very grateful to my wife, Verona, without whose encouragement, patience, and understanding this book could not have been written.

Fargo, North Dakota Aurel Carcoana

# Acknowledgments

The author gratefully acknowledges permission to use material from the following sources:

*Chapter 1, condensed text,* pp. 1–2, from *Petroleum Production Handbook,* T. C. Frick, ed. © 1962 Society of Petroleum Engineers (SPE-AIME).

*Tables 3-1a and 3-1b,* abridged tables from *Thermodynamic Properties of Steam,* by Keenan and Keyes, thirty-ninth printing, pp. 32, 38, copyright © 1965 Courtesy of John Wiley & Sons, Inc. *Table 3-2,* from *Oil and Gas Journal,* May 30, © 1960 PennWell Publishing Co., used with permission. *Table 3-4,* from *Fundamentals of Thermal Oil Recovery: How to Calculate Heat Transmission in Hot Fluids,* Ramey, J. H., Jr., p. 165, © 1965 Petroleum Engineer International Publishing Co. *Table 3-5,* from *Oil and Gas Journal,* March 1 and 8, © 1965 PennWell Publishing Co. *Figure 3-3,* from *Oil and Gas Journal,* February 15, © 1965 PennWell Publishing Co. *Figure 3-4,* from "Improved Oil Recovery," Ch. VII, in *Steam Flooding,* by Farouq Ali and Meldau, p. 319, © March 1983 The Interstate Oil Compact Commission. *Figure 3-6,* from Well-Bore Heat Loss, by Pacheco et al., *Journal of Petroleum Technology,* February 1972, © 1972 SPE-AIME. *Figure 3-7,* from "Improved Oil Recovery," Ch. VII, in *Steam Flooding,* by Farouq Ali and Meldau, p. 324, © March 1985 The

Interstate Oil Compact Commission. *Figure 3-8,* from "A Direct-Fired Downhole Steam Generator," by Schrimer et al., *Transactions,* Vol. 279 (1985), © 1985 SPE-AIME. *Figure 3-9,* from "Improved Oil Recovery," Ch. VII, in *Steam Flooding,* by Farouq Ali and Meldau, p. 314, © March 1983 The Interstate Oil Compact Commission. *Figure 3-10,* from "Steam Stimulation Heavy Oil Recovery," by Buckles, presented at the 49th California Regional Meeting, Ventura, California, April 18–20, 1979, © 1979 SPE-AIME. *Ch. 3, text condensed,* from "Feed Water Treatment," *API Recommendations,* API-RP 11T, © 1983 API.

*Table 4-1,* "Error Functions," by Marx et al., *Transactions,* Vol. 216 (1959), p. 312, © 1959 SPE-AIME. *Tables 4-3 and 4-4,* from "Thermal Recovery from Tar Sands," by Carrigy, presented at the 1988 California Regional Meeting, Long Beach, California, March 1988, © 1988 SPE-AIME. *Figure 4-3,* from "Improved Oil Recovery," Ch. VII, in *Steam Flooding,* by Farouq Ali and Meldau, p. 339, © March 1983 The Interstate Oil Compact Commission. *Figure 4-5,* from *Fundamentals of Thermal Oil Recovery,* by Ramey, H. J., Jr., p. 165, © 1965 Petroleum Engineer International Publishing Co. *Figures 4-6 through 4-9, and text condensed* (*chart utilization*), from "Correlations for Predicting Oil Recovery of Steamflood," by Gomaa, E. E., *Journal of Petroleum Technology,* February 1980, © 1980 SPE-AIME. *Figures 4-11 through 4-14 and condensed text,* from SPE-DOE 17380, "Kern River Steam Foam Pilots," by Patzek et al., presented at the Sixth Symposium on Enhanced Oil Recovery, Tulsa, Oklahoma, 1988, © 1988 SPE. *Figures 4-15 and 4-16,* adapted from "The 200 Steamflood Demonstration Project," by Gagner, M. J., © 1986 Department of Energy DOE/ET/12059-9. *Figures 4-17 through 4-20 and case history condensed text,* from SPE 18774 "Successful Conversion of the Pikes Peak Viscous Oil Cyclic Steam Project to Steamdrive," by Miller, K. A., et al., presented to 1989 California Regional Meeting, April 5–7, Bakersfield, California, © 1989 SPE-AIME.

*Figure 5-1,* from "Chemical Aspect of In-Situ Combustion," by Burger and Sahuquet, *Society of Petroleum Engineers Journal* (*SPEJ*), Vol. 12, 1972 © 1972 SPE-AIME. *Figure 5-2,* from "Laboratory Research on Wet Combustion," by Burger and Sahuquet, *Journal of Petroleum Technology,* Vol. 25, © 1973 SPE-AIME. *Figure 5-3,* from "Automation of an In-Situ Combustion Tube," by Vossoughi, S., et al., *SPEJ* August 1982, © 1982 SPE-AIME. *Figure 5-7,* from "Factors Affecting Fuel Availability and Composition," by Alexander, J. D., et al., *Journal of Petroleum Technology,* October 1962, © 1962 SPE-AIME. *Figure 5-10,* from *Combustion as a Primary Recovery Process,* by Gates, C. F., and Sklar, I., in 1985 ed. of SPE Reprint Series, No. 7, © 1985 SPE-AIME. *Figures 5-11; 5–18a, b; 5-19; 5-20; and case history,* from "Results and Difficulties of the World's Largest In-Situ Combustion Process," SPE/DOE paper 20248, by Carcoana, A., presented at the SPE/DOE Seventh Symposium on EOR, Tulsa, Oklahoma, April 1990, © 1990 SPE-AIME. *Figures 5-21 and 5-22,* from "Pressure Maintenance by In-Situ Combustion, W.

Heidelberg Unit," by Huffman et al., *Journal of Petroleum Engineers,* October 1983, © 1983 SPE-AIME. *Figure 5-24,* from "A Method for Engineering In-Situ Combustion Oil Recovery Projects," by Gates, C. F., and H. J. Ramey, Jr., *AIME Transactions,* © 1954 SPE-AIME. *Ch. 5, Example 5-1,* from *Thermal Oil Recovery,* by White, P. D., and Moss, G. T. (The Atomic H/C Ratio Calculation, pp. 72–73), © 1983 PennWell Publishing Co. *Ch. 5, Section 5–15* (*Figures 5-25 and 5-28*), from "How to Engineer an In-Situ Combustion Project," by Nelson, T. W., and McNiel, J. S., *Oil and Gas Journal,* June 5, 1961, © 1961 PennWell Publishing Co.

*Table 6-1,* adapted from "A Technical Survey of Polymer Flooding Projects," by Manning, R. K., et al., Report U.S. DOE/BC/10327-19, September 1983. *Figure 6-3,* from "Improved Secondary Recovery by Control of Water Mobility," by Pye, David, *Journal of Petroleum Technology,* August 1964, © 1964 SPE-AIME. *Figure 6-5,* from "Reservoir Stability of Polymer Solutions," by Knight B. L., *Journal of Petroleum Technology,* May 1973, © 1973 SPE-AIME. *Figure 6-6,* from "Factor Influencing Mobility Control by Polymer Solutions," by Jennings, R. R., et al., *Journal of Petroleum Technology,* March 1971, © 1971 SPE-AIME. *Figures 6-7 through 6-10,* from "Prediction of Oil Recovery by Waterflood," by Johnson, C. E., Jr., *AIME Transactions,* Vol. 207, 1956, © 1956 SPE-AIME. *Figure 6-11,* from *The Reservoir Engineering Aspects of Waterflooding,* by Craig, F. F., third printing, © 1980 SPE-AIME. *Figures 6-13 through 6-15, and condensed text,* from "Performance and Operation of a Successful Polymer Flood in the Sleepy Hollow Reagan Unit," by Christopher, C. A., et al., SPE/DOE paper 17395 presented at SPE/DOE Sixth Symposium on EOR, Tulsa, Oklahoma, April 1988, © 1988 SPE-AIME.

*Tables 7-1 and 7-2,* from "New Variations of Chemicals EOR: The ASP Process," by Ball, J., and Surkalo, H., *Proceedings of the Improved Oil Recovery Conference,* Abilene, Texas, September 1988, U.S. Department of Energy, © 1989 DOE/BC/14286-1. *Figures 7-2 through 7-4,* from "Field Trial of Caustic Flooding Process," by Graue, D. J., and Johnson, C. E., Jr., *Journal of Petroleum Technology,* December 1974, © 1974 SPE-AIME. *Figure 7-5 and text condensed,* from "An Evaluation of the Alkaline Waterflooding Demonstration Project Ranger Zone, Wilmington Field," by Dauben et al., in U.S. DOE/BC/10830-5, © 1987 US/DOE.

*Figure 8-2,* from *Applied Petroleum Reservoir Engineering,* by Craft, B. C., and Hawkins, M. F., p. 64, © 1959 Prentice-Hall, Inc. *Figure 8-3,* from "Carbon Dioxide Miscible Flooding: Past, Present, and Outlook for the Future," by Stalkup, F. I., *Journal of Petroleum Technology,* August 1978, © 1978 SPE-AIME. *Figure 8-6,* from *Miscible Displacement,* by Stalkup, F. I., p. 3, in SPE Monograph Series, Vol. 8, © 1984 SPE-AIME. *Figures 8-10 through 8-14 and condensed text,* from "Tertiary Application of a Hydrocarbon Miscible Flood: Rainbow Keg River 'B' Pool," by Mazzocchi et al., SPE/DOE 17355, presented at EOR Symposium, Tulsa, Oklahoma, April 1988, © 1988 SPE-

AIME. *Figures 8-15 through 8-17 and condensed text,* from "Assessment of a Vertical Hydrocarbon Miscible Flood in the Westpem Nisku D Reef," by Da Sie and Guo, SPE/DOE 17354 presented at EOR Symposium, Tulsa, Oklahoma, April 1988, © 1988 SPE-AIME. *Chapter 8 condensed text,* from *Miscible Displacement,* Ch. 2, pp. 6–16, and *Screening Criteria,* Ch. 5–7, by Stalkup, F. I., SPE Monograph Series Vol. 8, © 1984 SPE-AIME.

*Figures 9-3 through 9-5,* from "Commercial Scale Demonstration EOR by Micellar-Polymer Flood," Final Report prepared by Stover, D. F., Marathon Oil Co. for the U.S. DOE, AC-19-78 ET13077-130, Bartlesville, Oklahoma, 1988, © 1988 U.S./DOE. *Figure 9-6 and condensed text,* from "A Review of the Loudon Surfactant Flood Pilot Test," work performed by Dauben, D. L., under contract No. AC19-85BC 10830-8 for U.S./DOE Bartlesville, Oklahoma, 1988, © 1988 U.S./DOE. *Figure 9-7 and Figures 9-10 through 9-13, and condensed text,* from "Selection of Reservoir Amenable to Micellar Flooding—Critical Reservoir and Micellar Quantities; Screening and Evaluation," *Proceedings,* Part III and App. III, First Annual Report DOE/BC/00048-20, Bartlesville, Oklahoma, December 1980, © 1980 U.S./DOE. *Figures 9-16 and 9-17,* from "Selection of Reservoirs Amenable to Micellar Flooding," Final DOE Report performed by Goldburg, A., et al., US-DOE BC00048-29, August 1983, © 1983 U.S./DOE.

*Figure 10-1,* from *Fundamentals of Enhanced Oil Recovery,* by Van Poollen and Associates, p. 133, Tulsa, Oklahoma, © 1980 PennWell Publishing Co., *Figures 10-2 and 10-3,* from "Generalized Correlations of $CO_2$-Crude Oil System," by Simon, R., and Graue, D. J., *Journal of Petroleum Technology,* January 1965, © 1965 SPE-AIME. *Figures 10-4 and 10-8,* from "Mechanism of Oil Displacement by Carbon Dioxide," by Holm, L. W., and Josendal, V. A., *Journal of Petroleum Technology,* December 1974, lc 1974 SPE-AIME. *Figure 10-5,* adapted from *Miscible Displacement,* by Stalkup, F. I., SPE Monograph Series, Vol. 8, © 1984 SPE-AIME. *Figure 10-7,* from "Determination and Prediction of $CO_2$ MMP," by Yelling, W. R., et al., *Journal of Petroleum Technology,* January 1980, © 1980 SPE-AIME. *Figure 10-8,* from "Carbon Dioxide Flooding Fundamentals," by Mungan, N., *Journal of Canadian Petroleum Technology,* pp. 87–92, © 1981. *Figures 10-9 and 10-10 and Tables 10-3 and 10-4* (*adapted*), from "Performance Review of a Large-Scale $CO_2$-WAG Project, SACROC-Unit, Kelly-Snyder Field," by Kane, A. V., *Journal of Petroleum Technology,* February 1979, pp. 217–231, © 1979 SPE-AIME, *Figures 10-11 through 10-14 and SACROC case history,* from "Definitive $CO_2$ Flooding, Response in the SACROC Unit," Paper SPE/DOE 17321 presented by Langston, M. V., et al., at the SPE/DOE EOR Symposium, Tulsa, Oklahoma, April 1988, © 1988 SPE-AIME. *Figures 10-15, 10-16,* case history, and Table 10-5, from "Review of an Immiscible $CO_2$ Project, Tar Zone, Fault Blok V, Wilmington Field," by Spivak, A., et al, *SPE Reservoir Engineering,* May 1990, pp. 150–162, © 1990 SPE-AIME. *Example 10-2, condensed text,* from

"Carbon Dioxide Flooding," Ch. IV, in *Improved Oil Recovery,* by Mungan, N., pp. 143–145, © 1983 IOCC.

*Figure 11-1,* from "Improved Oil Mining: A Feasibility Study Final Report," NIPER-328 performed by Tham, M. K., et al., under Coop. Agreem. No. FC22-83FE60149, Bartlesville, Oklahoma, © 1988 U.S./DOE. *Figure 11-2,* from "Recovery of Oil from Underground Drillsites," Paper SPE 19344 presented by Streeter et al. at the 1989 SPE Eastern Regional Conference and Exhibition, October 1989, Morgantown, West Virginia, © 1989 SPE-AIME. *Figure 11-3 and condensed text,* from "Microbial-Enhanced Waterflood Field Experiment," Topical NIPER-356 Report prepared by Bryant, R. S., et al., for US-DOE under Coop. Agreem. No. FC22-83FE60149, Bartlesville, Oklahoma, © 1989 U.S./DOE. *Figures 11-4 through 11-6 and condensed text, case history,* from "Field Studies of Microbial EOR Paper SPE/DOE 20254, presented by Sheehy, A. J., at the SPE/DOE Seventh Symposium on EOR, Tulsa, Oklahoma, April 1990, © 1990 SPE-AIME.

*Figure 12-3,* from "Free World Oil Discovery Index," by Ivanhoe, L. B., *Oil & Gas Journal,* November 21, 1983, pp. 88–90, © 1983 PennWell Publishing Co. *Figure 12-5 and Tables 10-1 and 10-2,* from "Review of EOR Projects, Trends and Thermal EOR Technology—Topical Report," by Pautz, J. F., et al., from IIT Research Institute, NIPER-461, under Coop. Agreem. No. FC22-83FE60149, U.S. DOE Bartlesville, Oklahoma, 1990, © 1990 U.S./DOE.

# Nomenclature

| | |
|---|---|
| $A$ | Area, acres, $ft^2$ |
| $a$ | Geothermal gradient, °F/ft |
| $a$ | Well-to-well spacing, ft (Eq. 5-19) |
| $a_s$ | Active surfactant retention, mg/g rock (Eq. 9-7) |
| $B$ | Formation volume factor, bbl/STB |
| $b$ | Surface geothermal temperature, °F (Eq. 3-10) |
| $C$ | Constant (Eq. 3-6) |
| $C$ | Specific heat capacity, Btu/lbm × °F or J/kg × °C. |
| $C_a$ | Amount of air required to burn through a cubic foot of reservoir rock, $scf/ft^3$ |
| $C_u$ | Amount of coke deposited or fuel content, $lbm/ft^3$ |
| $C'_p$ | Injection pressure gradient, psi/ft (Eq. 9-1) |
| $C_s$ | Concentration of active surfactant in the injected slug (Eq. 9-7) |
| $C_{1,S_l}$ | Volume fraction of pseudocomponent 1 in phase $S_l$ |
| $C_{TS}$ | Surfactant requirements, bbl or $m^3$ |

| | |
|---|---|
| $C_{TP}$ | Polymer requirements, bbl or $m^3$ |
| $D$ | Depth, ft |
| $D$ | Thermal diffusivity of the cap rock, $ft^2$/hr (Eqs. 4-1 and 4-5) |
| $D_e$ | Effective diffusion coefficient for $CO_2$/oil or $N_2/CO_2$, $cm^2$/s (Eq. 10-7) |
| $D_s$ | Surfactant retention, dimensionless (Eqs. 9-6, 9-14) |
| $d$ | Tubing inside diameter, in. (Eq. 10-13) |
| $E$ | Efficiency, fraction or percent |
| $E$ | Expenses, $ |
| $E_D$ | Micellar-polymer displacement efficiency, fraction |
| $E_r$, $E_R$ | Overall oil recovery factor, percent |
| $F$ | Fluid volume, bbl |
| $f$ | Fraction (such as the fraction of a flow stream consisting of a particular phase) |
| $f$ | Friction factor, fraction (Eq. 10-13) |
| $f_{hv}$ | Vertical heat loss, fraction (Eq. 4-11) |
| $f_s$ | Steam quality, percent |
| $f(t_D)$ | Dimensionless transient heat conduction time function (Eq. 3-10) |
| $f_{opk}$ | Oil cut or the peak oil rate, volume percent (Eq. 9-20) |
| $H$ | Heat of combustion, Btu/$lb_m$ or J/kg |
| $H_o$ | Heat injection rate, Btu/hr |
| $h$ | Formation thickness, ft |
| $h$ | Enthalpy, Btu/$lb_m$ or J/kg |
| $h$ | Differential pressure, inches of water (Eq. 3-6) |
| $h_f$ | Enthalpy of saturated liquid, Btu/$lb_m$ or J/kg |
| $h_g$ | Total enthalpy, Btu/$lb_m$ or J/kg |
| $h_{fg}$ | Enthalpy of vaporization, Btu/$lb_m$ or J/kg |
| $K$ | Thermal conductivity of the cap rock, Btu/ft $\times$ hr $\times$ °F |
| $k$ | Permeability, md |
| $k$ | Thermal conductivity of the earth, Btu/day $\times$ ft $\times$ °F |
| $\overline{k}$ | Mean permeability, md (Eq. 6-6) |
| $k_s$ | Permeability at 84.1 percent of the cumulative samples, md |
| $l_v$ | Specific latent heat (enthalpy) of vaporization, Btu/$lb_m$ or J/kg |
| $l_n$ | Natural logarithm |
| log | Base 10 logarithm |
| $M$ | Mobility ratio, mixture |
| $M_e$ | Empirical function (Eq. 9-17) |

| | |
|---|---|
| $M_s$ | Heat capacity of steam saturated rock, Btu/ft$^3$ × °F |
| $m$ | Mass, $lb_m$ or kg |
| $N$ | Oil in place, bbl |
| $N_c$ | Capillary number |
| $N_e$ | Reynolds number |
| $n$ | Tubing roughness, in. (Eq. 10-13) |
| $n$ | Atomic H/C ratio |
| $P, p$ | pressure, psi |
| $Q$ | Total heat amount, Btu or joule |
| $Q$ | Total air injected, scf |
| $Q_D$ | Total injection volume, pore volumes |
| $Q_f$ | Net amount of heat available to formation, Btu or joule |
| $Q_g$ | Steam generator heat loss, Btu or joule |
| $Q_s$ | Sensible heat, Btu or joule |
| $Q_s$ | Heat lost on surface lines, Btu or joule |
| $Q_v$ | Latent heat of vaporization, Btu or joule |
| $Q_w$ | Heat loss rate in wellbore, Btu/day (Eq. 3-10) |
| $q$ | Rate, bbl/day |
| $R$ | Resistance factor, ratio |
| $R$ | Revenue, $ |
| $r$ | Radius, ft |
| $S$ | Saturation, fraction |
| $S$ | Saturation phase (relative amount), fraction |
| $S_o(S')$ | Oil price (base), $/STB |
| $T$ | Temperature, reservoir temperature, °F or °C |
| $t$ | Time (injection), days, hours |
| $t_{Dob}$ | Oil breakthrough time, porous volume (Eq. 9-18) |
| $t_{D_S}$ | Time of peak oil rate, porous volume (Eq. 9-19) |
| $U$ | Air flux density, scf/ft$^2$ × hr |
| $U$ | Overall heat transfer coefficient, Btu/day × ft × °F |
| $U_o$ | Idem, Btu/hr × ft × °F |
| $u$ | Superficial or actual velocity, ft/day |
| $V$ | Volume, bbl |
| $V, v$ | Velocity, ft/day |
| $V_{DP}$ or $V$ | Permeability variation (Dykstra-Parsons) |
| $V_b$ | Rate of the burning front advance, ft/day |
| $V_o$ | Rate at which oil is displaced, bbl/day (Eq. 4-4) |

| | |
|---|---|
| $V_{gs}$ | Volume occupied by gases in reservoir after pressurization, bbl (Eq. 10-3) |
| $\nu$ | Specific volume of saturated liquid, $ft^3/lb_m$ |
| $\nu_g$ | Specific volume of saturated vapor, $ft^3/lb_m$ |
| $W$ | Flow rate of wet steam, gal/min (Eq. 3-6) |
| $W_d$ | Density of dry steam, $lb_m/ft^3$ (Eq. 3-6) |
| $W_o$ | Heat injected lost to adjacent strata, fraction (Eq. 4-2) |
| $W_p$ | Cumulative water produced, bbl |
| $X$ | Length of diffusion zone, ft |
| $Y$ | Mole fraction in combustion gases |
| $Z$ | Gas deviation factor |
| $Z$ | Formation depth, ft (Eq. 3-10) |
| *DA* | Developed area, acres |
| $\Delta N$ | Cumulative oil produced during an interval, bbl; reserves (recoverable), bbl |
| $\Delta T$ | Temperature difference, °F |
| *MD* | Measured depth, ft |
| *MP* | Micellar-polymer |
| *PV* | Porous volume, bbl |
| *RO* | Recoverable oil, bbl |
| *SG* | Specific gravity |
| *TO* | Target oil, bbl |
| *AOR* | Air-oil ratio |
| *API* | American Petroleum Institute |
| *EOR* | Enhanced oil recovery |
| *FPV* | Floodable pore volume, bbl |
| *GOR* | Gas-oil ratio |
| *NRP* | Number of repeated patterns |
| *OSR* | Oil-steam ratio |
| *SOR* | Steam-oil ratio |
| *STB* | Stock tank barrels |
| *TDS* | Total dissolved solids, ppm |
| *TDV* | True vertical depth, ft |
| *WOR* | Water-oil ratio |
| *ppg* | Parts per gallon |
| *ppm* | Parts per million |
| CFPM | Chemical flood predictive model |

| | |
|---|---|
| CRMQ | Critical reservoir and micellar quantities |
| HCPV | Hydrocarbon pore volume, bbl |
| OOIP | Original oil in place, bbl |
| PECφN | Preliminary Economic Evaluation Model |
| $\phi$ | Porosity, fraction or percent |
| $\lambda$ | Mobility, md/cp |
| $\mu$ | Viscosity, cp |
| $\rho$ | Density, $lb_m/ft^3$ or $g/cm^3$ |
| $\tau$ | Interfacial tension (IFT), dyne/cm |

# Subscripts

| | |
|---|---|
| a | air, actual, areal |
| A | areal |
| b | burning |
| c | combustion, caprock |
| cons | consumed |
| cr | critic |
| d | dry, diffusion |
| D | dimensionless, displacement |
| e | effective, external |
| ext | exterior |
| feedw | feedwater |
| g | gas |
| h | heated |
| i | initial, injection |
| I | invasion, vertical |
| inj | injected |

| | |
|---|---|
| int | interior |
| l | liquid |
| MB | mobility buffer |
| max | maximum |
| min | minimum |
| o | initial, oil |
| ob | oil bank |
| of | oil formation |
| om | mobil oil |
| or | residual oil |
| orc | micellar-polymer swept zone residual oil |
| orw | residual oil after waterflooding |
| ov | overburden |
| p | polymer, produced, pattern |
| pp | pressurization phase |
| r | relative, rock, residual |
| s | surfactant, specific, solid, soluble, steam |
| T | total |
| tf | tubing flowing (pressure) |
| ts | tubing static (pressure) |
| u | unburned zone |
| v | volumetric |
| w | water, wet, well |
| wb | water bank |
| wf | bottom well flowing (pressure) |
| ws | well static (pressure) |

# Conversion Factors

acre = $4046.856 \text{ m}^2$

acre-ft = $1233.482 \text{ m}^3$

atm = 101.325 kPa = 0.101325 MPa = 14.696 psi

bbl = 42 U.S. gal = $5.614583 \text{ ft}^3$ = $0.158987 \text{ m}^3$

bbl/acre-ft = $0.128893 \text{ m}^3/\text{m}^3$

Btu = 10505.056 J = 251.996 cal

Btu/ft$^3$ = $37.259 \text{ kJ/m}^3$

Btu/hr = 0.293071 W

Btu/lb$_m$ = 2.326 kJ/kg

$\text{Btu} \times \text{lb}_m^{-1} \times {}^\circ\text{F}^{-1} = 1 \text{ kcal} \times \text{kg}^{-1} \times \text{K}^{-1} = 4.186800 \text{ kJ} \times \text{kg}^{-1} \times \text{K}^{-1}$

cal = 4.186800 J

cp = 1 m Pa·s

Darcy = $0.986923 \ \mu\text{m}^2$

dyn/cm = 1 m$N$/m

ft = 0.304800 m

ft$^2$ = $9.290304 \times 10^{-2} \text{ m}^2$

$ft^3 = 2.831685 \times 10^{-2}\ m^3$
$ft^3/bbl = 0.178108\ m^3/m^3$
gal (U.S.) = 3.785412 $dm^3$ or liters = $3.785412 \times 10^{-3}\ m^3$
in. = 2.540 cm = 0.025400 m
$lb_m = 0.453592$ kg
$lb_m/ft^3 = 16.018463\ kg/m^3$
psi = 6.894757 kPa = 6,894.757 Pa = $6.894757 \times 10^{-3}$ MPa
ton = $10^3$ kg

Chapter 1

# Hydrocarbon Classification and Oil Reserves

## 1-1 HYDROCARBON CLASSIFICATION

There is a large number of hydrocarbon compounds with molecules composed of the chemical elements hydrogen and carbon in various proportions.

Hydrocarbons will exist in a fluid phase (gas or liquid) or as solids in a reservoir, depending upon changes in temperature or pressure.

Fluid hydrocarbons or petroleum are normally produced through wells and are subdivided into liquid hydrocarbons and natural hydrocarbon gases. The recommended classification and nomenclature of the fluid hydrocarbons (Arps, 1962; SPE, 1981) is provided in Figure 1-1, with SPE letter symbols standard (SPE, 1986).

Liquid hydrocarbons are subdivided into the following categories:

*Crude Oil*: "a mixture of hydrocarbons that exists in the liquid phase in natural underground reservoirs and remains liquid at atmospheric pressure after passing through surface separating facilities."

*Natural Gas Liquids*: "those portions of reservoir gas that are liquefied at the surface in lease separators, field facilities, or gas processing plants.

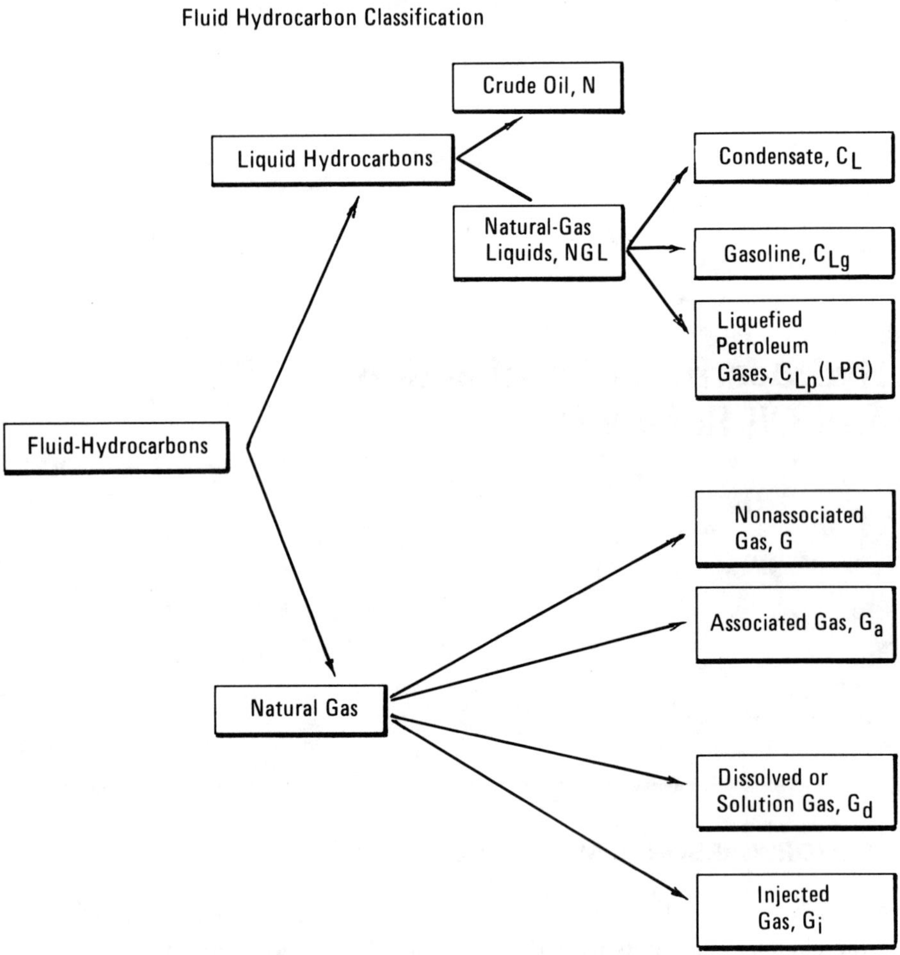


**Fig. 1-1** Fluid hydrocarbon classification

Natural gas liquids include but are not limited to ethane, propane, butanes, pentanes, natural gasoline, and condensate."

*Natural Gas*: "a mixture of hydrocarbons and varying quantities of non-hydrocarbons that exists either in the gaseous phase or in solution with crude oil in natural underground reservoirs." Natural gas is subdivided into the following categories:

*Nonassociated Gas*: natural gas that is in reservoirs that do not contain significant quantities of crude oil.

*Associated Gas*: natural gas, commonly known as gas-cap gas, which overlies and is in contact with crude oil in the reservoir.

*Dissolved Gas*: natural gas in solution with crude oil in the reservoir.

*Injected Gas*: gaseous hydrocarbons that have been injected in underground reservoirs for pressure maintenance or storage purposes.

The scope of this book is limited to crude oil as it occurs in the natural liquid phase under reservoir temperature and pressure and the problem of how to recover more of the original oil in place.

## 1-2 OIL RESERVES CLASSIFICATION

The oil reserves classification takes into consideration the following criteria:

- Recovery possibilities
- Degree of proof
- Development and producing status
- Energy source

### Recovery Possibilities

The recovery possibilities refer to the fact that the amount of oil produced from original oil in place (OOIP) is limited by existing recovery mechanisms, efficiency of known reservoirs, and economic conditions.

### Degree of Proof

The OOIP which initially saturates the porous space of the rock reservoir is difficult to determine exactly in the beginning (the exploration phase), when minimal information is available. Knowledge of the amount of original oil in place is improved by volumetric or material balance calculations at the start of development and during the exploitation of the oil reservoir. However, cumulative oil production obtained and measured at surface conditions can be accurately determined.

If $N$ is the amount of original oil in place (barrels, bbl, of oil) and $\Delta N$ is the cumulative oil produced (bbl) at a given time, the ratio $E_R = \Delta N \times 100/N$ % is the oil recovery factor at that time or the actual oil recovery factor.

The petroleum engineer is mainly interested in knowing from the beginning the ultimate oil recovery from a reservoir, in other words, the product $N \times E_{R_{final}}$, where $E_{R_{final}}$ is the ultimate recovery factor. During the early stages when few data are available, but when important decisions regarding the development of the reservoir must be made, and during the life of a reservoir the term "recoverable reserves" or simply

$$\text{reserves} = \Delta N = N \times E_{R_{final}}$$

should be estimated with more and more accuracy.

J. J. Arps (1962) shows three periods in the life of an oil reservoir. In the *first* period, before any wells are drilled, a general estimation of reserves in barrels per acre is made based on experience. In the *second* period, the first wells drilled are produced. The amount of original oil in place, $N$, is calculated on a volumetric or material balance basis and $E_{R_{final}}$ is estimated knowing the principal recovery mechanism. The reserves are expressed in barrels per acre-foot or barrels. In the *third* period a performance decline trend curve can be extrapolated or a mathematical model made matching the past performance of the reservoir. The reserves are estimated in barrels.

It is evident that the estimated reserves must have a degree of proof or certainty specific to each phase when the estimation is made.

## Development and Producing Status

It is also necessary to consider development status to distinguish between reserves recoverable through existing wells and "reserves" under undeveloped spacing units. These are "so close and so related to developed spacing units that they may be assumed with the same certainty to be produced when drilling" (Arps, 1962).

Producing status is the highest status of reserves because the oil reserves are produced through the existing wells and/or are expected to be recovered from existing wells. Producing reserves are the result of the natural energy source inherent in the reservoir or supplemented by artificial means.

Taking into consideration *recovery possibilities*, the *degree of proof*, the *development status*, and the *producing status* criteria, a classification of oil reserves is given in Figure 1-2.

The United States has already produced 139 billion bbl of oil. This represents an average actual recovery factor of 28 percent from a total of 492 billion bbl of OOIP discovered (Brashear, 1988).

Estimation of currently proven reserves of approximately 28 billion bbl raises the average oil recovery factor from 28 to 34 percent. This represents an ultimate value as the result of current primary and conventional recovery methods.* We can see now that proven reserves are reserves recoverable using different recovery methods or mechanisms. These are developed by the energy source which displaces the oil through the reservoir to the producers. Thus it is important that oil reserves be classified also by energy source criteria.

## Energy Source

In an oil reservoir, production results from a mechanism which utilizes existing pressure. This is the source of driving energy. The reservoir having primarily a natural recovery mechanism that uses principally the liberation and expan-

* Approximately 5 percent of the oil produced per year in the United States is however the result of thermal recovery and other enhanced oil recovery (EOR) methods.

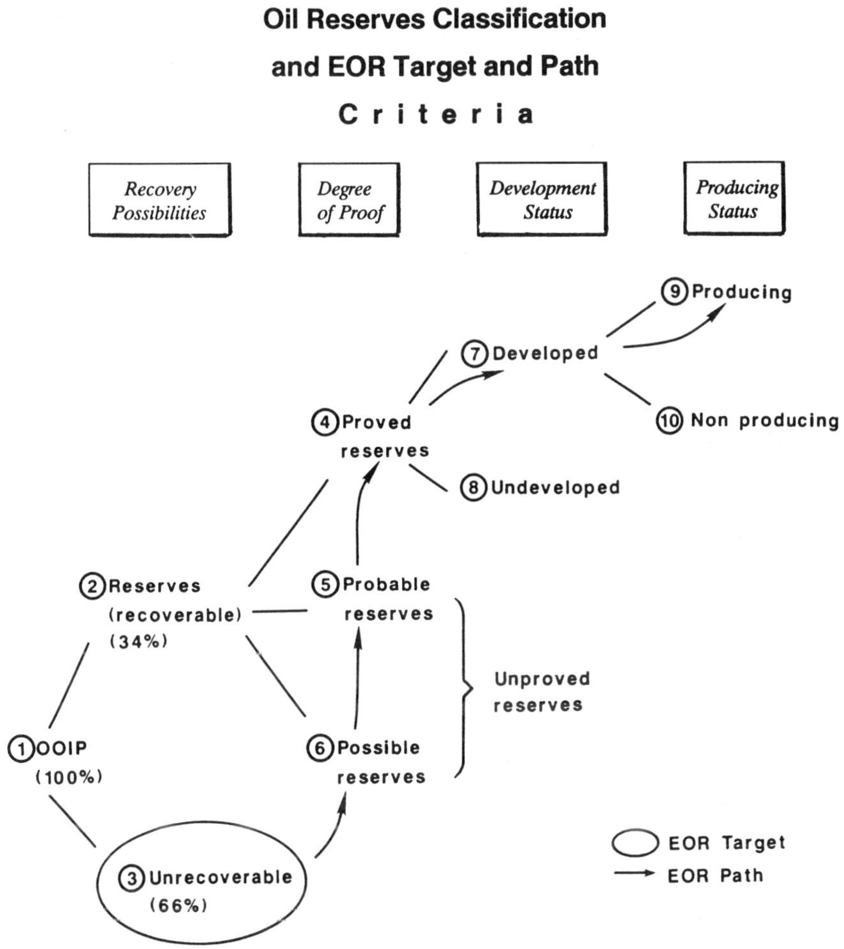


LEGEND

(1) = Original Oil in place, N

(2) = Estimated volumes of hydrocarbon anticipated to be commercially recoverable from known accumulations

(3) = 1 − 2

(4) = Are recoverable reserves under current economic conditions

(5) = Are less certain than proved reserves, more likely to be recovered than not

(6) = Are less certain than probable reserves and more likely not to be recovered

(7) = Are expected to be recovered from existing wells

(8) = 4 − 7

(9) = Are expected to be recovered from completion intervals open at the time of the estimate and producing

(10) = 7 − 9

**Fig. 1-2** Oil reserves classification and EOR target and path

sion of dissolved gas is termed a "dissolved gas drive reservoir," one that uses principally the expansion of a cap of free gas over the oil zone is termed a "gas-cap drive reservoir," and one that uses principally the influx of natural water is a "water drive reservoir." Driving energy may be derived also from gravity and from combinations of these mechanisms. An oil reservoir with a primary energy source produces oil using one or more of the primary recovery mechanisms just defined.

When oil recovery involves the introduction of energy into a reservoir by injecting gas or water under pressure, the oil is produced by secondary or so-called conventional recovery methods. Waterflooding has been and continues to be very successful and improves the recovery of oil from known reservoirs.

Despite the improvement made in the technology and methods used for development and production of oil reservoirs, a substantial amount of oil, nearly 325 billion bbl or 66 percent of OOIP, remains as droplets trapped in the pores of reservoir rock or as films partly coating the pore walls. Entrapment of the remaining oil is due mainly to capillary forces and interfacial tensions and to a partial sweep of the reservoir by injected fluids. These remaining or nonrecoverable reserves are the target of more sophisticated and expensive so-called enhanced oil recovery (EOR) methods.

Considering the *energy source* criteria for oil reserve classification, we can conclude that

*Primary reserves* are reserves recoverable commercially with current equipment and under current economic conditions as a result of primary recovery methods using the natural energy inherent in the reservoir (Arps, 1962).

*Secondary reserves* are reserves recoverable commercially under current economic conditions, in addition to the primary reserves, as a result of supplementing by conventional methods (water and/or gas injection) the natural energy inherent in the reservoir (Arps, 1962).

*Tertiary reserves* are reserves beyond those proved recoverable by conventional methods but recoverable by EOR methods.

Each of these categories could be subdivided into *proved*, probable, and possible; *developed* and undeveloped; and *producing* and nonproducing, depending on the information available at the specific time when the reserve estimation is made.

Changes in reserves take place each year and are assigned to newly discovered petroleum reservoirs, extensions of existing reservoirs, and revisions (Lovejoy and Homan, 1965). This last category results from improved knowledge of the geological characteristics of the reservoir and of the implementation of conventional and EOR methods. The goal of the geologist and the petroleum engineer is to promote more proved, developed, and producing reserves. The proved, developed, and producing reserves assure the oil production of the United States.

## QUESTIONS AND EXERCISES

**1-1** In what phases may the fluid hydrocarbons *N*, *NGL*, and *G* exist in reservoir conditions?

**1-2** How are OOIP, $E_R$, and $E_{R_{final}}$ defined?

**1-3** Calculate the "reserves" of an oil reservoir knowing the OOIP and the ultimate oil recovery factor.

**1-4** Assuming a cumulative oil production of 853,000 bbl, the amount of original oil in place $7.6 \times 10^6$ bbl and the ultimate oil recovery 28 percent, calculate the remaining reserves.

**1-5** Define oil reserves by energy source criteria.

## REFERENCES

ARPS, JAN J., *Petroleum Production Handbook*, T. C. Frick, ed. (Richardson, TX: Society of Petroleum Engineers, AIME, 1962), Volume XI, Chapter 37, p. 37-1.

BRASHEAR, J. P., A. B. BECKER, and K. H. BIGLARBIGI, "Incentives, Technology, and EOR: Potential for Increased Oil Recovery at Lower Oil Prices," SPE 17454, California Regional Meeting, Long Beach, CA, March 23–25, 1988.

LOVEJOY, W. F., and P. T. HOMAN, *Methods of Estimating Reserves of Crude Oil, Natural Gas and Natural Gas Liquids* (Baltimore, MD: Johns Hopkins University Press, 1965).

SPE, "Society Adopts Proved Reserves Definitions," *Journal of Petroleum Technology*, (November 1981), pp. 2113–14.

SPE, *SPE Letter and Computer Symbols Standard* (Richardson, TX: Society of Petroleum Engineers, 1986).

# Chapter 2

# Producing Reserves

## 2-1 OIL RECOVERY METHODS

Close examination of producing reserves is revealing and helpful in classifying the oil recovery methods by energy source criteria.

### Primary Recovery Methods

Producing reserves are called "primary" when they are produced by primary recovery methods using the natural energy inherent in the reservoir. The driving energy may be derived from the liberation and expansion of dissolved gas, from the expansion of the gas cap or of an active aquifer, from gravity drainage, or from a combination of these effects.

### Improved Recovery Methods

The producing reserves are called "improved recovery reserves" when they are produced by improved recovery methods, in addition to the primary reserves, using additional energy. Improved recovery methods are subdivided into

*Conventional methods* (secondary methods), which involve the injection of gas and/or water into the reservoir, and

*Enhanced oil recovery methods* (tertiary methods), of which thermal, chemical, and miscible methods are generally recognized as the most promising.

Primary and improved recovery methods are presented in Figure 2-1.

The object of this book is to cover the processes involved in enhanced oil recovery. The EOR target and path shown in Figure 1-2 is better understood now. It is a difficult and complex path

- from research and lab experiments, which can improve the *possible* reserves,
- through field pilot tests, which can increase the *probable* reserves,
- to *proved* and *producing* reserves when the commercial development of the reservoir under one of the EOR methods is in place.

## 2-2 ENHANCED OIL RECOVERY METHODS

Since the early 1950s, a significant amount of laboratory research and field testing has been undertaken, and some of the resulting findings have been developed on a commercial scale. The intent of enhanced recovery methods (Latil et al., 1980) is to

- improve sweep efficiency by reducing the mobility ratio between injected and in-place fluids,
- eliminate or reduce the capillary and interfacial forces and thus improve displacement efficiency, and
- act on both phenomena simultaneously.

The basic principles of the most promising EOR methods used are given in Table 2-1.

Other processes, such as bacterial activity, electrical heating of the reservoir, and so on have been proposed, but their potential for adding to proved oil reserves must be demonstrated.

*Chemical methods* of enhanced oil recovery are characterized by the addition of chemicals to water in order to generate fluid properties or interfacial conditions that are more favorable for oil displacement. Polymer flooding, using polyacrylamides or polysaccarides, is conceptually simple and inexpensive, and its commercial use is increasing despite the fact that it raises potential production by only small increments. Surfactant flooding is complex, requiring detailed laboratory testing to support field project design. It is also expensive

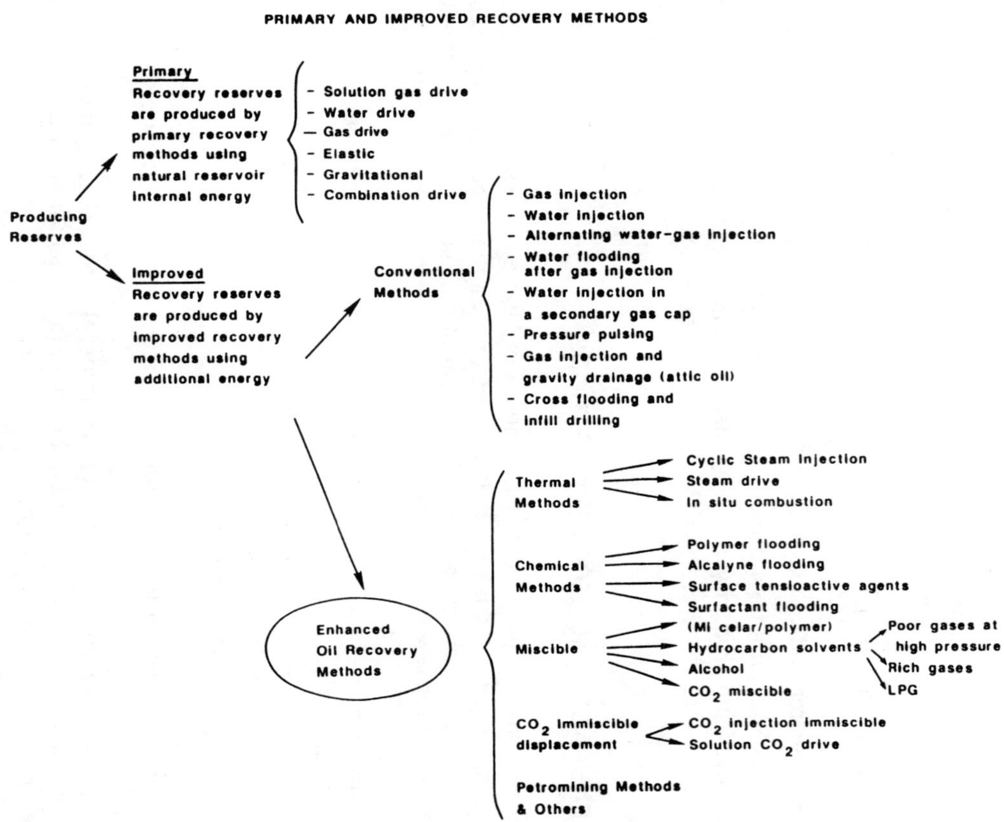


**Fig. 2-1** Primary and improved recovery methods

**TABLE 2-1.** Methods of Enhanced Recovery

| | Methods Used | Basic Principle |
|---|---|---|
| Chemical methods | Polymer-augmented waterflooding; surfactant flooding; alkaline flooding; $CO_2$-augmented waterflooding; immiscible $CO_2$ displacement. | Improvement of sweep efficiency; improvement of displacement efficiency. |
| Miscible methods | Miscible fluid displacement using $CO_2$, nitrogen, alcohol, LPG or rich gas, dry gas. | Improvement of displacement efficiency. |
| Thermal methods | Cyclic steam injection; steam drive; in situ combustion | Improvement of both sweep efficiency and displacement efficiency |

and is used in few large-scale projects. Alkaline flooding has been used only in those reservoirs containing specific types of high-acid-number crude oils.

*Miscible methods* have their greatest potential for enhanced recovery of low-viscosity oils. Among these methods, $CO_2$ miscible flooding on a large scale is expected to make the greatest contribution to miscible enhanced recovery in the future.

*Thermal methods* provide a driving force and add heat to the reservoir to reduce oil viscosity and/or vaporize the oil. This makes the oil more mobile, so that it can be more effectively driven to producing wells. Steam injection has been commercially applied in California since the early 1960s and is the most advanced of all EOR methods in terms of field experience. The performance of steam injection can be estimated with less uncertainty than other methods. In situ combustion is normally applied to reservoirs containing low-gravity oil, but has been field tested under a wide variety of reservoir conditions. Only a few projects have proven economical enough to advance to a commercial scale. To date, in situ combustion has been most effective for the recovery of viscous oils in moderately thick reservoirs (National Petroleum Council, 1984).

## 2-3 OIL RECOVERY FACTOR

The oil reserves obtained as a result of EOR methods in addition to the primary or conventional reserves may be expressed as the percentage of original oil in place (OOIP).

To estimate how much EOR methods can add to oil reserves, the recovery

**TABLE 2-2.** Conventional and EOR Methods Efficiency ($E_R$)

| $E_R$ BY PRIMARY AND CONVENTIONAL RECOVERY | | 5–10% | 10–25% | 25–40% | 40–55% |
|---|---|---|---|---|---|
| ADDITIONAL OIL, % OF OOIP TO FINAL RECOVERY / EOR METHODS | | ADDITIONAL OIL, $\frac{\%\ OOIP}{E_{R_{final}}}$ | ADDITIONAL OIL, $\frac{\%\ OOIP}{E_{R_{final}}}$ | ADDITIONAL OIL, $\frac{\%\ OOIP}{E_{R_{final}}}$ | ADDITIONAL OIL, $\frac{\%\ OOIP}{E_{R_{final}}}$ |
| IN SITU COMBUSTION | | $\frac{+35\%}{40\%-45\%}$ | $\frac{+30\%}{40\%-55\%}$ | $\frac{+15\%}{40\%-55\%}$ | $\frac{+10\%}{50\%-65\%}$ |
| STEAM INJECTION | | $\frac{+25\%}{30\%-35\%}$ | $\frac{+20\%}{30\%-45\%}$ | $\frac{+10\%}{35\%-50\%}$ | |
| POLYMER INJECTION | | — | $\frac{+25\%}{35\%-50\%}$ | $\frac{+15\%}{40\%-55\%}$ | — |
| SOLVENT INJECTION | DRY AND RICH GAS | — | — | $\frac{+12\%}{37\%-52\%}$ | $\frac{+8\%}{48-63\%}$ |
| | ALCOHOL, LPG | — | $\frac{+25\%}{35\%-50\%}$ | $\frac{+15\%}{40\%-55\%}$ | — |

| | | | | | |
|---|---|---|---|---|---|
| SURFACTANT FLOODING | | — | +30% / 40%–55% | +15% / 40%–55% | +10% / 50%–65% |
| $CO_2$ INJECTION | IMMISCIBLE DISPLACEMENT | — | +20% / 30%–45% | +10% / 35%–50% | — |
| | MISCIBLE DISPLACEMENT | — | — | +15% / 40%–55% | +10% / 50%–65% |
| IMPROVED CONVENTIONAL METHODS | INFILL DRILLING | | | +2–4% | — |
| | ALTERNATING WATER-GAS INJECTION | | +7% | +5% | — |
| | WATERFLOODING AFTER GAS INJECTION | | | — | +5% |
| | WATER INJECTION IN SECONDARY GAS CAP | | | +5% | — |
| | PRESSURE PULSING | | +3–5% | +3% | — |
| | GAS INJECTION AND GRAVITATIONAL DRAINAGE (ATTIC OIL) | | | +5% | — |
| | CROSS-FLOODING | | | — | +5% |

potential of the reservoir has to be known. This is defined by the reservoir's characteristics and prior recovery mechanism.

For instance, the ultimate oil recovery factor of individual reservoirs under primary and/or conventional recovery methods may range from 5 percent of OOIP for the poorest reservoir characteristics or for viscous oil, to as high as 55 or 60 percent of OOIP for the best reservoir characteristics or for light oil.

To achieve this desideratum the oil reservoirs are classified by several models according to the average of the ultimate oil recovery $E_{R_{final}}$, expressed as a percentage of OOIP, possibly attained by the respective recovery mechanism, as follows:

| $E_{R_{final}}$ | |
|---|---|
| 5–10% | Tight oil reservoirs, slightly fractured or heavy oil reservoirs |
| 10–25% | Oil reservoirs produced mainly by solution gas drive |
| 25–40% | Oil reservoirs producing under partial water drive, gas injection, or gravity drainage |
| 40–55% | Oil reservoirs produced by conventional waterflood |

A possible estimation (Carcoana and Aldea, 1976) of the additional oil reserves, percentage of OOIP, that could be recovered as the result of EOR processes is shown in Table 2-2. These values can be safely used for a quick evaluation of the improved oil reserves and for the selection of methods that should be applied. For instance, for oil reservoirs where $E_R$ = 25–40 percent of OOIP obtained by primary or conventional methods, the implementation of the EOR methods would increase these values to 35–55 percent of OOIP. However, the selection of a certain process aimed at enhancing oil recovery must be made only after detailed investigations of the pertinent oil reservoir data (especially residual oil saturation), laboratory tests, and field pilots. The National Petroleum Council (NPC) study shows that the potential exists to add 11 billion bbl, approximately 40 percent of current proven reserves, to U.S. supply with existing EOR technology.

## QUESTIONS AND EXERCISES

**2-1** Define producing reserves by energy source criteria.

**2-2** Show the target and path of EOR.

**2-3** Explain why the ultimate recovery factor is different for different oil reservoirs.

**2-4** Estimate the additional oil, expressed as a percentage of OOIP, possible to obtain by steam injection and by surfactant flooding when applied to oil reservoirs developed by primary and conventional methods.

## REFERENCES

CARCOANA, A., and GH. ALDEA, *Marirea Factorulni Final de Recuperare la Zacamintele de Hidrocarburi* (Bucharest, RO: Ed. Technica, 1976).

LATIL, M., et al., *Enhanced Oil Recovery* (Houston, TX: Gulf, 1980), p. 206.

NATIONAL PETROLEUM COUNCIL, *Enhanced Oil Recovery* (Washington, D.C.: U.S. Department of Energy, June 1984).

# Chapter 3

# Steam: A Heat Carrier Agent

Thermal recovery of oil using steam injected into the formation is similar to other hot fluid injection methods using water or gas. However, since steam is a better heat carrier, a higher oil displacement efficiency can be obtained with steam than with hot water or gas. For example, if water at 350 °F is injected into a reservoir with a 130 °F temperature, the heat content added to the reservoir is 224 $Btu/lb_m$. However, when steam at 350 °F is injected into a reservoir with a 130 °F temperature, the added heat content is 1194 $Btu/lb_m$. In addition to the higher heat content, the steam front's high temperature also generates other favorable effects such as vaporization and condensation. For these reasons steam has been preferred as an injection agent instead of hot water or gas.

## 3-1 LIQUID TO VAPOR PHASE CHANGE

Before steam is injected into the oil reservoir, it should be produced in the field using steam generators. We know how water is transformed into steam, but it

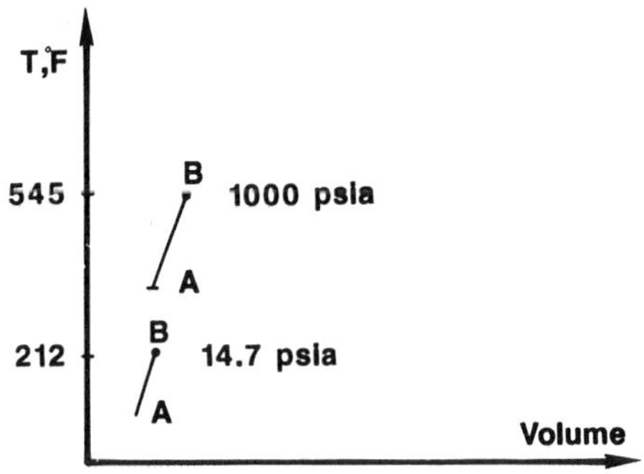


**Fig. 3-1a** Water to boiling point, B

is useful to consider some details regarding the liquid to vapor phase change (Bleakley, (a), 1965).

### Boiling Point

By the application of heat, water gains internal energy, and the temperature increase can be illustrated as line A-B on a temperature-volume diagram (Figure 3-1a). When the boiling point B is reached, some liquid molecules have enough kinetic energy to escape through the liquid surface tension as vapor. At atmospheric conditions (14.7 psia) the boiling temperature of water is 212 °F. At higher pressures the boiling temperature increases. For instance, at 1000 psia the boiling temperature of water increases to 545 °F. The amount of heat required to raise the water temperature until the boiling point B is reached (change in temperature without a phase change) is called sensible heat, $Q_s$, and is expressed in British thermal units (Btu) or joules (J).

$$Q_s = mc\Delta T \tag{3-1}$$

where

$m$ = mass of water, lb or kg

$c$ = specific heat capacity, Btu/(lb$_m$ °F) or J/(kg °C) (heat required to produce a unit temperature change in a unit mass)

$$\Delta T = T_1 - T_0 \text{ °F or °C}$$

where $T_1$ is the final temperature and $T_0$ is the initial temperature of the mass $m$.

### Vaporization Point

Continued application of heat causes the water to boil and vaporize at a constant temperature and pressure (line B-V, Figure 3-1b). When vaporization

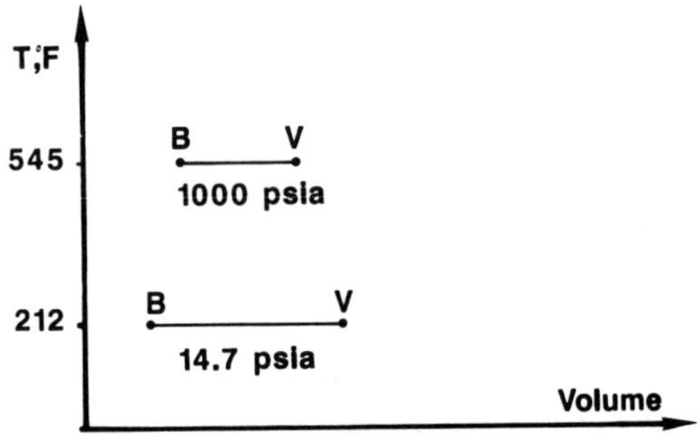


**Fig. 3-1b** Water and steam: the vaporization point, V

point V is reached all water in the liquid phase changes to vapor. The amount of heat required to change the phase from liquid (water) to vapor (steam) at a constant temperature and pressure is called latent heat of vaporization, $Q_v$ (Btu or J),

$$Q_v = ml_v \tag{3-2}$$

where $l_v$ is the specific latent heat of vaporization (enthalpy of vaporization) and is expressed in units of Btu/$lb_m$ or J/kg. Further addition of heat causes the temperature of steam to rise without an increase in pressure. This steam is said to be superheated.

## Water-Steam Pressure-Volume Diagram

The line joining points B with different pressure values is called the bubble point line or saturated liquid line (Figure 3-1c).

When the liquid is saturated (one phase region) and heat is added, the temperature remains constant, the vaporization process begins and liquid

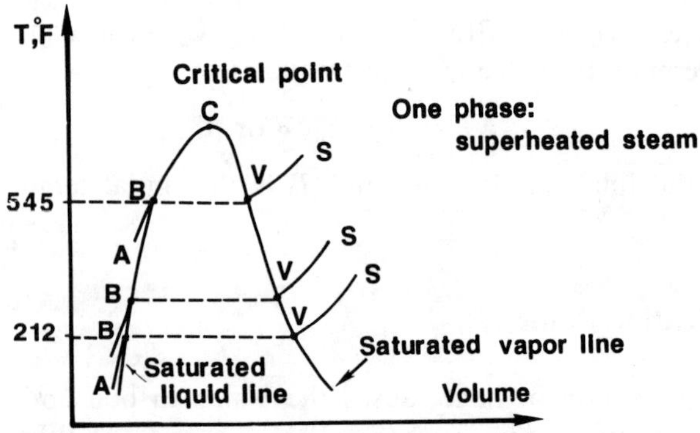


**Fig. 3-1c** Water-steam temperature-volume diagram

phase changes to vapor phase (vaporization line B-V, two phases). The line joining points V is called the dew point line or saturated vapor line. When the vapor is saturated and heat is added, the temperature increases (line V-S, one phase) and the steam is superheated. Inversely, when heat is taken from the superheated steam system, temperature decreases until point V is reached on the saturated vapor line. If more heat is taken out, the temperature remains constant, vapor phase changes to liquid phase (condensation line V-B, two phases) until point B is attained, where all vapor is changed to liquid (one phase). Critical point C (3206.2 psia and 705.4 °F) is the point where the saturated liquid line and the saturated vapor line converge.

Each A-B-V-S line is also a constant-pressure line. The conversion of liquid water to steam occurs along a constant-pressure and constant-temperature line (line B-V). For a given temperature there is only one pressure at which water and steam can both be present. Field steam generators operate at constant pressure, higher than reservoir pressure, and in the two-phase region. The temperature of the output steam is also constant. In the two-phase region the steam temperature cannot be increased or decreased without respective increase or decrease in pressure. The two-phase region where both water and steam coexist is the area of importance in oil field steam operations.

## 3-2 THE HEAT CONTENT OF STEAM

The total amount of heat $Q_T$ absorbed in the process of converting water into steam is given by

$$Q_T = Q_s + Q_v = mc\Delta T + ml_v \tag{3-3}$$

### Steam Enthalpy

The amount of heat per unit of mass is called enthalpy, $h$, and is expressed in Btu/lb$_m$ or J/kg. It is arbitrarily assumed that saturated water has a zero enthalpy value at 32 °F or 0 °C temperature and 0.08854 psia or 0.006 atmosphere ($h_{DATUM} = 0$).

The amount of heat necessary for a mass unit of water at 32 °F and 0.08854 psia to reach the boiling point B on the saturation liquid curve is given by the enthalpy of saturated liquid $h_f$, Btu/lb$_m$ or J/kg.

A Cartesian plot of the amount of heat (enthalpy) as a function of pressure is shown in Figure 3-2.

The total enthalpy $h_g$ necessary to vaporize all liquid and to reach the point V on vaporization line is given by

$$h_g = h_f + h_{fg} \tag{3-4}$$

where $h_{fg}$ is the amount of absorbed heat needed to vaporize the water and is

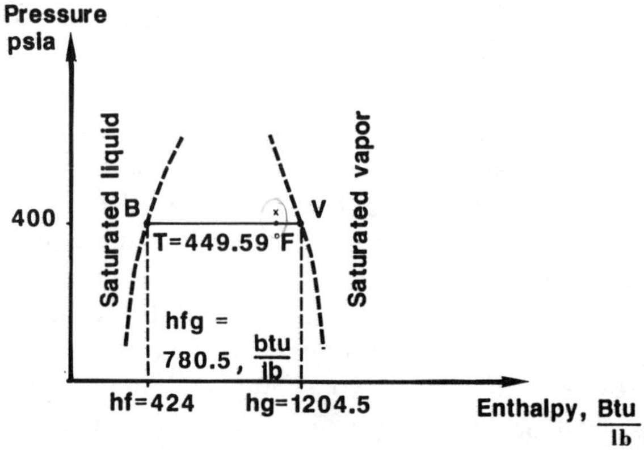


**Fig. 3-2** Water-steam enthalphy curve at $P$ = 400 psia and $T$ = 449.59 °F

called enthalpy of vaporization at a given saturation temperature and the corresponding pressure. It must be pointed out that when the saturation temperature is 212 °F or 100 °C and the corresponding pressure is 14.696 psia or 1 atm, the enthalpy of vaporization $h_{fg}$ coincides with the specific latent heat of vaporization $l_v$.

**Example 3-1.** Find the total enthalpy of 1 $lb_m$ of steam at $p$ = 400 psia and $T$ = 444.59 °F, given

- the enthalpy of saturated liquid (sensible heat)

$$h_f = 424 \text{ Btu/lb}_m$$

- the enthalpy of vaporization (latent heat of vaporization)

$$h_{fg} = 780.5 \text{ Btu/lb}_m$$

SOLUTION The total enthalpy or the enthalpy of steam when all water is transformed into steam (point V, Figure 3-2) is

$$h_g = h_f + h_{fg} = 424 + 780.5 = 1204.5 \text{ Btu/lb}_m$$

In the field steam generators do not convert all water into steam. Only a certain mass fraction, $f_s$, of water is converted. Output from the generator is a mixture of saturated water (liquid) and dry steam. This is called "wet steam" of quality $f_s$. The enthalpy or the total heat of wet steam is given by

$$h_{gw} = h_f + f_s h_{fg} \tag{3-5}$$

Oil field steam generators produce wet steam usually of 80 to 85 percent quality. This means 80 to 85 percent of the water mass is vaporized and 10 to 15 percent is still in the liquid phase (point X in Fig. 3-2). The liquid phase retains in solution the dissolved solids present in the feedwater. However, it is important not to exceed the solubility limits of the liquid phase. For this reason the feedwater's initial concentration of dissolved solids should be reduced by a water treatment unit.

**Example 3-2.** Find the enthalpy of steam and the change in enthalpy necessary to convert 83 percent of the water into steam under the same conditions of pressure and temperature as in Example 3-1.

SOLUTION The generator produces wet steam of quality $f_s = 83\%$.

The enthalpy or heat content of the wet steam per unit of mass, $h_{gw}$, is given by Eq. 3-5:

$$\begin{aligned} h_{gw} &= h_f + f_s h_{fg} \\ &= 424 + 0.83 \times 780.5 \\ &= 1071.8 \text{ Btu/lb}_m \end{aligned}$$

The change in enthalpy during the conversion of 1 $\text{lb}_m$ of water at 400 psia and 444.59 °F into wet steam of quality $f_s = 83\%$ is

$$\begin{aligned} h_{gw} - h_f &= f_s h_{fg} \\ &= 0.83 \times 780.5 \\ &= 647.8 \text{ Btu/lb}_m \end{aligned}$$

It is important to point out that the net enthalpy of steam available to reservoir oil (assuming there are no heat losses) is that given by Eq. 3-5 minus the water enthalpy at reservoir temperature. For instance, if wet steam with 1071.8 $\text{Btu/lb}_m$ enthalpy is injected into an oil reservoir whose temperature is $T_R = 185$ °F, the net enthalpy of steam available for heating the reservoir rock and fluid content is

$$1071.8 \text{ Btu/lb}_m - 152.93 \text{ Btu/lb}_m = 918.87 \text{ Btu/lb}_m$$

where 152.93 $\text{Btu/lb}_m$ is the water enthalpy at 185 °F (saturation temperature) from steam tables.

### Steam Tables and Charts

The properties of steam are tabulated (Keenan and Keyes, 1967) for wide range of saturation temperatures and corresponding pressures in step increments of 1 °F between 32 °F and 200 °F, 2 °F between 200 °F and 400 °F, and 5 °F between 400 °F and 700 °F. There are also tables with saturation pressures and corresponding temperatures for different step increments. For instance, between 350 psia and 500 psia, the step increment is 10 psia, and between 500 psia and 1000 psia, the step increment is 20 psia. The properties of interest in oil field steam operations are

- Enthalpy of saturated liquid (sensible heat), $h_f$

**TABLE 3-1a.** Properties of Steam

| Temp., °F $T$ | Abs. Press., psia $p$ | Specific Volume ($ft^3/lb_m$) of Saturated: Liq. $v_f$ | Specific Volume ($ft^3/lb_m$) of Saturated: Vap. $v_g$ | Enthalpy, $Btu/lb_m$: Saturated Liquid $h_f$ | Enthalpy, $Btu/lb_m$: Change in Enthalpy $h_{fg}$ | Enthalpy, $Btu/lb_m$: Saturated Vapor $h_g$ |
|---|---|---|---|---|---|---|
| 425 | 325.92 | 0.01902 | 1.4226 | 402.77 | 801.2 | 1203.5 |
| 430 | 343.72 | 0.01910 | 1.3499 | 407.79 | 796.0 | 1203.8 |
| 435 | 362.27 | 0.01918 | 1.2815 | 413.34 | 790.8 | 1204.1 |
| 440 | 381.59 | 0.01926 | 1.2171 | 418.90 | 785.4 | 1204.3 |
| 445 | 401.608 | 0.01935 | 1.1565 | 424.49 | 780.0 | 1204.5 |

- Enthalpy of saturated vapor (total heat), $h_g$
- The change in enthalpy (latent heat), $h_{fg} = h_g - h_f$
- Specific volume of saturated liquid, $v_f$, $ft^3/lb_m$
- Specific volume of saturated vapor, $v_g$, $ft^3/lb_m$

Tables 3-1a and 3-1b show enthalpy and specific volume values of water and steam at saturation temperatures from 425 to 445 °F (Table 3-1a) and at saturation pressures between 400 and 440 psia (Table 3-1b).

The enthalpies and specific volumes of saturated water and saturated steam at different pressures and temperatures can also be found from charts especially prepared. Figure 3-3 is a pressure-enthalpy chart for steam.

Figure 3-4 represents the variation of sensible ($h_v$), latent ($h_{fg}$), and total heat of steam ($h_g$) with pressure. It is interesting to observe that, starting at approximately 470 psia, the total heat of steam decreases with an increase in pressure. The reason for this can be understood by observing how the two-phase area (latent heat) is reduced when the saturated vapor line and saturated liquid line converge at the critical point (Fig. 3-4). The decrease in the latent heat content of steam becomes larger than the increase of the sensible heat with pressure.

**TABLE 3-1b.**

| Abs. Press., psia $p$ | Temp., °F $T$ | | | | | |
|---|---|---|---|---|---|---|
| 400 | 444.59 | 0.0193 | 1.1613 | 424.0 | 780.5 | 1204.5 |
| 410 | 447.01 | 0.0194 | 1.1330 | 426.8 | 777.7 | 1204.5 |
| 420 | 449.39 | 0.0194 | 1.1061 | 429.4 | 775.2 | 1204.6 |
| 430 | 451.73 | 0.0194 | 1.0803 | 432.1 | 772.5 | 1204.6 |
| 440 | 454.02 | 0.0195 | 1.0556 | 434.6 | 770.0 | 1204.6 |

Abridged tables are from Joseph H. Keenan and Frederick G. Keyes, *Thermodynamic Properties of Steam*, thirty-ninth printing (New York: John Wiley & Sons, © 1967), pp. 32, 38.

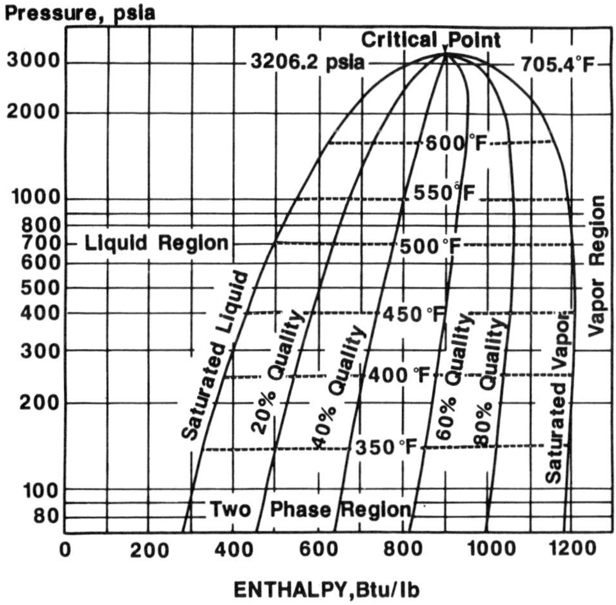


**Fig. 3-3** Pressure-enthalpy chart for steam (From Bleakley, 1965)

In other words, if the steam's injection pressure is just enough to displace the reservoir fluids, it will have more heat content than at higher pressures.

### Steam Quality

Figure 3-3 also shows steam quality lines or lines connecting the points with the same steam quality values. Steam quality can be determined by different methods such as the separator method, conductivity-meter method, chloride method, and the orifice-meter measurement (Bleakley, (c), 1965).

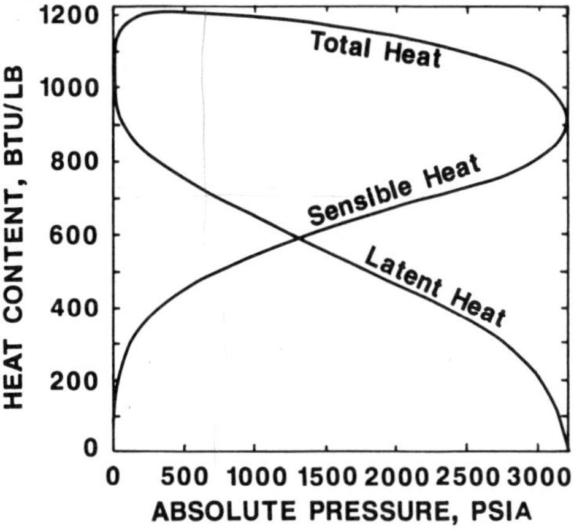


**Fig. 3-4** Variation of sensible, latent, and total heat of steam with pressure (From Farouq Ali, 1970)

**Saturation Method.** The mass rates of the liquid phase and of the dry steam separated in an insulated vessel are measured under pressure over a short period of time. The steam quality is given by the ratio mass rate of vapor flow to mass rate of flow of the vapor and liquid steam.

**Conductivity-Meter Method.** This method is based on the resistance to flow of electrical current. More salt dissolved leads to less resistance to flow, which leads to higher electrical conductivity. Steam quality can be calculated by measuring the electrical conductivity of the feedwater and liquid phase of the wet steam. For instance, if the conductivity of the liquid phase of steam is six times higher than that of the feedwater, five-sixths of the feedwater has been vaporized and the steam quality is $\frac{5}{6} = 0.833$.

**Chloride Method.** The chloride method measures only the chloride ion, $C_l$, contained in both stream flows instead of measuring the conductivity resulting from the total salt content.

**Orifice-Meter Method.** This method measuring superheated and high-quality steam flow rates has been adapted by Pryor (1966), to determine quality $x$ of wet steam used in injection steam operations:

$$x = \left(\frac{C\sqrt{W_d h}}{W}\right)^{1/2} \tag{3-6}$$

where

$W$ = flow rate of wet steam, gallons per minute (gpm), is determined by measuring the inlet water using an orifice meter, since oil field generators are designed for steady flow

$C$ = combination of flow constant for the size of pipe, orifice diameters, and orifice plate expansion; units conversion is given for average conditions in tables such as Table 3-2 (abridged)

**TABLE 3-2.** *C*, Constant Values (bellows type, W.C. meter)

| | 2.626-in. ID meter run (W in gpm) | | |
|---|---|---|---|
| Orifice plate | 100 in. | 200 in. | 400 in. |
| 2.125 | 2.640 | 3.720 | 5.280 |
| 2.000 | 2.196 | 3.092 | 4.392 |
| 1.875 | 1.844 | 2.600 | 3.688 |
| 1.750 | 1.524 | 2.148 | 3.048 |
| 1.625 | 1.272 | 1.796 | 2.544 |
| 1.500 | 1.048 | 1.480 | 2.096 |

**TABLE 3.2.** *(continued)*

| 3.068-in. ID meter run (W in gpm) | | | |
|---|---|---|---|
| Orifice plate | 100 in. | 200 in. | 400 in. |
| 2.375 | 3.144 | 4.44 | 6.288 |
| 2.250 | 2.692 | 3.80 | 5.384 |
| 2.125 | 2.312 | 3.260 | 4.624 |
| 2.000 | 1.980 | 2.788 | 3.960 |
| 1.875 | 1.688 | 2.380 | 3.376 |
| 1.750 | 1.424 | 2.004 | 2.848 |
| 1.625 | 1.208 | 1.704 | 2.416 |
| 1.500 | 1.016 | 1.432 | 2.032 |
| 1.375 | 0.844 | 1.192 | 1.688 |
| 1.250 | 0.692 | 0.976 | 1.384 |

| 3.438-in. ID meter run (W in gpm) | | | |
|---|---|---|---|
| Orifice plate | 100 in. | 200 in. | 400 in. |
| 2.750 | 4.380 | 6.20 | 8.76 |
| 2.625 | 3.820 | 5.40 | 7.64 |
| 2.500 | 3.320 | 4.692 | 6.64 |
| 2.375 | 2.872 | 4.060 | 5.74 |
| 2.250 | 2.504 | 3.544 | 5.008 |
| 2.125 | 2.192 | 3.100 | 4.384 |
| 2.000 | 1.892 | 2.672 | 3.784 |
| 1.875 | 1.624 | 2.300 | 3.248 |
| 1.750 | 1.400 | 1.980 | 2.80 |
| 1.625 | 1.192 | 1.684 | 2.384 |
| 1.500 | 1.000 | 1.416 | 2.00 |
| 1.375 | 0.836 | 1.180 | 1.668 |
| 1.250 | 0.688 | 0.972 | 1.376 |
| 1.125 | 0.556 | 0.784 | 1.108 |
| 1.000 | 0.436 | 0.616 | 0.872 |

| 4.026-in. ID meter run (W in gpm) | | | |
|---|---|---|---|
| Orifice plate | 100 in. | 200 in. | 400 in. |
| 3.000 | 4.84 | 6.86 | 9.68 |
| 2.875 | 4.32 | 6.116 | 8.64 |
| 2.750 | 3.828 | 5.42 | 7.652 |
| 2.625 | 3.40 | 4.808 | 6.80 |
| 2.500 | 3.008 | 4.26 | 6.012 |

From Pryor (1966).

$h$ = differential pressure across the orifice in inches of water (from recorder)
$W_d$ = density of the dry saturated steam determined from steam tables, $lb_m/ft^3$

**Example 3-3.** Calculate the wet-steam quality of a 150 $m^3$/day feedwater steam generator working at 1000 psia saturation pressure given a 2.75-in.-diameter orifice plate in 4.026-in. ID meter run and 100 in. W.C. bellows meter recorder.

SOLUTION From recorder

$$\sqrt{h} = 3.8 \quad \text{or} \quad 28.9 \text{ in. of water}$$

From steam tables

$$W_d = \frac{1}{0.4456} \text{ lb}_m/\text{ft}^3 \qquad \text{the density of saturated vapor at 1000 psia}$$

From Table 3-2

$$C = 3.828$$

$$x^2 = \frac{3.828\sqrt{2.244} \times 3.8}{27.5} = 0.89 \text{ or } 89.0\%$$

where

$$150 \text{ m}^3/\text{d} \times \frac{1}{0.159} \frac{\text{bbl}}{\text{m}^3} \times 42 \text{ gal/bbl} \times \frac{1}{24 \times 60 \text{ min/day}} = 27.5 \text{ gpm}$$

## 3-3 WET-STEAM GENERATORS

In the field the steam needed for injecting through wells in the oil reservoir is produced by wet-steam generators. Field steam generators are oil- or gas-fired units designed for automatic operation, with forced and continuous water circulation. Air pollution regulations are very strict regarding the control of emissions from oil-fired generators, especially when the fuel used to fire the generator contains sulphur.

A wet-steam generator is aquatubular, having water-filled tubes with the flame and hot gases surrounding the tubes. The tubes can be coil shaped, and water is pumped through them at high velocity and turbulence, contrary to the flow direction of the hot gases as shown in Figure 3-5a. The water-filled tubes can also be straight, running back and forth along the length of the generator. In this case the unit has an economizer to preheat the water (Figure 3-5b). A comparison between these two types of generators is shown in Table 3-3.

Steam generators are furnished as mobile units on a skid, trailer mounted (truck mounted), or as permanent installations on a pad. The recommended practice for installation and operation of wet-steam generators was established by the American Petroleum Institute (API, 1983).

Wet-steam generators are usually rated in millions of Btus per hour of

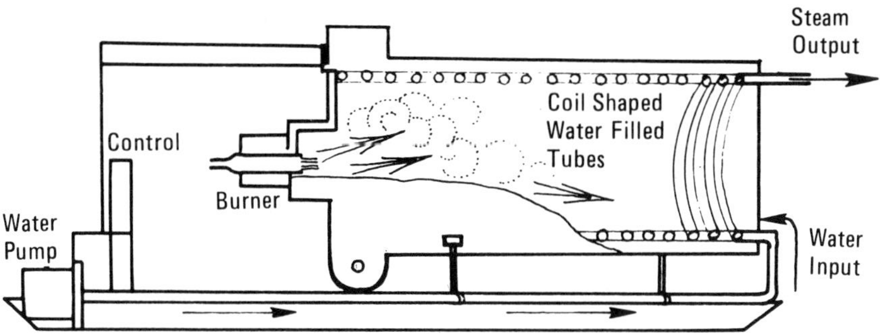


**Fig. 3-5a** Steam generator with coil-shaped water filled tubes

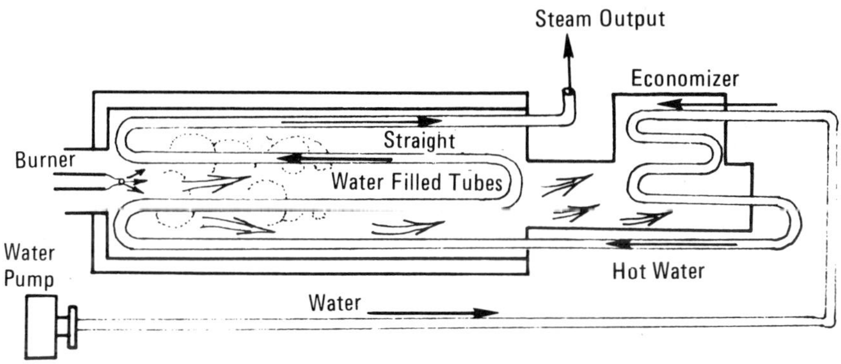


**Fig. 3-5b** Steam generator with straight water-filled tubes

heat absorbed. Those used in enhanced oil recovery range from 12- to 50-MM Btu/hr steam output. They can produce steam with a saturation pressure of up to 2000–2500 psia and a quality frequently between 80 and 85 percent. The steam saturation temperature corresponds to the respective saturation pressure.

**Example 3.4.** Find the capacity in tons of steam per hour and the saturation temperature of a 24-MM Btu/hr wet-steam generator operating at 1560 psia saturation pressure and producing steam with $f_s = 80\%$ quality.

SOLUTION Using the steam tables we find that the given condition of 1560 psia saturation pressure is not listed. The problem requires interpolation between 1500 and 1600 psia values. At a saturation pressure of 1560 psia (105 atm), the heat content of the wet steam

$$h_{gw} = h_f + f_s h_{fg} = 619.1 + 0.8\ (543.3)$$

$$= 1053.7 \text{ Btu/lb}_m$$

**TABLE 3-3.** Steam Generator Water-Filled Tubes Comparison

| Water-Filled Tubes | Coil | Straight |
|---|---|---|
| Advantages | Requires no preheated water. Reduced refractory surface and better portability. Coils easily cleaned by acid solution circulation and rinsing. | Strongly built and fewer chances of damages. Easier replacement of the straight tubes. |
| Disadvantages | Higher temperature per unit of heated area and more chances of damages. Difficult repair work for the first row (external) of coils. | More refractory material. Heavier. Elbow shock loads at both ends of the straight tubes. Hot gas carbon deposits over the economizer tubes. |

If a 1-lb mass of steam at 1560 psia saturation pressure has 1053.7 Btu, then 1000 kg or 1 ton of steam has

$$\frac{1000 \text{ kg} \times 1053.7 \text{ Btu/lb}_m}{0.4536 \text{ kg/lb}_m} = 2.323 \times 10^6 \text{ Btu}$$

and

$$24\text{MM Btu/hr} = \frac{24 \times 10^6 \quad \text{Btu/hr}}{2.323 \times 10^6 \text{ Btu/ton}} = 10.33 \text{ tons/hr of steam}$$

at 1560 psia saturation pressure and 601.43 °F saturation temperature. A rough but faster estimation can be made using the pressure enthalpy chart for steam (Figure 3-3) to read the water and steam enthalpy values at 1560 psia.

## 3-4 FEEDWATER TREATMENT

The water used in the field to feed a wet-steam generator should be of good quality to avoid scaling, tube corrosion, and suspended solids in the effluent. The American Petroleum Institute (API-RP, 1983) recommends that the following factors be considered in the treatment of feedwater:

| | |
|---|---|
| Total hardness | Less than one part per million (ppm). |
| Iron concentration | Less than 0.1 ppm. |
| Total dissolved solids (TDS) | "Levels of TDS become a cause of concern only when liquid phase concentrations approach solu- |

| | |
|---|---|
| | bility. Therefore a unit producing 80% quality steam should be able to tolerate feedwater dissolved solids in concentrations approaching 20 percent of their solubility limits." |
| Suspended solids | Below 5 ppm and preferably below 1 ppm. |
| Suspended oil content | Below 1 ppm. |
| Oxygen | Less than 0.01 ppm and preferably 0.0 ppm. |
| Alkalinity | Moderate alkalinity levels help reduce corrosion and maintain silica solubility. Bicarbonate alkalinity levels of over 2000 ppm should be avoided. |
| Silica | Control consists of maintaining solubility which is strongly affected by alkalinity. Alkalinity should be maintained at a level at least three times that of the silica content. |
| pH | From 7 to 12. |

Different types of steam generators are provided with water treatment units. To avoid high maintenance costs and low generator efficiency, the source and chemical composition of the water must be analyzed and the water treatment unit must be adjusted to the specific conditions determined by the analysis.

## 3-5 HEAT LOSSES

The steam generated by a wet-steam generator is the heat carrier agent injected into the reservoir. It raises the temperature of the rock and fluids it contains and displaces the oil. Not all the heat carried by the steam reaches the reservoir fluid and stays in the reservoir. Some of the heat is lost at the surface, some is lost into the wellbore, and some is lost to the adjacent formations. Heat can be transmitted away by conduction, convection, radiation, or combinations of all three means. Also, part of the heat reaching the reservoir is lost through produced fluids. Detailed information regarding the heat loss calculations and heat transmission were presented by Ramey (1962), Pacheco and Farouq Ali (1972), Prats (1982), and White and Moss (1983), among others.

The amount of formation heated depends on the amount of heat lost

- in the steam generator
- on the surface transmission lines
- from the wellbore
- to adjacent formations

### Steam Generator Heat Loss

The heat lost in the steam generator, $Q_g$, is given by a material balance between the heat released through the fuel-burning process and the heat gained by steam. The total heat, $Q$, liberated by the direct combustion of fuel is

$$Q = Hm \tag{3-7}$$

where $H$ is the heat of combustion or the heat evolved when a unit mass $m$ (or volume) of fuel is completely burned, Btu/$lb_m$ or J/kg. The total heat absorbed or the enthalpy, $h_{gw}$, of wet steam is given by Eq. 3-5 minus the enthalpy of the feedwater. Therefore, the steam generator heat loss is

$$Q_g = Q - (h_f + f_s h_{fg}) - f_{\text{feedw}} \tag{3-8}$$

**Example 3-5.** A steam generator produces steam of 85 percent quality at 1000 psia saturation pressure, consuming 911 $lb_m$/hr fuel oil with 19,800 Btu/$lb_m$ heat of combustion. The feedwater rate is 150 $m^3$/day at 60 °F. Find the heat loss and the efficiency of the generator.

SOLUTION Total heat produced is

$$Q = 19{,}800 \text{ Btu/lb}_m \times 911 \text{ lb}_m\text{/hr} = 18.04 \times 10^6 \text{ Btu/hr}$$

Wet-steam enthalpy at 1000 psia saturation pressure (from steam tables) is

$$h_{gw} = h_f + f_s h_{fg} = 542.4 + 0.85(649.4) = 1094.4 \text{ Btu/lb}_m$$

The change in enthalpy from water to wet steam is

$$1094.4 \text{ Btu/lb}_m - 28.06 \text{ Btu/lb}_m = 1066.34 \text{ Btu/lb}_m$$

where 28.06 Btu/$lb_m$ is $h_{\text{feedw}}$ or the enthalpy of feedwater at 60 °F saturation temperature (steam tables).

Total heat gained by steam is

$$150 \text{ m}^3\text{/day} \times 1000 \text{ kg/m}^3 \times 2.204 \text{ lb}_m\text{/kg} \times \frac{1 \text{ day}}{24 \text{ hr}} \times 1066.34 \text{ Btu/lb}_m = 14.689 \times 10^6 \text{ Btu/hr}$$

The heat lost is

$$Q_g = (18.04 - 14.689) \times 10^6 = 3.35 \times 10^6 \text{ Btu/hr or } 18.6\%$$

and is mostly due to flue gas emissions.

The generator efficiency is

$$E = 1 - \left(\frac{3.35 \times 10^6}{18.04 \times 10^6}\right) \times 100 = 81.4\%$$

### Heat Loss on the Surface Transmission Lines

The surface transmission lines conduct steam from the generator to the wellhead and into the wellbore. The heat lost, $Q_s$, by conduction and radiation on surface lines is

$$Q_s = AU_o \Delta T \tag{3-9}$$

where

$A$ = the surface area of steam pipelines, $ft^2$
$U_o$ = overall heat transfer coefficient, Btu/hr × $ft^2$ × °F (an average of various transfer coefficients of all the exposed surfaces that make up the steam transmission lines)
$\Delta T = T_{int} - T_{ext}$, temperature difference, °F

The heat losses are minimized to 3 to 5 percent if the surface steam pipelines are insulated or buried and are higher if lines are bare and/or the climate is cold.

### Heat Loss from the Wellbore

Wellbore heat loss is a factor seriously limiting the use of steam injection to shallow wells. As the wet steam flows through tubing down the wellbore to the reservoir, the tubing is heated to the steam temperature. The tubing loses heat with time by transferring it through the annulus to the casing and through the cement behind the casing to the ground. The problem of heat transmission in the wellbore is complex, and since it is also important, a number of authors have treated it in detail. Ramey (1962) considered the geothermal gradient and radiation conditions, and Willhite (1967) explained the overall heat transfer coefficient in steam injection. More recently Pacheco and Farouq Ali (1972) and Farouq Ali (1981) improved a mathematical model for wellbore steam injection under various flow regimes. Also, White and Moss (1983) published details of heat transfer in the wellbore and examples of the calculation procedures.

Figure 3-6 shows wellbore heat loss as a function of injection rate.

As we observe, increasing the injection rate causes the steam pressure to decline due to higher friction along the tubing string. Correspondingly, at a lower saturation pressure, there is a lower temperature and more hot liquid will vaporize. The steam quality increases and the heat loss as a percentage of total heat can be reduced.

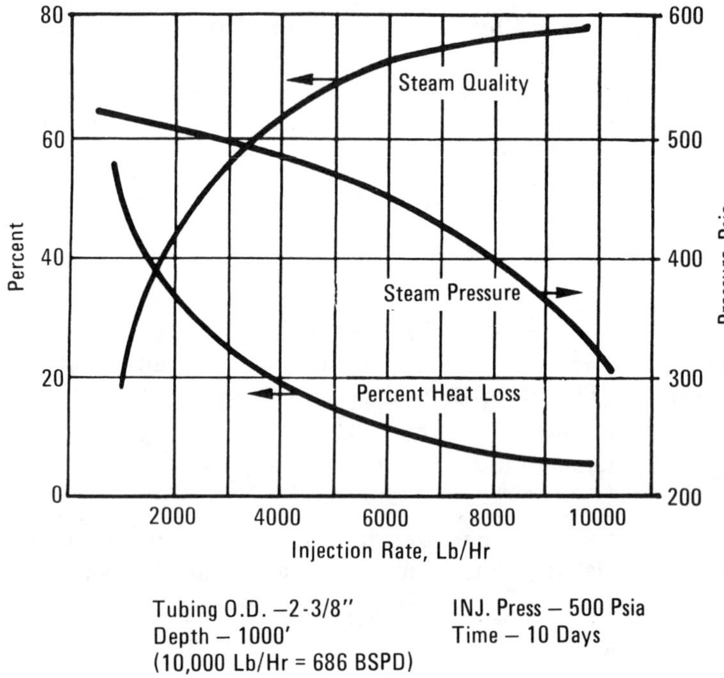


**Fig. 3-6** Wellbore heat loss as a function of injection rate (From Pacheco and Farouq Ali, 1972)

For a given steam injection rate, field methods to reduce heat loss from the wellbore (Farouq Ali and Meldau, 1983) include

- insulated tubing
- casing-tubing annulus vented to the atmosphere
- concentric tubing strings with insulating material between
- crude oil gel placed in the annulus
- high-pressure gas pack in the annulus

All these methods increase the resistance to heat flow from the wellbore. The effect of reducing the heat loss from the wellbore is illustrated in Figure 3-7. The heat lost in the wellbore ranges from 5 percent of total heat input, if it is well insulated, to 25 percent without insulation. The comparison is made for the working conditions listed in Figure 3-7.

The total heat loss from the wellbore when steam is injected down tubing

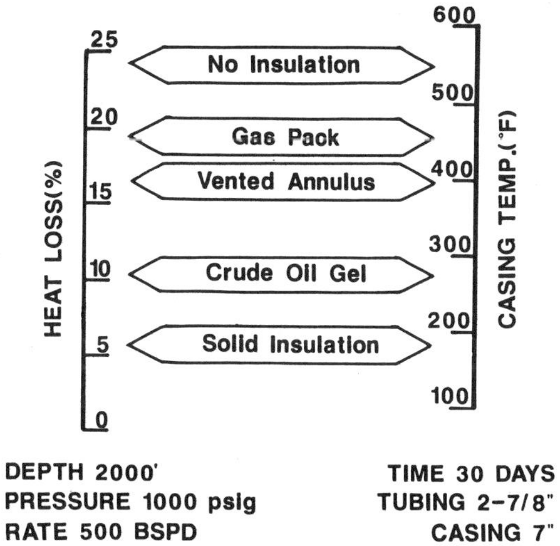


**Fig. 3-7** Insulated tubing heat loss comparison (From Farouq Ali and Meldau, 1983)

can be estimated in the field using Ramey's (1965) equation for the heat loss rate $Q_w$, Btu/day:

$$Q_w = \frac{2\pi r_1 Uk}{k + r_1 Uf(t_D)}\left[(T_o - b)Z - \frac{aZ^2}{2}\right] \tag{3-10}$$

where

$r_1$ = inside tubing radius, ft
$U$ = overall heat transfer coefficient between inside of tubing and outside of casing, Btu/day × ft$^2$ × °F, and where area is based on $r_1$
$k$ = thermal conductivity of the earth, Btu/day × ft × °F
$f(t_D)$ = dimensionless transient heat conduction time function
$T_o$ = saturation temperature of steam at prevailing pressure, °F
$b$ = surface geothermal temperature, °F
$Z$ = formation depth, ft
$a$ = geothermal gradient, °F/ft

Under the particular conditions listed in Table 3-4, Ramey obtained the heat loss rate values after 100 days' steam injection time.

For different casing sizes and injection times, and assuming that the thermal diffusivity of earth is a constant = 0.96 ft$^2$/day, the dimensionless transient heat conduction time function $f(t_D)$ values are given in Table 3-5.

**TABLE 3-4.** Estimated Wellbore Heat Loss Rate for Steam Injection at 100 Days' Injection Time

Conditions:
- Geothermal gradient, $a$ = 0.02 °F/ft
- Geothermal surface temperature = 70 °F
- Overall heat-transfer coefficient, $U$ = 30 Btu/day × ft² × °F
- Tubing size = 2 in.; $r_1 = \frac{1}{12}$
- Casing size = 7 in.
- Thermal conductivity of earth, $k$ = 33.6 Btu/day × ft × °F
- Thermal diffusivity of earth = 0.96 ft²/day

| Formation depth, ft | Heat Loss Rate, MM Btu/day for steam temperature at 300 °F | 400 °F | 500 °F | 600 °F |
|---|---|---|---|---|
| 500 | 1.36 | 1.97 | 2.57 | 3.18 |
| 1000 | 2.67 | 3.88 | 5.10 | 6.30 |
| 1500 | 3.91 | 5.72 | 7.55 | 9.36 |
| 2000 | 5.08 | 7.50 | 9.92 | 12.32 |
| 2500 | 6.21 | 9.25 | 12.26 | 15.30 |
| 3000 | 7.28 | 10.90 | 14.54 | 18.19 |

Source: Ramey (1965).

**Example 3-6.** Calculate the percentage of heat loss in a well-insulated wellbore when the wet steam produced by a generator (Example 3-5) reaches the wellhead and is injected through 3-in. tubing to a 2000-ft depth. The wellbore conditions are those outlined by Ramey in Table 3-5, and the injection time is 100 days.

SOLUTION The wet-steam temperature is 544.61 °F when saturation pressure is 1000 psia (steam tables). The dimensionless function $f(t_D)$ for 7-in. casing size and 100 days' injection time is 3.98 (Table 3-5).

**TABLE 3-5.** Values of $f(t_D)$ for Different Casing Sizes and Injection Time

| Casing Size, in. | Days: 5 | 25 | 50 | 75 | 100 |
|---|---|---|---|---|---|
| $4\frac{1}{2}$ | 2.96 | 3.81 | 4.08 | 4.37 | 4.48 |
| $5\frac{1}{2}$ | 2.89 | 3.56 | 3.99 | 4.08 | 4.27 |
| 7 | 2.64 | 3.32 | 3.64 | 3.90 | 3.98 |
| $8\frac{5}{8}$ | 2.46 | 3.10 | 3.42 | 3.64 | 3.81 |

From Bleakley (1965).

The heat loss rate is

$$Q_w = \frac{2\pi \frac{1.0\text{ in.}}{12\text{ in./ft}}(30\text{ Btu/day} \times \text{ft}^2 \times {}^\circ\text{F})(33.6\text{ Btu/day} \times \text{ft} \times {}^\circ\text{F})}{(33.6\text{ Btu/day} \times \text{ft} \times {}^\circ\text{F}) + \frac{1.0}{12}\text{ft}(30\text{ Btu/day} \times \text{ft}^2 \times {}^\circ\text{F})3.98}$$

$$\times \left[(544.61 - 70)\,{}^\circ\text{F} \times 2000\text{ ft} - \frac{0.02\,{}^\circ\text{F/ft}}{2}(2000)^2\text{ ft}^2\right]$$

$$= \frac{527.78}{43.550}(949{,}220 - 40{,}000) = 11.02 \times 10^6\text{ Btu/day}$$

or 459,116 Btu/hr. This represents 3.12 percent heat lost from the total heat gained by the steam ($14.689 \times 10^6$ Btu/hr, Example 3-5). The heat loss can increase four to five times, and mechanical problems may occur if the wellbore completion is not provided with insulation.

### Downhole Steam Generator

The failure conditions to which tubular goods are subjected in steam injection wells and excessive heat losses can be avoided by generating the steam downhole. Mechanical problems and heat losses were the main incentives for developing a downhole steam generator. Schrimer and Eson (1985), among others, investigated the concept of using a direct-fired downhole steam generator (DFDSG) in steam injection operations.

In this type of generator, fuel and air are injected separately into the wellbore reaching a downhole combustion chamber placed in front of the productive formation. After the fuel is ignited with an electrical torch igniter, water injected into the combustion chamber comes in contact with the burner flame and vaporizes into steam (Figure 3-8).

Use of both fuel and air transported continuously downhole to the steam generator creates safety problems not experienced in conventional surface steam generators. The hazards that should be avoided are the leakage of fuel (gas) or air and the mixture conditions favorable to explosion in the wellbore.

The main advantages over conventional steaming methods are described as

- reduction in heat losses
- reduction in air pollution
- deeper steaming potential
- offshore potential (smaller-size facilities and use of sea water)
- reservoir pressurization (higher pressure for steam drive around the well)

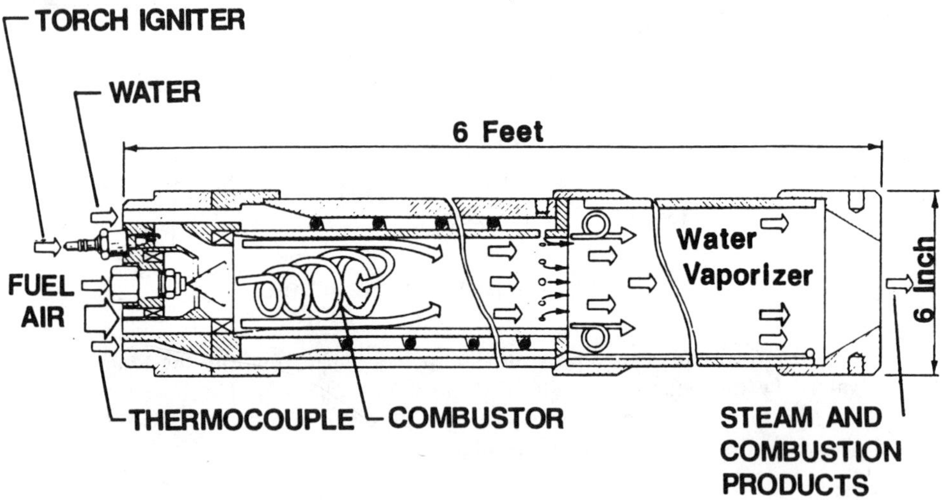


**Fig. 3-8** Schematic of flow patterns for DFDSG (From Schrimer and Eson, 1985)

### Reservoir Heat Loss to Adjacent Formations

Before the steam reaches the heat front, heat is also lost to adjacent formations. When steam is injected continuously (steam drive) for a long period of time, the heated area of the reservoir rock is large. Therefore, the cumulative heat loss to adjacent formations is also large. The reservoir volume heated by a sustained steam injection project is determined by the heat lost to the base and to the cap rock and by the amount of steam required to heat the formation.

Details regarding the heat lost to adjacent strata are given in Chapter 4, where oil displacement by a steam mechanism and the evaluation of a heat injection project are presented.

## 3-6 THE HEAT EFFECT ON RESERVOIR OIL VISCOSITY

The mobility of a fluid in the porous space of an oil reservoir is expressed as the ratio of the effective permeability to fluid viscosity. By reducing the viscosity a higher mobility is developed and the fluid flows much more easily and/or is displaced through the reservoir. Oil production is also inversely proportional to viscosity. Raising the reservoir temperature has the effect of decreasing the crude oil's viscosity, sometimes dramatically, and large increases in the production rate can be predicted. A useful correlation of oil viscosity as a function of temperature and API gravity has been made by Farouq Ali and Meldau (1983) using viscosity-temperature data of 60 heavy crude oil samples. (See Figure 3-9.)

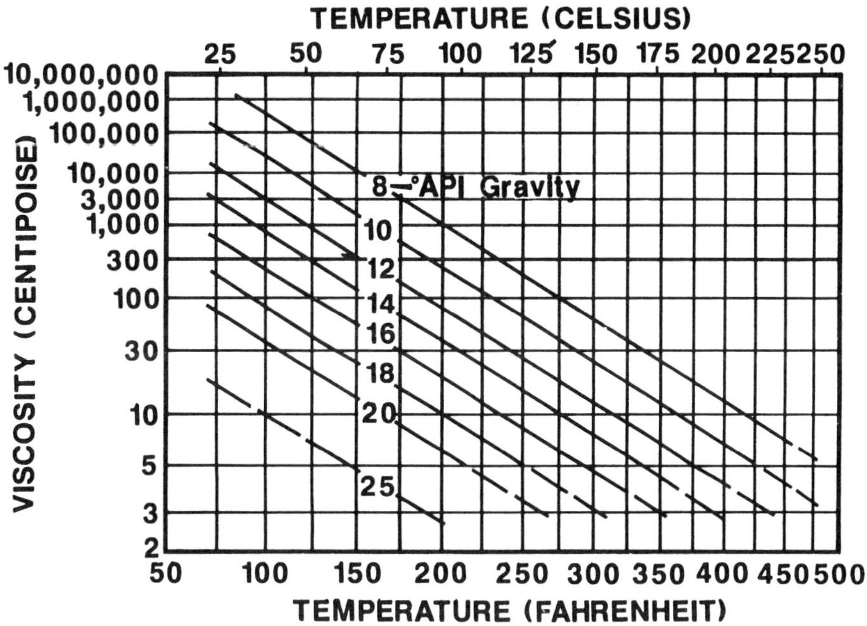


**Fig. 3-9** Oil viscosity as a function of temperature and gravity (From Farouq Ali and Meldau, 1983)

At the same temperature the viscosity of the crude increases with its density on a doubled logarithmic scale. As a result, for the same increase in temperature the reduction of the crude oil viscosity is more evident for heavier crudes. Data from some typical heavy oil reservoirs in the US and Canada are given in Figure 3-10 (Buckles, 1979).

For example, raising the reservoir fluid temperature to 300 °F (147 °C) will decrease the viscosity of the Cold Lake "crude" (10 °API) from 100,000 cp to approximately 10 cp, or 10,000 times. The same increase in temperature reduces the viscosity of the Kern River "A" oil reservoir (14 °API) only 500 times, from 2000 cp to 4 cp.

Two important observations can be made regarding the effect of heat on reservoir oil viscosity. First, to reduce the oil viscosity in a large area of a reservoir, there needs to be a flow of hot fluids within the reservoir. In other words, an effective permeability of the rock needs to exist for the heat carriers (steam and condensed water). For reservoirs that contain low-gravity, high-viscosity oil (≈10 °API or less), there is a lack of mobility of the crude bitumen. To get it moving, steam has to be injected above fracture pressure. Therefore the upper viscosity range of steam applications can be extended by inducing horizontal parting at the base of the formation, by electrical preheat, or by horizontal wells.

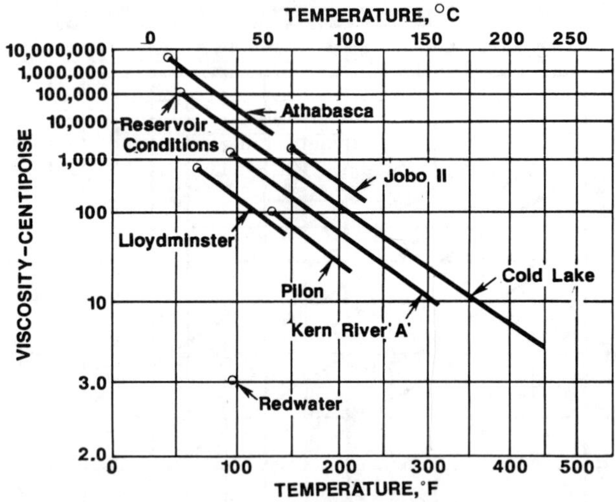


**Fig. 3-10** Typica heavy oil viscosity-temperature relationships (From Buckles, 1979)

The second observation is that reduction of the crude oil's viscosity is the main but not the only recovery mechanism developed in the reservoir by steam injection. There are several other displacement mechanisms that occur such as steam vapor drive, distillation of the lighter fractions, and thermal expansion of the oil. Therefore steam injection may also be applied to reservoirs with medium or light crude oils to which these effects are prevailing.

## QUESTIONS AND EXERCISES

**3-1** Explain the water-steam (temperature-volume) phase diagram.

**3-2** What is the main characteristic of the two-phase region on a steam pressure-enthalpy diagram?

**3-3** Is it possible to increase the temperature of steam coming from a generator and still operate at about the same steam quality? If so, how is it possible?

**3-4** A mass of 5000 $lb_m$ of water has to be vaporized at 212 °F saturation temperature. Find

**(a)** the sensible heat content (the enthalpy of saturated liquid)
**(b)** the total heat content after vaporization
**(c)** the change in enthalpy
**(d)** the saturation pressure

**3-5** How much heat (in joules and Btu) is produced when 1000 $lb_m$ of steam are condensed at 150°C?

**3-6** What is the amount of heat per hour required to generate steam of 82 percent quality at 860 psia if the 70 $m^3$/day feedwater temperature is 45 °F?

**3-7** Find the wet-steam quality of a 75 $m^3$/day feedwater steam generator working at 420 psia saturation pressure, given 2.5-in. orifice plate in 3.438-in. meter run and 400-in. W.C. bellows meter recorder; the meter reading is 1.6 (5.12 in. of water).

**3-8** A 50-MM Btu/hr steam generator operates at 400 psia saturation pressure and produces steam with 82 percent quality. Find the capacity of stream in tons per hour and the saturation temperature of the steam.

**3-9** Find the efficiency of a steam generator and the heat loss when the working conditions are

| | |
|---|---|
| Waterfeed rate | 140 $m^3$/day |
| Saturation temperature | 445 °F |
| Waterfeed temperature | 70 °F |
| Wet-steam quality | 84% |
| Fuel heat of combustion | 19,800 Btu/$lb_m$ |
| Fuel consumed | 890 $lb_m$/hr |

**3-10** Calculate, after 100 days' injection time, the amount of heat transported by wet steam of 83 percent quality injected through $2\frac{1}{2}$-in. tubing and reaching the production formation at a depth of 1400 feet. Assume a wet-steam saturation temperature of 430 °F and the wellbore conditions given in Table 3-5. The water-feed rate is 550 bbl/day at 60 °F. Assume the heat loss from surface transmission lines is 5 percent of the heat gained by steam in the generator.

## REFERENCES

API, *Recommended Practice for Installation and Operation of Wet Steam Generators* (Dallas, TX: American Petroleum Institute, Production Department, March 1983).

API, *Recommended Practice—Wet Steam Generator Feed Water Quality*, Appendix A, 26 (Dallas, TX: American Petroleum Institute, 1983).

BLEAKLEY, W. B., "The How and Why of Steam," Parts 1 and 2, *Oil and Gas Journal* (January 11 and 18, 1965), pp. 75, 102. (a)

BLEAKLEY, W. B., "Finding Enthalpy and Specific Volume from Charts," Part 5, *Oil and Gas Journal* (February 15, 1965), p. 121. (b)

BLEAKLEY, W. B., "How to Find Steam Quality," Part 7, *Oil and Gas Journal* (March 1 and 8, 1965), pp. 76–77. (c)

BUCKLES, R. S., "Steam Stimulation Heavy Oil Recovery at Cold Lake Alberta," SPE 7994 presented at the 49th California Regional Meeting, Ventura, California, April 18–20, 1979.

FAROUQ ALI, S. M., *Oil Recovery by Steam Injection* (Bradford, PA: Producers, 1970).

FAROUQ ALI, S.M., "A Comprehensive Well-bore Steam-Water Flow Model for Steam Injection and Geothermal Applications," *Society of Petroleum Engineering Journal* (October 1981).

FAROUQ ALI, S.M., and R. F. MELDAU, "Steam Flooding," Chapter VII in *Improved Oil Recovery* (Oklahoma City, OK: Interstate Oil Compact Commission, March 1983), pp. 314, 324, 339.

KEENAN, J., and F. KEYES, *Thermodynamic Properties of Steam* (New York: John Wiley & Sons, 1967), pp. 32, 38.

PACHECO, E. F., and S. M. FAROUQ ALI, "Well-Bore Heat Losses and Pressure Drop in Steam Injection," *Journal of Petroleum Technology* (February 1972).

PRATS, M., *Thermal Recovery*, SPE Monograph Series (New York, Dallas: American Institute of Mining, Metallurgical and Petroleum Engineers, 1982).

PRYOR, J. A., "Orifice Meter Measures Steam Quality," *Oil and Gas Journal* (May 30, 1966), p. 86.

RAMEY, H. J., JR., "Well-bore Heat Transmission," *Petroleum Transactions* (AIME, 1962), pp. 225, 427.

RAMEY, H. J., JR., "How to Calculate Heat Transmission in Hot Fluid Injection," in *Fundamentals of Thermal Oil Recovery* (Dallas, TX: Petroleum Engineer, 1965), p. 165.

SCHRIMER, R. M., and R. L. ESON, "A Direct-Fired Downhole Steam Generator—From Design to Field Test," *Transactions of the SPE*, Vol. 279 (1985).

WILLHITE, G. P., "Overall Heat Transfer Coefficients in Steam Injection and Hot Water Wells," *Journal of Petroleum Technology* (May 1967), pp. 607–615.

WHITE, P. D., and J. T. MOSS, *Thermal Recovery Methods* (Tulsa, Oklahoma: PennWell, 1983), Chapter 6.

# Chapter 4

# Steam Injection

## 4-1 GENERAL

Steam injection is the thermal process which supplies the heat needed to increase reservoir temperature and the energy to displace oil. Steam is the heat carrier agent provided by the steam generator. Continuous steam injection is called steam drive. Steam drive uses a pattern flood with injector and producers. In a single well operation, injecting steam and then producing oil from the same well, steam injection is called cyclic steam injection, steam soak, or "huff-and-puff."

The main purpose of steam drive is to increase the ultimate recovery factor. Cyclic steam injection's principal effect is to stimulate the formation to produce at a higher rate. Only when specific reservoir characteristics are present (shallow oil reservoirs with high dip and high lateral permeability) does cyclic steam injection stimulate gravity and drainage effect and increase the recovery factor through interwell communication.

The first report of a continuous steam injection field experiment was published in 1934 by Stovall. The Woodson-Texas field test was performed in an inverted nine-spot pilot by injecting steam at 200 psia pressure for 235 days in a formation 18 ft thick at a depth of 380 ft. The production rate of the wells around the injection well increased eight times.

Cyclic steam injection was implemented in the United States in 1933 and 1936. After a gap of 25 years this method resumed and developed, especially in California. The partially depleted California oil reservoirs (with depths of less than 3000 ft, a thickness between 40 and 300 ft, and still high oil saturations) have been appropriate candidates for steam injection. The number of cyclic steam injection operations increased from 4 in 1962 to 111 in 1964 and to more than 420 in 1967. This "boom" in the use of cyclic steam injection can be credited to the method's immediate effect of increasing wells' oil production rates and to the availability of high-performance steam generators. Efficiency was increased by an average of 200 bbl/day additional oil produced by each steam generator in operation.

Commercial application of steam processes has been underway since the early 1960s, and the wealth of field experience with respect to their use makes them the most advanced of all EOR methods.

## 4-2 PROCESSES DESCRIPTION AND RECOVERY MECHANISMS

The steam injected into the reservoir transmits heat to the formation and the fluids it contains. The heat lowers the oil's viscosity and increases its mobility. As steam cools, it condenses into water. In a steam drive process, as steam is continually injected into the injection wells, the condensed water is pushed further through the reservoir to the producing wells. In a cyclic steam injection process, the condensed water is produced along with the oil stream by the same producing well.

### Steam Drive Process

A cross section of a steam injection well and a producer in a reservoir formation with uniform permeability is shown in Figure 4-1.

As we observe, because of density differences, the steam segregates gravitationally and overrides. This tendency favors the early breakthrough of steam into the producers. As a consequence, only the upper one-third of the reservoir is swept by steam. To reduce the override tendency around the well, the perforated interval should be placed at the bottom of the formation. Before breakthrough, during the steam drive process, different zones are formed which advance through the reservoir:

*Zone 1* is the condensing zone with hot water essentially at steam temperature.

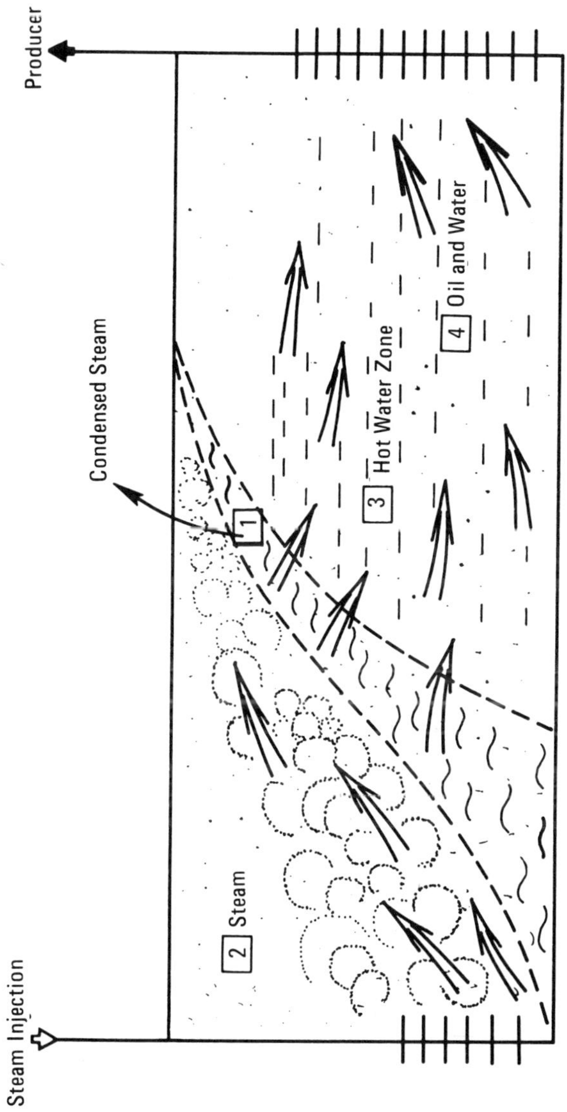


**Fig. 4-1** Steam drive process scheme and zones formed in the reservoir

*Zone 2* is the steam-saturated zone in which the oil saturation is reduced to less than 0.15, depending on the oil viscosity at reservoir temperature and on the steam temperature in the formation.

*Zone 3* is the hot water transition zone with a decreasing temperature from hot water to water near reservoir temperature.

*Zone 4* is the oil bank pushed and/or drawn to the producers by the water bank.

Beyond the effect of the oil displacement by the water bank at reservoir temperature (zone 4), the steam drive process develops other mechanisms which contribute to increased oil recovery.

In zone 3 (the transition zone), thermal expansion of the oil takes place and the heated oil increases in volume so that the residual oil saturation is reduced. In the steam-saturated zone (zone 2), oil is displaced by gas (steam) drive, and a steam distillation process starts to develop. If the reservoir oil contains light components in the condensing zone (zone 1) and beyond, in the hot water zone, the light oil vapor condenses and enriches the existing oil. The effect is that of a solvent slug reducing the residual oil saturation. Overall, the viscosity reduction improves the efficiency of the oil displacement by increasing its mobility.

### Cyclic Steam Injection Process

Cyclic steam injection is a single well process and involves the injection of steam for 2 to 6 weeks into a producing well. After a short soak period of 3 to 6 days, the well produces at a higher rate for several months to a year (Figure 4-2).

A typical reservoir response to three cycle steaming is well illustrated in Figure 4-3 by Farouq Ali and Meldau (1983).

As we observe, production response to first-cycle steaming is 8 to 10 times higher than the cold production rate, and cycle duration is also longer. The second- and third-cycle durations are shorter and have less effect on the production rate, which still is higher than the cold production rate. The reason first-cycle steaming results in better reservoir response is explained by the two main effects of steam stimulation: reduction of viscosity and wellbore cleanup (asphaltic and/or parafinic deposits are accumulated around the wellbore reducing the formation permeability). A third effect is the pressure gradient caused by steam injection, which is the same for the next successive cycles while viscosity reduction and cleanup effects diminish.

Steam stimulation is a very useful method for increasing oil production, especially when the reservoir rock is discontinuous. The method is also used in other recovery processes to stimulate the producers and to clean up the formation around the wellbore. Wells treated in this manner are better pre-

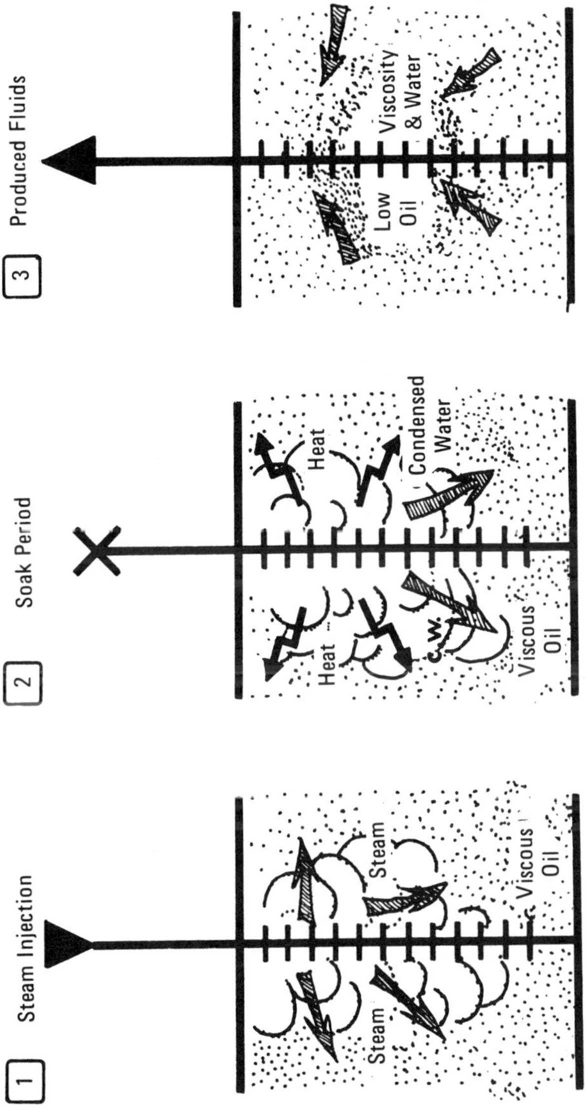


**Fig. 4-2** Cyclic steam injection process scheme

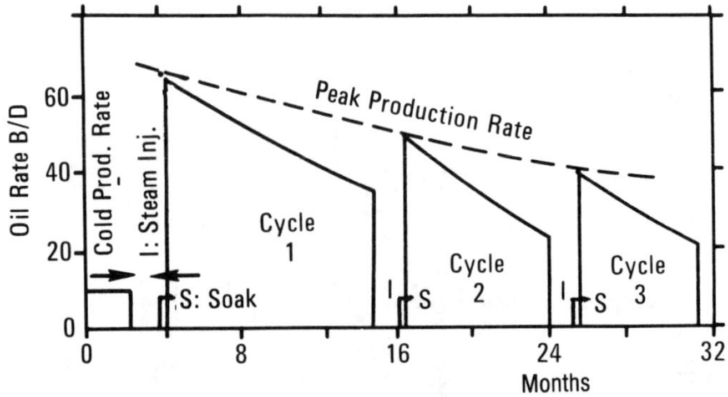


**Fig. 4-3** Typical cyclic steaming performance (From Farouq Ali and Meldau, 1983)

pared for picking up the fluid bank displaced by a burning front in a process of in situ combustion, for instance. When the productive formation is continuous, some operators convert steam stimulation to steam drive, after a couple of cycles, to increase the oil recovery. Sometimes, when shallow oil reservoirs have a thick productive formation or high dip and good lateral permeability, the cyclic steam injection operation results in a higher oil recovery. In these cases the steam cycles repeated at the beginning, in closely spaced wells, transmit the heat through the oil beyond the normal drainage area of the well (Figure 4-4). The new drainage areas of two neighboring steam-stimulated producers overlap.

Interwell communication determines the oil to flow under the influence of the gravitational effect (Seba and Perry, 1969). This very efficient mechanism results in high recoveries.

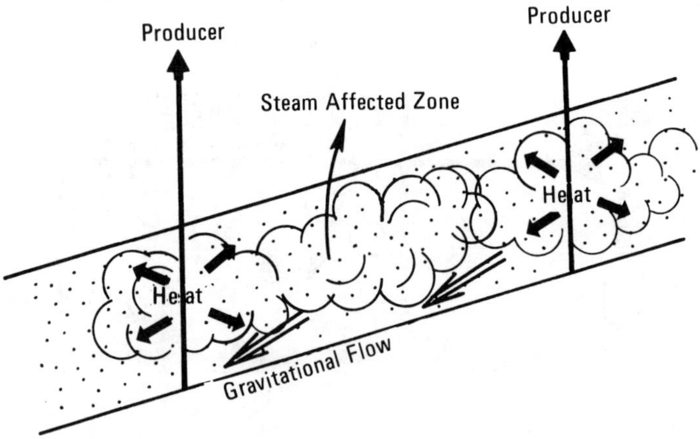


**Fig. 4-4** Cyclic steam stimulated producers with drainage area overlapped and gravitational effect in place

## 4-3 HEAT AMOUNT TO THE FORMATION

As pointed out in Chapter 3, the heat losses in a steam injection system begin at the steam generator and continue to the surface injection lines, to the wellbore, and to the adjacent strata. The remaining heat is given up to the formation itself, and some of it is transported with the fluids produced.

The beneficial effect of steam injection on oil displacement and recovery depends on the amount of heat transferred to the formation itself and on the porous volume swept by the hot fluids. To evaluate this effect the amount of heat loss to adjacent strata must be known. The first results of mathematical heat models of heat transfer and heat loss in the formation were published by Lawerier (1955), Marx and Langenheim (1959), Rubinstein (1959), and Willman et al. (1961). Gates and Ramey (1964) and Ramey (1965) have made a graphical comparison of these results by plotting on a Cartesian scale the fraction of total heat injected lost to adjacent strata, $W_o^*$ versus the logarithm of the dimensionless time function, $t_D$ (Figure 4-5). The dimensionless time is defined as

$$t_D = \frac{4D \times t}{h^2} \qquad \text{dimensionless} \tag{4-1}$$

where

$t$ = time, days
$D$ = thermal diffusivity of the cap rock, ft²/day
$h$ = formation thickness, ft

The amount of heat lost to adjacent strata varies directly with time of injection and inversely with formation thickness. So the heat lost can be a high percentage of the cumulative heat injected in a steam drive process's lifetime. To reduce the amount of heat lost to adjacent strata, a process of heat scavenging is accomplished by displacing the hot fluids "slug" with cold water injection. The process is also useful for saving energy when the formation is heated and the steam breaks through into the producers.

Gates and Ramey (1964) expressed the cumulative heat loss resulting from injection of hot fluid from time zero to time $t_o$, followed by injection of cold fluid from time $t_o$ to time $t$ as:

$$(W_o)_t = \frac{t}{t_o}\left[(W_o^*)_t - \left(\frac{t - t_o}{t}\right)(W_o^*)_{t-t_o}\right] \tag{4-2}$$

where

$(W_o)_t$ = fraction of total heat injected in "slug" up to time $t$ that is lost
$t$ = total time of injection (hot and cold fluid), days
$t_o$ = time heat was injected, days
$W_o^*$ = is obtained from Figure 4-5 at times $t$ and $t - t_o$.

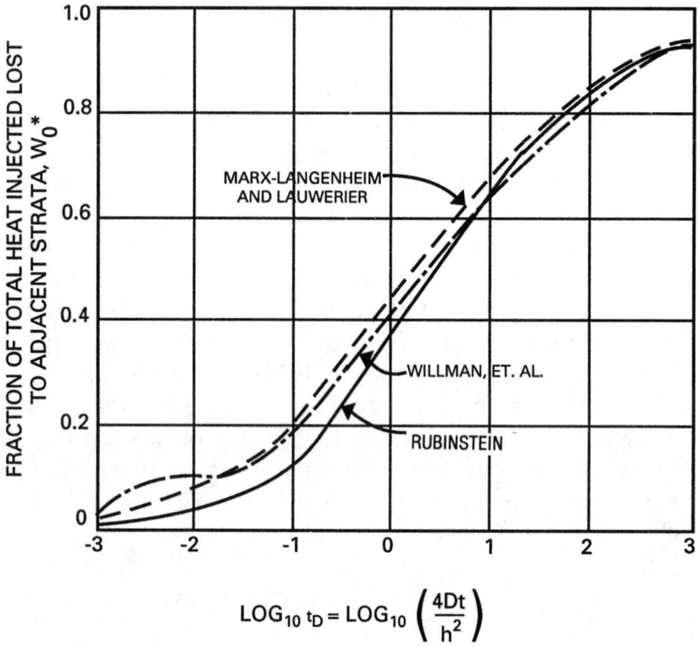


**Fig. 4-5** Vertical heat loss $W_o^*$ versus the logarithm of the dimensionless time function, $t_D$ (From Ramey, 1965)

**Example 4-1.** Heat at an amount of 14 MM Btu/hr is injected as wet steam into a formation 70 ft thick for a period of 400 days followed by cold water injection for another 500 days. Calculate the vertical heat loss to adjacent strata if the thermal diffusivity $D = 0.96$ ft²/day.

SOLUTION

$$t = 400 + 500 = 900 \text{ days}$$

$$t_o = 400 \text{ days}$$

At time $t$,

$$\log t = \log \frac{4Dt}{(h)^2}$$

$$\log \frac{(4)(0.96)(900)}{(70)^2} = -0.15$$

$(W_o^*)_t = 0.3$ from Figure 4-5 (Rubinstein curve)

At time $t - t_o$,

$$\log t = \frac{(4)(0.96)(500)}{(70)^2} = -0.4$$

$(W_o^*)_{t-t_o} = 0.22$ and the vertical heat loss is given by Eq. (4-2).

$$(W_o)_t = \frac{900}{400}\left[0.30 - \frac{(900-400)}{900}(0.22)\right] = 0.40$$

or 40 percent of the total heat injected into the formation is lost to the overburden. Before applying the percentage of losses to the total amount of heat injected, we must consider the water enthalpy at reservoir temperature (see Example 4-2).

## 4-4 HEATED RADIUS

The amount of heat lost to adjacent formations is subtracted from the total amount of heat injected as wet steam into the formation. The remaining heat raises the temperature of the rock and fluids content above the reservoir temperature.

The heated radius $r_h$, assuming radial and uniform propagation, can be calculated using Eq. 4-3:

$$r_h = \sqrt{\frac{Q_f}{\pi \times M_s(T_s - T) \times h}} \text{ ft} \tag{4-3}$$

where

$Q_f$ = net amount of heat available to formation, Btu
$M_s$ = heat capacity per cubic feet of steam saturated rock, Btu/ft$^3$ × °F (32 to 38 Btu/ft$^3$ × °F)
$T_s$ = steam temperature, °F
$T$ = reservoir temperature, °F
$h$ = formation thickness, ft

**Example 4-2.** Assuming the formation temperature in Example 4-1 is 100 °F, steam quality is 72 percent, and steam injection pressure is 760 psia at sand face, estimate the net heat gained by the formation, the heat lost to the overburden, and the heated radius.

SOLUTION From steam tables, at $p_i$ = 760 psia and 72% steam quality,

- the wet-steam enthalpy is 502.6 + 0.72(697.1) = 1004.5 Btu/lb$_m$
- the water enthalpy at 100 °F is 67.9 Btu/lb$_m$
- the net enthalpy of the wet steam injected into the formation is

$$1004.5 - 67.9 = 936.6 \text{ Btu/lb}_m$$

Steam injection rate is

$$q_s = \frac{14 \times 10^6 \text{ Btu/hr} \times 24 \text{ hr/day}}{350 \text{ lb}_m\text{/bbl} \times 1004.5 \text{ Btu/lb}_m} = 955.7 \text{ bbl/day}$$

available heat above reservoir temperature is

$$\frac{955.7 \times 350 \times 936.6}{24} = 13 \text{ MM Btu/hr}$$

Heat lost to the overburden is

$$Q_{ov} = 0.40 \times 13(10^6) = 5.2 \text{ MM Btu/hr}$$

Net heat to the formation is

$$Q_f = (13 - 5.2)10^6 \text{ Btu/hr}(24 \text{ hr/day})(400 \text{ days}) = 74.88 \times 10^9 \text{ Btu}$$

The rock has a reservoir temperature of 100 °F.
The steam temperature at 760 psia is 512.3 °F (from steam tables).
The heat capacity of the rock is given as 36 Btu/ft³ x °F.
The heated radius (Eq. 4-3) is

$$r_h = \sqrt{\frac{74.88 \times 10^9 \text{ Btu}}{\pi \times 36 \text{ Btu/ft}^3 \times {}^\circ\text{F}(512.3 - 100)\ {}^\circ\text{F} \times 70 \text{ ft}}} = 151.5 \text{ ft}$$

## 4-5 STEAM DRIVE DISPLACEMENT

Several analytical and empirical models have been developed and are available in the literature describing the displacement of oil by continuous steam injection (Myhill and Stegemeyer, 1978; Williams et al., 1980; Farouq Ali, 1981; Prats, 1982; Aydelotte and Pope, 1983; Peake, 1989). Most of the investigators improved and refined the frontal displacement model of Marx and Langenheim (1959), modified by Ramey (1965), to evaluate steam drive projects. Correlation charts for predicting steamflood oil recovery as a function of reservoir characteristics were also developed (Gomaa, 1980).

Besides the original works a detailed description of some models and calculation examples are given by van Poollen (1980), White and Moss (1983), and Lake (1989), among others.

### Oil Displacement Rate

The rate $V_o$ at which oil is displaced from the reservoir is given by the Marx–Lagenheim equation:

$$V_o = 4.273\left[\frac{H_o\,\phi(S_o - S_{or})}{M_s\,\Delta T}\right] \times (e^{x^2} \operatorname{erfc} x) \text{ bbl/day} \qquad (4\text{-}4)$$

where

$H_o$ = heat injection rate, Btu/hr
$\phi$ = formation's effective porosity
$S_o$ = oil saturation
$S_{or}$ = residual oil saturation
$M_s$ = heat capacity, Btu/ft$^3$ x °F
$\Delta T$ = temperature difference $(T_s - T_f)$, °F

The error function is the group of terms in the parentheses on the right-hand side of Eq. (4-4) and is given in Table 4-1 for different values of

$$x = \left(\frac{2K}{M_s h\sqrt{D}}\right)t^{0.5} \qquad \text{dimensionless}$$

where

$K$ = thermal conductivity of the cap rock, Btu/ft × hr × °F
$D$ = thermal diffusivity of the cap rock, ft$^2$/hr
$t$ = injection time, days
$h$ = formation thickness, ft

The thermal diffusivity of the cap rock (overburden) is defined as

$$D = \frac{K}{\rho_c C_c} \tag{4-6}$$

where

$\rho_c$ = cap rock density, lb$_m$/ft$^3$
$C_c$ = specific heat of the cap rock, Btu/lb$_m$ × °F

Marx and Langenheim also provide an equation for the cumulative heated area at time $t$:

$$A(t) = \left(\frac{H_o M_s hD}{4K^2 \Delta T}\right)\left(e^{x^2} \operatorname{erfc} x + \frac{x}{\sqrt{\pi}} - 1\right) \text{ft}^2 \tag{4-7}$$

and the cumulative oil displaced:

$$\frac{A(t) \times h}{43{,}560} \times 7758(\phi)(S_o - S_{or}) \text{ bbl} \tag{4-8}$$

The heated area will expand continuously with time, and, assuming an isotropic and homogeneous formation, the expansion will be radial. Since the heat loss to the overburden increases with time and since the injection rate is constant, the results obtained using Eqs. 4-5 and 4-7 are correct as long as the heat injection rate is greater than the heat consumption rate. When the

**TABLE 4-1.** Error Function Values

| $x$ | $e^{x^2} \operatorname{erfc} x$ | $e^{x^2} \operatorname{erfc} x + \frac{2x}{\sqrt{\pi}} - 1$ | $x$ | $e^{x^2} \operatorname{erfc} x$ | $e^{x^2} \operatorname{erfc} x + \frac{2x}{\sqrt{\pi}} - 1$ | $x$ | $e^{x^2} \operatorname{erfc} x$ | $e^{x^2} \operatorname{erfc} x + \frac{2x}{\sqrt{\pi}} - 1$ |
|---|---|---|---|---|---|---|---|---|
| 0.00 | 1.00000 | 0.00000 | 1.00 | 0.42758 | 0.55596 | 4.50 | 0.12248 | 4.20019 |
| .02 | .97783 | .00039 | .05 | .41430 | .59910 | .60 | .11994 | .31048 |
| .04 | .95642 | .00155 | .10 | .40173 | .64295 | .70 | .11749 | .42087 |
| .06 | .93574 | .00344 | .15 | .38983 | .68746 | .80 | .11514 | .53136 |
| .08 | .91576 | .00603 | .20 | .37854 | .73259 | .90 | .11288 | .64194 |
| 0.10 | 0.89646 | 0.00929 | 1.25 | 0.36782 | 0.77830 | 5.00 | 0.11070 | 4.75260 |
| .12 | .87779 | 0.1320 | .30 | .35764 | .82454 | .20 | .10659 | 4.97417 |
| .14 | .85974 | .01771 | .35 | .34796 | .87127 | .40 | .10277 | 5.19602 |
| .16 | .84228 | .02282 | .40 | .33874 | .91847 | .60 | .09921 | .41814 |
| .18 | .82538 | .02849 | .45 | .32996 | .096611 | .80 | .09589 | .64049 |
| 0.20 | 0.80902 | 0.03470 | 1.50 | 0.32159 | 1.01415 | 6.00 | 0.09278 | 5.86305 |
| .22 | .79318 | .04142 | .55 | .31359 | .06258 | .20 | .08986 | 6.08581 |
| .24 | .77784 | .04865 | .60 | .30595 | .11136 | .40 | .08712 | .30874 |
| .26 | .76297 | .05635 | .65 | .29865 | .16048 | .60 | .08453 | .53184 |
| .28 | .74857 | .06451 | .70 | .29166 | .20991 | .80 | .08210 | .75508 |
| 0.30 | 0.73460 | 0.07311 | 1.75 | 0.28497 | 1.25964 | 7.00 | 0.07980 | 6.97845 |
| .32 | .72106 | .08214 | .80 | .27856 | .30964 | .20 | .07762 | 7.20195 |
| .34 | .70792 | .09157 | .85 | .27241 | .35991 | .40 | .07556 | .42557 |
| .36 | .69517 | .10139 | .90 | .26651 | .41043 | .60 | .07361 | .64929 |
| .38 | .68280 | .11158 | .95 | .26084 | .46118 | .80 | .07175 | 7.87311 |
| 0.40 | 0.67079 | 0.12214 | 2.00 | 0.25540 | 1.51215 | 8.00 | 0.06999 | 8.09702 |
| .42 | .65912 | .13304 | .05 | .25016 | .56334 | .20 | .06830 | .32101 |
| .44 | .64779 | .14428 | .10 | .24512 | .61472 | .40 | .06670 | .54508 |

| | | | | | | | | |
|---|---|---|---|---|---|---|---|---|
| .46 | .63679 | .15584 | .15 | .24027 | .66628 | .60 | .06517 | .76923 |
| .48 | .62609 | .16771 | .20 | .23559 | .71803 | .80 | .06371 | 8.99344 |
| 0.50 | 0.61569 | 0.17988 | 2.25 | 0.23109 | 1.76994 | 9.00 | 0.06231 | 9.21772 |
| .52 | .60588 | .19234 | .30 | .22674 | .82201 | .20 | .06097 | .44206 |
| .54 | .59574 | .20507 | .35 | .22255 | .87424 | .40 | .05969 | .66645 |
| .56 | .58618 | .21807 | .40 | .21850 | .92661 | .60 | .05846 | 9.89090 |
| .58 | .57687 | .23133 | .45 | .21459 | 1.97912 | .80 | .05727 | 10.11539 |
| 0.60 | 0.56780 | 0.24483 | 2.50 | 0.21081 | 2.03175 | 10.00 | 0.05614 | 10.33993 |
| .62 | .55898 | .25858 | .60 | .20361 | .13740 | | | |
| .64 | .55039 | .27256 | .70 | .19687 | .24350 | | | |
| .66 | .54203 | .28676 | .80 | .19055 | .35001 | | | |
| .68 | .53387 | .30117 | .90 | .18460 | .45690 | | | |
| 0.70 | 0.52593 | 0.31580 | 3.00 | 0.17900 | 2.56414 | | | |
| .72 | .51819 | .33062 | .10 | .17372 | .67169 | | | |
| .74 | .51064 | .34564 | .20 | .16873 | .77954 | | | |
| .76 | .50328 | .36085 | .30 | .16401 | .88766 | | | |
| .78 | .49610 | .37624 | .40 | .15954 | 2.99602 | | | |
| 0.80 | 0.48910 | 0.39180 | 3.50 | 0.15529 | 3.10462 | | | |
| .82 | .48227 | .40754 | .60 | .15127 | .21343 | | | |
| .84 | .47560 | .42344 | .70 | .14743 | .32244 | | | |
| .86 | .46909 | .43950 | .80 | .14379 | .43163 | | | |
| .88 | .46274 | .45571 | .90 | .14031 | .54099 | | | |
| 0.90 | 0.45653 | 0.47207 | 4.00 | 0.13700 | 3.65052 | | | |
| .92 | .45047 | .48858 | .10 | .13383 | .76019 | | | |
| .94 | .44455 | .50523 | .20 | .13081 | .87000 | | | |
| .96 | .43876 | .52201 | .30 | .12791 | 3.97994 | | | |
| .98 | .43311 | .53892 | .40 | .12514 | 4.09001 | | | |

From Marx and Langenheim (1959).

so-called critical time period is reached, steam drive projects are evaluated using more complex models.

**Example 4-3.** An isotropic and homogeneous oil reservoir is subjected to a steamdrive process in which 1200 bbl/day wet steam of 80 percent quality is injected at 660 psia into 65 ft of productive formation. Calculate the oil displaced by steam drive and the cumulative heated area if the injection period is 5 years and the reservoir rock characteristics are as follows:

| | |
|---|---|
| Porosity | $\phi = 26\%$ |
| Oil saturation at start of the process | $S_o = 0.60$ |
| Residual oil saturation in steam zone | $S_{or} = 0.12$ |
| Reservoir temperature | $T = 130$ °F |
| Heat capacity | $M_s = 32$ Btu/ft$^3$ × °F |
| Thermal diffusivity | $D = 0.70$ ft$^2$/day |
| Thermal conductivity | $K = 0.95$ Btu/ft × hr × °F |

SOLUTION

The argument of the error function, $x$, is

$$x = \left(\frac{2 \times 0.95}{32 \times 65\sqrt{0.70/24}}\right) t^{0.5} = 0.0053 t^{0.5}$$

Heat injection rate, $H_o$ is

$$H_o = C q_s (h_f + f_s h_{fg} - h_{f\text{water}}) \text{ Btu/hr} \tag{4-9}$$

where

$q_s$ = steam injection rate, bbl/day
$f_s$ = steam quality, decimal fraction
$h_{fg}$ = enthalpy of vaporization (tables)
$h_f$ = enthalpy of saturated liquid (tables)
$h_{f\text{water}}$ = enthalpy of water at initial reservoir temperature (tables)

$$C = \frac{350 \text{ lb}_\text{m}/\text{bbl}}{24 \text{ hr/day}} = 14.583 \frac{\text{lb}_\text{m} \times \text{day}}{\text{bbl} \times \text{hr}}$$

$$H_o = 14.583 \frac{(\text{lb}_\text{m}) \text{ day}}{\text{bbl (hr)}} \times 1200 \text{ bbl/day} \times 960.54 \text{ Btu/lb}_\text{m} = 16.808 \text{ MM Btu/hr}$$

Rate of oil displaced, $V_o$ (bbl/day) is

$$V_o = 4.275 \frac{\text{hr bbl}(10 \times 10^6)(0.26)(0.60 - 0.12) \text{ Btu/hr}}{\text{day ft}^3 \text{ 35 Btu/ft}^3 \times \text{°F} (496.58 - 130) \text{ °F}} (e^{x^2} \operatorname{erfc} x)$$

$$V_o = 698.5(e^{x^2} \operatorname{erfc} x)$$

where

$$x = (0.0053)\sqrt{t}$$

and $t$ is introduced in hours.

**TABLE 4-2.** Oil Displaced by Steam Drive

| | | | | Oil Displaced | | |
|---|---|---|---|---|---|---|
| Steps | Time (days) | x | $e^{x^2}$ erfc $x$ * | Barrels per Day | Barrels per Step | Cumulative Oil per Barrel |
| 1 | 1 | 0.0259 | 0.9715 | 678.6 | 678.6 | 679 |
| 2 | 50 | 0.1836 | 0.8224 | 574.4 | 30,698 | 31,377 |
| 3 | 100 | 0.2596 | 0.7632 | 533.1 | 27,687 | 59,064 |
| 4 | 200 | 0.3672 | 0.6907 | 482.4 | 50,775 | 109,839 |
| 6 | 600 | 0.6360 | 0.5521 | 385.6 | 173.600 | 283,439 |
| 8 | 1000 | 0.8211 | 0.4819 | 336.6 | 144,440 | 427,879 |
| 10 | 1400 | 0.9715 | 0.4338 | 303.0 | 127,920 | 555,799 |
| 12 | 1825 | 1.1092 | 0.3995 | 279.0 | 123,675 | 679,474 |

* By interpolating the error function values (Table 4-1).

The oil displaced by steam drive is given in Table 4-2 at different time increments.

The cumulative heated area after 5 years of steam injection is given by Eq. 4-7.

$$A(t) = \frac{16.808 \times 10^6 \text{ Btu/hr}(32 \text{ Btu/ft}^3 \times {}^\circ\text{F})65 \text{ ft}(0.70/24) \text{ ft}^2\text{/hr}}{4(0.95 \text{ Btu/ft} \times \text{hr} \times {}^\circ\text{F})^2(496.58 - 130)\,{}^\circ\text{F}}$$

$$\left(e^{x^2} \text{erfc } x + \frac{2x}{\sqrt{\pi}} - 1\right)$$

$$= 770{,}531 \text{ ft}^2(0.65114) = 501{,}723 \text{ ft}^2$$

The cumulative oil displaced (Eq. 4-8) is

$$\frac{501{,}723 \text{ ft}^2 \times 65 \text{ ft}}{43{,}560 \text{ ft}^3\text{/acre}} \times 7758 \text{ bbl/acre}(0.26)(0.60 - 0.12) = 724{,}860 \text{ bbl}$$

The cumulative oil displaced corresponds to the entire thickness of the formation. In reality, as we saw, less oil is displaced since the steam segregates gravitationally and overrides. However, oil displaced does not mean oil produced. Areas of reservoir rock between producers or the lower half of the formation thickness can receive some of the displaced oil. In a real oil reservoir, nonuniformity of displacement and bypass regions of high oil saturation do occur.

### Oil Recovery, Oil-Steam Ratio

Correlation charts for predicting steamflood oil recovery and oil-steam ratio have been developed by Gomaa (1980) using reservoir simulation.

The first step in estimating the oil recovery using Gomaa's charts

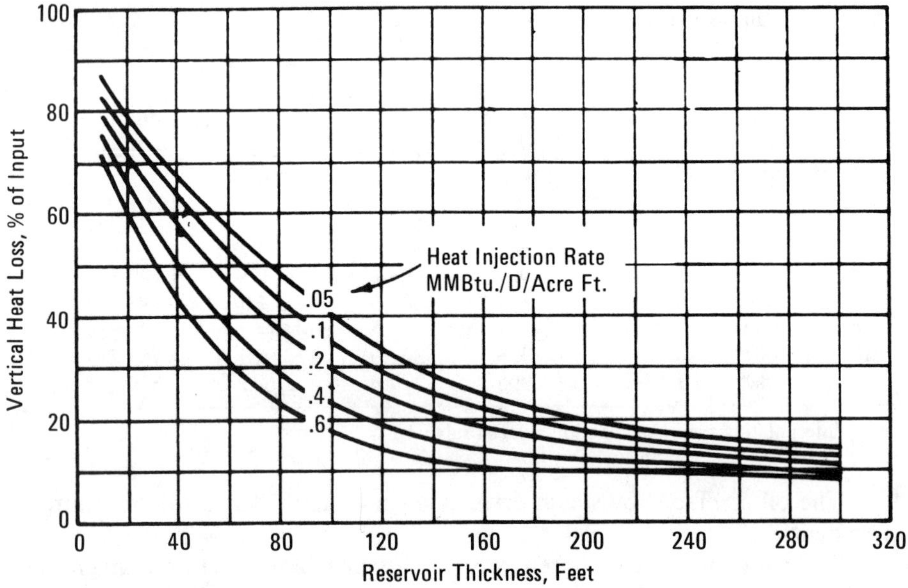


**Fig. 4-6** Heat loss to overlying and underlying strata (From Gomaa, 1980)

and equations is to read the vertical heat loss ($f_{hv}$) to adjacent strata from Figure 4-6.

For instance, for a gross reservoir thickness of 100 ft and heat injection rate of 0.05 MM Btu/day × acre × ft, the heat loss to vertical adjacent strata is 40 percent of input.

The next step is to read the heat utilization factor $Y$ from Figure 4-7.

The heat utilization factor is a function of steam quality and is defined by Gomaa as the ratio of the effective heat injected to the net heat injected for each steam quality value:

$$Y = \frac{Q_e}{Q_{inj}} \tag{4-10}$$

The effective heat injected ($Q_e$) is the fraction of the net heat injected ($Q_{inj}$) utilized effectively in the reservoir.

The net heat injected for each time increment $\Delta t$ is given by

$$Q_{inj} = 0.128 q_s h(1 - f_{hv})\Delta t \text{ MM Btu/day acre-ft} \tag{4-11}$$

where

$q_s$ = injection rate, bbl/day acre-ft
$h$ = wet-steam enthalpy, Btu/lb$_m$ (tables)
$f_{hv}$ = vertical heat loss, fraction
$\Delta t$ = time increments, years

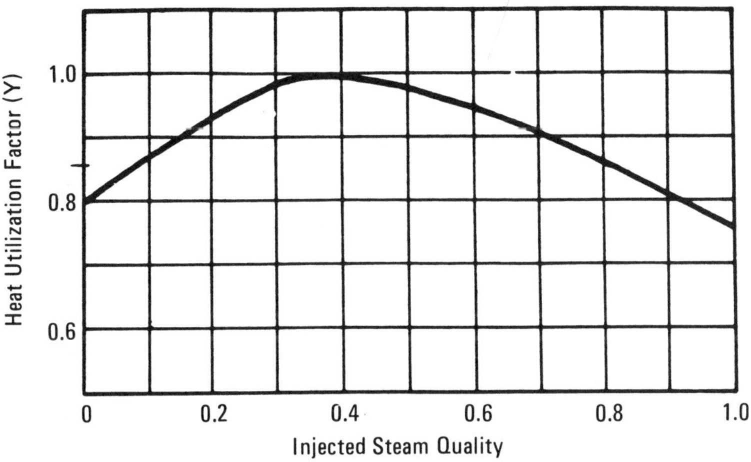


**Fig. 4-7** Heat utilization factor as a function of steam quality (From Gomaa, 1980)

Now, the effective heat injected can be calculated and used to read the oil recovery factor from Figure 4-8, as a function of different initial values of mobile oil saturation defined by Gomaa (1980) as

$$S_{om} = S_{oi} - S_{or} \tag{4-12}$$

where

$S_{oi}$ = initial oil saturation prior to steam drive operation
$S_{or}$ = residual oil saturation in the steam zone.

**Example 4.4.** Estimate the oil recovery factor after 4 years of wet-steam injection of 0.8 quality with a constant rate of 4900 bbl/day at 820 psia sand-face injection pressure. Other data are as follows:

| | |
|---|---|
| Reservoir productive area | 100 acres |
| Reservoir thickness | |
| Gross | 43 ft |
| Effective | 25 ft |
| Oil saturation at start of process | 0.58 |
| Residual oil saturation in steam zone | 0.08 |
| Reservoir temperature (initial) | 95 °F |

SOLUTION

The wet-steam enthalpy is given by Eq. 3-5 and by subtracting the water enthalpy at initial reservoir temperature: $h$ = 513.2 Btu/lb$_m$ + 0.8(684.8) Btu/lb$_m$ − 62.98 Btu/lb$_m$ = 998.06 Btu/lb$_m$, and the heat injected per day per acre-ft is

$$4900 \text{ bbl/day} \times 350 \text{ lb}_m\text{/bbl} \times \frac{998.06 \text{ Btu}}{10.0 \text{ acres} \times 43 \text{ ft} \times \text{lb}_m}$$

$$= 398{,}063 \qquad \text{or} \qquad 0.39806 \text{ MM Btu/day} \times \text{acre} \times \text{ft}$$

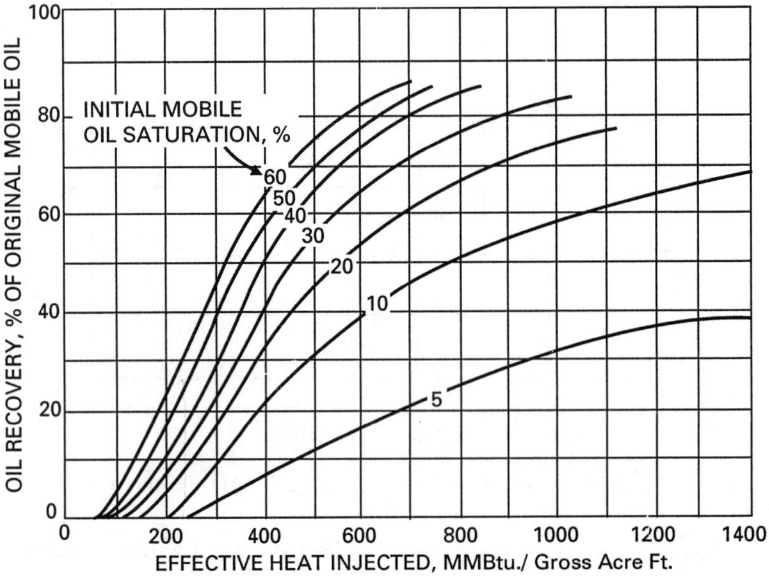


**Fig. 4-8** Steamflood oil recovery as a function of effective heat injected and mobile oil saturation (From Gomaa, 1980)

From Figure 4-6 corresponding to a gross thickness of 43 ft we read a vertical heat loss $f_{hv}$ approximately = 52%.

The net heat injected in 4 years is

$$Q_{\text{inj}} = 0.128 \frac{4900 \text{ bbl/day}}{100 \text{ acres} \times 43 \text{ ft}} (998.06 \text{ Btu/lb}_m)(1 - 0.52)\, 4$$

$$= 279.5 \text{ MM Btu/acre} \times \text{ft}$$

or

$$0.39806 \times 10^6 \text{ Btu/day} \times \text{acre} \times \text{ft} \times (1 - 0.52) \times 365 \text{ days/year}$$
$$\times\, 4 \text{ years} = 279 \text{ MM Btu/acre} \times \text{ft}$$

For a wet-steam quality of 0.8, we read from Figure 4-7 the heat utilization factor $Y = 0.86$.

The effective heat injected is

$$Q_e = 0.86(279) \text{ MM Btu/acre} \times \text{ft} = 240 \text{ MM Btu/acre} \times \text{ft}$$

The oil recovery after 4 years of steam drive is obtained from Figure 4-8 as 28 percent of the oil in place at the start of the process. Oil recovery will increase by continuing the steam injection or by starting to inject water. The final oil recovery is the total amount of oil produced from the time the oil reservoir was discovered, expressed as a percentage of the original oil in place (OOIP).

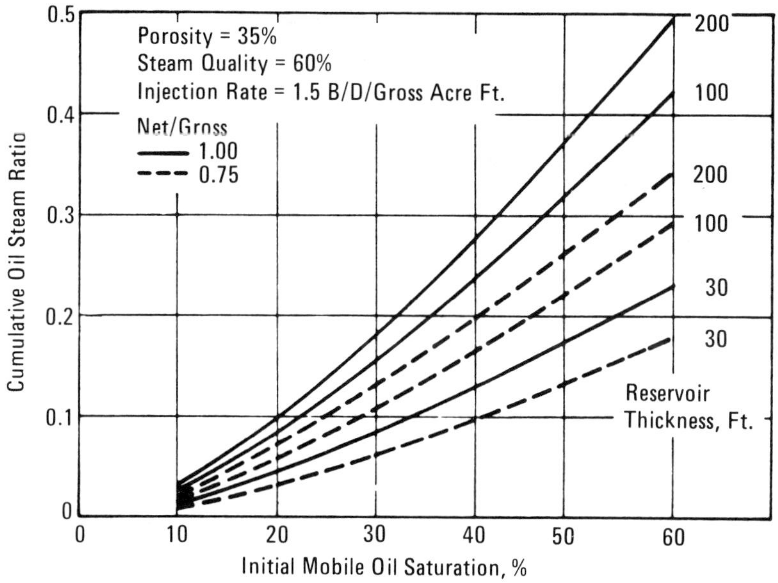


**Fig. 4-9** Effect of oil saturation, reservoir thickness, and net-gross ratio on cumulative oil-steam ratio (From Gomaa, 1980)

**Oil-steam ratio.** This is the ratio of stimulation, less primary oil production, to the cumulative steam injected, expressed as barrels of condensate.

The oil-steam ratio is influenced by reservoir thickness, oil saturation, and the net/gross ratio. Gomaa (1980) plotted the oil-steam ratio function of these parameters (Figure 4-9) for 35 percent porosity, 60 percent steam quality, and 1.5 bbl/day × acre × ft. The correlation is useful for reservoirs with similar conditions.

## 4-6 CYCLIC STEAM INJECTION

As pointed out earlier, cyclic steam injection is a method of stimulating well production to obtain higher oil rates, primarily from the first 3 to 4 steam cycles. Only when the productive formation is thick, and/or the reservoir is dipped with good permeability along the strata, is the producing mechanism that develops due to gravity. Cyclic steam injection in these conditions also increases the oil recovery.

A very simple solution for estimating the reservoir response to cyclic steam injection, taking into consideration only its effect on viscosity, was given by Smith (1985). Assuming the radial system of flow as illustrated in Figure 4-10, the heated reservoir extends a distance $r_h$ from the wellbore. The effect

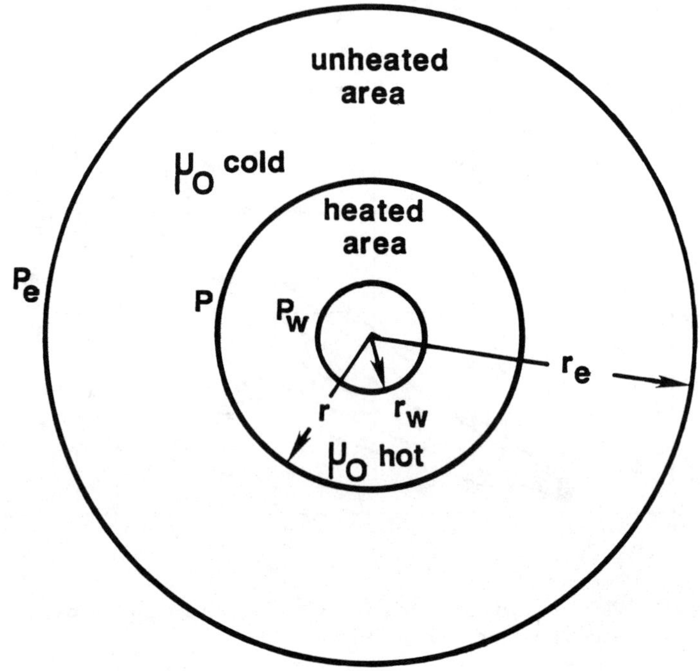


**Fig. 4-10** Idealized sketch of heated area around a cycling steam injection well

of the heated zone on well productivity can be understood by picturing a system of two concentric hollow cylinders of radii $r$ and $r_e$, with a pressure drop given by

$$p_e - p_w = (p_e - p) + (p - p_w) \tag{4-13}$$

where

$p_e - p_w$ = pressure drop before the first steam injection cycle (reservoir cold)

$(p_e - p) + (p - p_w)$ = pressure drop after the steam cycle injection

Using Darcy's law to express the oil rate before ($q_{o\text{cold}}$), and after ($q_{o\text{hot}}$), the ratio $q_{o\text{hot}}/q_{o\text{cold}}$ is

$$\frac{q_{o\text{hot}}}{q_{o\text{cold}}} = \frac{\mu_{o\text{cold}} \ln(r_e/r_w)}{\mu_{o\text{hot}} \ln(r_h/r_w) + \mu_{o\text{cold}} \ln(r_e/r_h)} \tag{4-14}$$

where $\mu_o$ is the oil viscosity value in both heated and unheated areas. The heated radius $r_h$ can be calculated using Eq. 4-3.

**Example 4-5.** Calculate the productivity increase of a well which produces oil from Kern River reservoir with 1100 cp initial viscosity, assuming after the first cycle of steam injection that

| | |
|---|---|
| Heated radius | 47 ft |
| Reservoir temperature | 100 °F |
| Reservoir temperature of heated area | 300 °F |
| Drainage radius | 700 ft |
| Wellbore radius | 3.5/12 ft |

SOLUTION The temperature-viscosity relationship for the Kern River oil reservoir (Fig. 3-10) shows that oil viscosity in the heated zone decreases to 10 cp. The increase in the well's productivity is

$$\frac{q_{o^{\text{hot}}}}{q_{o^{\text{cold}}}} = \frac{1100 \text{ cp} \ln(700/0.29)}{10 \ln(47/0.29) + 1100 \ln(700/47)} = 2.83 \text{ times}$$

and is due only to the decrease of the oil viscosity. The productivity of the well is also improved by the steam's wellbore cleanup effect, which increases the rock permeability around the wellbore.

## 4-7 FIELD DEVELOPMENT AND RESULTS

The field development of steam injection, as cyclic steam injection and as steam drive, has made an impact on 1988 EOR oil production in the United States. About 80 percent of total EOR production—or 455,484 bbl/day additional oil—can be attributed to 133 active processes using these methods (Lake, 1989).

The development of steam injection field projects covers oil reservoirs with a variety of characteristics in different exploitation phases. The evolution and current status of steamflooding has been presented by many investigators: Farouq Ali and Meldau (1979), Matthews (1983), and Chu (1985), among others. The method has been applied to reservoirs having

- depth between 200 ft (Charco Redondo, Texas) and 5000 ft (Brea, California)
- formations of sand and sandstones
- average gross thickness between 30 ft (Slocum, Texas) and 550 ft (Brea, California)
- dip between 0° (Huntington Beach and Inglewood, California) and 70° (Midway Sunset, California—Tenneco)
- porosity between 20 percent (Shiells, California) and 38 percent (Tiajuana, Mexico)

- absolute permeability between 70 md (Brea, California) and 15,000–24,000 md (Mount Poso, California)

Most steam injection operations have been applied to heavy crude oil reservoirs with densities between 12 and 18 °API and viscosities between 600 to 6000 cp in reservoir conditions. The main objectives were to increase oil production by reducing oil viscosity and to increase oil recovery by steam displacement. Steam injection has also been applied to reservoirs with light and intermediate crude oils with 20 to 40 °API densities and low viscosities, that is, Brea, California, with 24 °API and 6 cp. El Dorado, Kansas, with 37 °API and 4 cp; and Shiells Canyon, California, with 34 °API density and 6-cp viscosity (Chu, 1985). Steam injection in these applications is known as steam distillation drive. Its primary objective is to reduce the residual oil saturation below that obtainable by waterflooding and to increase the oil recovery. Other field tests and processes of steam injection, especially cyclic steam injection, have been developed, particularly in Canada, to recover bitumen from tar sands.

Characteristics and results of steam injection are presented through four examples of field applications:

- The *Kern River* steam foam pilots project, in which foam is utilized to retard steam override
- The "*200*" *Sand* steamflood project, which is a typical heavy oil reservoir which had unfavorable response to cyclic stimulation
- The *Pikes Peak* very viscous oil reservoir, which is an example of conversion from cyclic steam to steam drive.
- The steam injection for recovery of bitumen from *tar sands*.

## Kern River Steam Foam Pilots, California, United States

Two large-scale steam foam pilots (one on the Mecca Lease and the other on the Bishop Fee) were conducted in the Kern River Field, and the evaluated data and results were presented by Patzek and Koinis (1988).

**The Reservoir.** The Kern River oil field located near Bakersfield, California, has three productive sand intervals, named "J," "M," and "Q" sands. Figure 4-11 presents the type log of "M" and "Q" sands where the Mecca and Bishop pilots took place, respectively. The two pilots each consisted of four inverted five-spot patterns with a total area of 11.6 acres at Mecca and 14 acres at Bishop. The formation is 3° SW dip and has 30 percent porosity, 70 percent

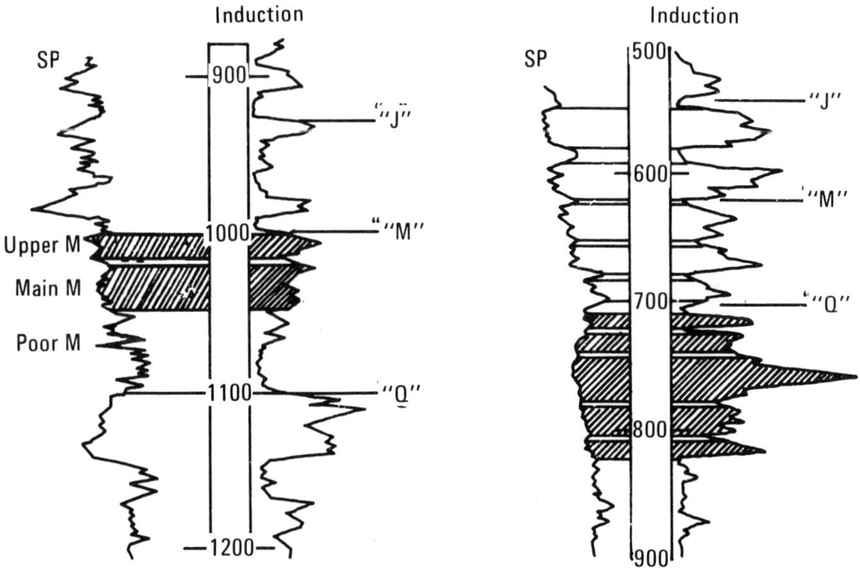


**Fig. 4-11** Mecca and Bishop type logs (From Patzek and Koinis, 1988)

initial oil saturation, and 13 °API gravity oil. The formation's gross thickness differs, being 83 ft at Mecca and 99 ft at Bishop, and is situated at 1000 ft depth and at 600 ft depth, respectively.

**Performance.** The production and injection history of the Mecca pilot prior to foam injection is represented by a recovery of 6.8 percent of OOIP in primary, with 15.0 percent of OOIP additional oil as a result of 9 years of cyclic steam injection, and with another 28.7 percent of OOIP additional oil as result of a 10-year steam drive period.

The Bishop pilot area has a recovery of 8 percent of OOIP in primary, with 55.2 percent of OOIP additional oil as a result of 19 years' cyclic steam injection and with 1.9 percent of OOIP additional oil after only 1 year of steam drive.

**Steam foam pilots.** The purpose of steam foam pilots was to retard steam override, increase vertical sweep, and increase oil production and oil recovery consequently.

Foams are dispersions of gas bubbles, air, or only nitrogen in water (steam) with surfactant. Foam forms downhole after the injection of chemicals. The presence of foam in the reservoir's pore space reduces the relative permeability to steam, and steam foam fluid has a lower mobility than steam.

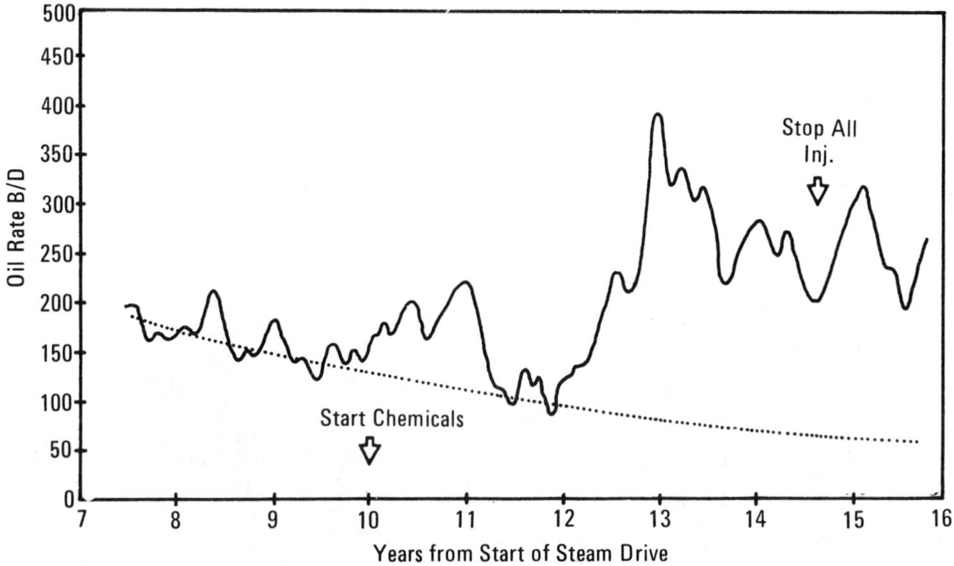


**Fig. 4-12** Mecca steam foam pilot (From Patzek and Koinis, 1988)

**Results.** The increase in oil production is illustrated in Figure 4-12 for the Mecca pilot and represents 14 percent of OOIP additional oil after 5 years of steam foam injection.

The increase in oil production for the Bishop pilot was 8.5% of OOIP additional oil after 3.7 years of steam foam injection with infill drilling and cyclic steam stimulation.

The increase in vertical sweep is illustrated by Patzek and Koinis in Figures 4-13 and 4-14 using simultaneous temperature surveys and gamma-ray–neutron logs run in the observation wells.

The steam foam presence in the porous space means an increase in gas saturation detected by the neutron log. As we can observe, steam override was retarded and vertical sweep increased.

The residual oil saturation when using steam foam was 10 percent in both pilots. This value is practically the same as that when using steam. Steam foam injection's effect of increasing the oil recovery is consequently due to the increase of sweep efficiency and not to the reduction of residual oil saturation.

**Comments.** The experience gained using steam foam injection in the field is very useful and shows that

- Steam foam injection retards steam override and increases vertical sweep, even after a long period of steam drive.

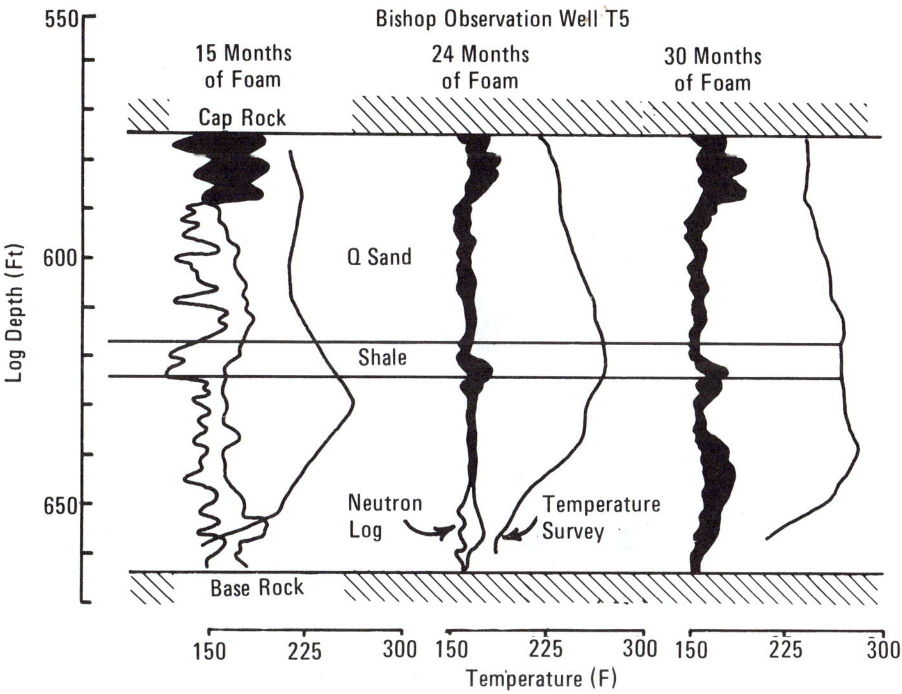


**Fig. 4-13** Improved vertical sweep by steam foam 70 ft from injector (From Patzek and Koinis, 1988)

- Oil production is increased and oil recovery is enhanced by better vertical sweep efficiency.
- Infill drilling is necessary to improve the injection-production balance if the economic conditions for drilling new wells are acceptable.
- Cyclic steam injection is still used to clean old wellbore wells or to link new cold producers with the effect of injectors.

### The "200" Sand, Midway Sunset Steamflood, California, United States

Steam injection has been applied with good results in Midway Sunset Field reservoirs. However, there were still shallow heavy oil reservoirs in this field with poor cyclic steam performance and, for this reason, not adequately developed.

The so-called "demonstration project" was initiated in the "200" Sand reservoir "to demonstrate the operational, recovery, and economic aspects of steamflooding" (Gagner, 1986).

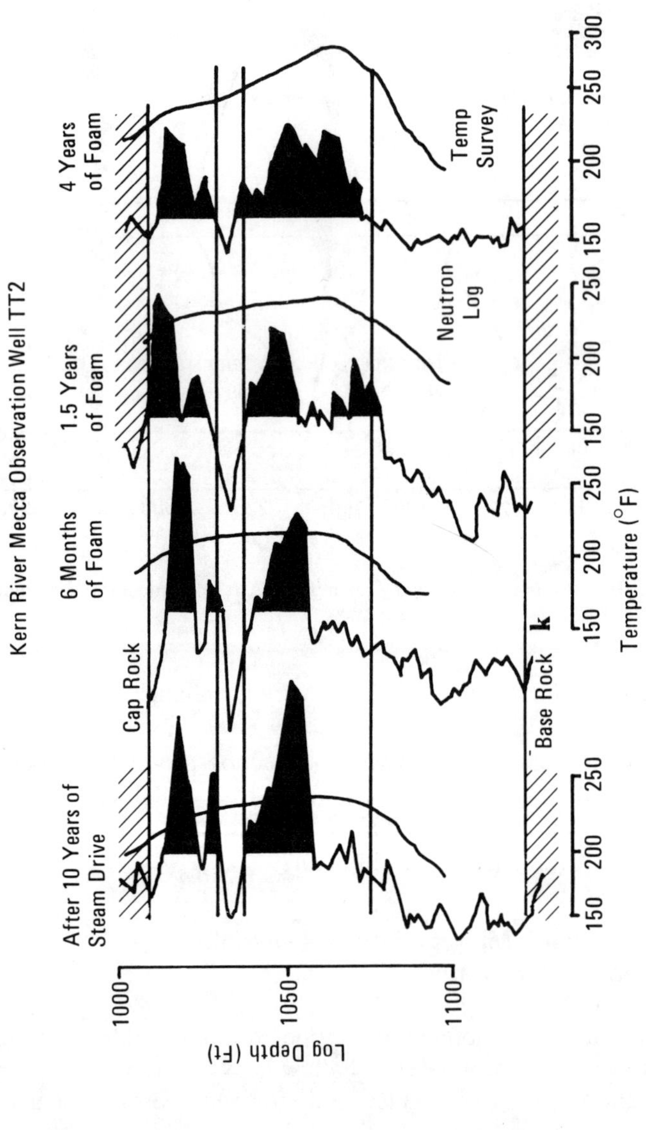


**Fig. 4-14** Improved vertical sweep by steam foam 90 ft from injector (From Patzek and Koinis, 1988)

**The reservoir.** The "200" Sand Midway Sunset oil reservoir consists of interbedded sands, conglomerates, silts, and diatomites, discontinuous laterally. The reservoir has a depth of 400 to 700 ft, a productive area of 250 acres, and a gross thickness of 200 ft (150 ft effective). The 12 °API oil gravity has 6500-cp viscosity at reservoir temperature (90 °F). The viscosity is reduced to 60 cp when the temperature is increased to 210 °F. The oil in place corresponding to a porosity of 30 percent, and average oil saturation of 59 percent is 51.4 million bbl.

**Performance.** The "200" sand has a very good permeability, with values between 1000 and 3400 md. However, even with the sharp decrease of oil viscosity with temperature increase and favorable flowing conditions, the primary oil production was far from satisfactory, and the reservoir response to cyclic steam stimulation was poor. A logical explanation was that the reservoir had a low average pressure value of 40 psig and consequently lacked enough energy to assure adequate production rates. Indeed, the "200" Sand Midway Sunset reservoir was the right choice to demonstrate steamflood performance.

**The project.** The pilot area started with 4 × 2.35-acre inverted seven-spot steam drive patterns and was expanded in 1980–1981 to 14 fully developed patterns (Figure 4-15).

The project has been monitored and evaluated monthly through

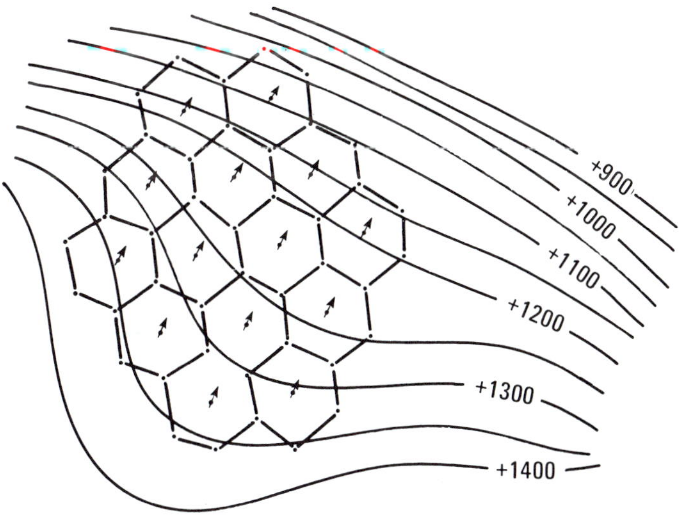


**Fig. 4-15** Structural map at "200" sand steamflood project (Adapted from Gagner, 1986)

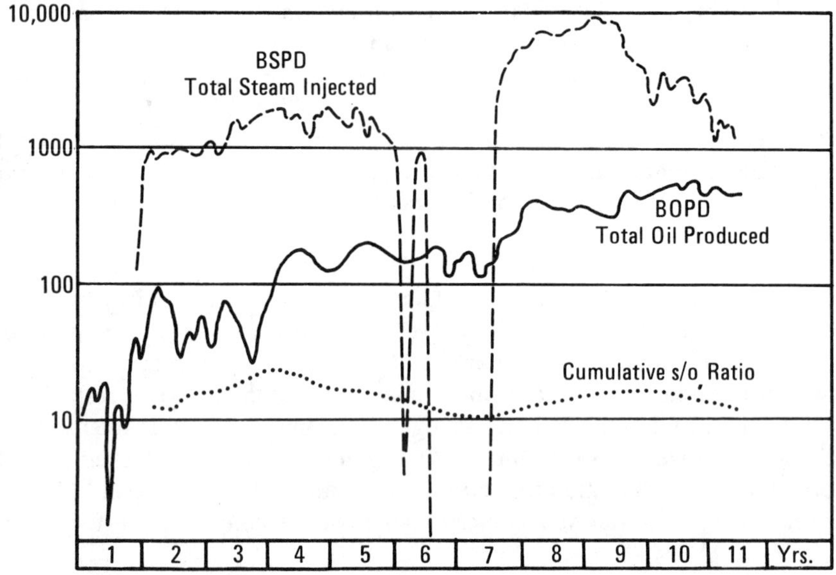


**Fig. 4-16** Production and injection history "200" sand steamflood project (Adapted from Gagner, 1986)

production-injection data, through temperature and injection profiles run in observation wells, and through reports that have been prepared periodically. The project production and injection history reported by Gagner (1986) is illustrated in Figure 4-16.

**Results.** As we observe, the total oil production rate increased to an average of 150 bbl/day (and 500 bbl/day water) after the first three years of continuous steam injection at a rate of 1000–1800 bbl/day. Due to the heat energy accumulated in the formation, the recorded oil rate was over 100 bbl/day even when the injection rate was stopped for a period of time. When the steam rate was increased to 7000 bbl/day, oil production reached 400 bbl/day (with 1700 bbl/day water). Up to this point the reservoir responded to steam rate increases in only one way: The higher the injection rate, the higher the production. However, there was an injection rate value limit when steam breakthrough to producers occurred. Indeed a further increase of the injected steam rate to over 9000 bbl/day had a consequence of channeling the injected steam to several producing wells and causing the decline of oil production to 280 bbl/day (with 270 bbl/day water). To reduce steam channeling, the injection rate was reduced. At this time, as a direct result, the oil production rate went up to 460 bbl/day (with 1920 bbl/day water). The cumulative steam-oil ratio has decreased to 12 and is predicted to approach lower values as oil production increases.

**Comments.** The experience gained from the "200" Sand steamflood project shows that

- The 14 injection wells completed with casing strings cemented to total depth are jet perforated at intcrvals rcstrictcd to 50 ft at thc basc of thc formation for the first 4 wells and to 20 ft for the next 10 wells. Limited-entry perforations in a heterogeneous formation with high-permeability stringers can cause severe channeling when the steam injection rates are high.
- The 42 producing wells (except 1) were completed with casing cemented at the top of the formation and with slotted liner and gravel packing along the entire formation. This configuration did not restrict production to the lower part of the formation to avoid steam override effect, since the formation is heterogeneous and there is no control of steam drive profile.
- The steam generated at 420 °F with 80 percent quality entered the formation at 350 °F sand-face temperature and 72 percent quality.
- The oil production was very sensitive to the back pressure on the formation. Wells had to be kept in pumped-off condition and cyclic steam injection had to be utilized to clean liner slots and the formations around the wellbore.
- The poor response of the downstructure producers was due to the poor formation characteristics in the corresponding injectors as revealed by high injection pressures. A more detailed geology and a well pattern adapted to the reservoir nonuniformity are important features to be considered in similar applications.

The "200" Sand Midway Sunset project demonstrated that shallow heavy oil reservoirs with poor cyclic steam performances could be developed by steamflooding.

## Pikes Peak, High-Viscosity Oil Reservoir, Canada

**The reservoir.** Pikes Peak reservoir (Canada) is closer to being defined as a tar sand reservoir than an oil reservoir due to 25,000 cp viscosity of the oil (or bitumen) at 64°F reservoir temperature. The reservoir, located in Waseca formation of the Lloydminster area, has 1640-ft depth, an average net pay of 49 ft with 32 to 36 percent porosity and 4500 to 10,000 md permeability. The reservoir description, the cyclic steam injection evaluation, and the conversion to steamflood are described by Miller and others (1989).

**The project.** The Pikes Peak oil reservoir consists of a homogeneous and highly permeable lower unit in communication with a sand/shale "A" interbeds unit. The reservoir was developed during a period from 1981 to 1987 by cyclic steam stimulation and 92 thermal wells drilled in a seven-spot pattern (Figure 4-17).

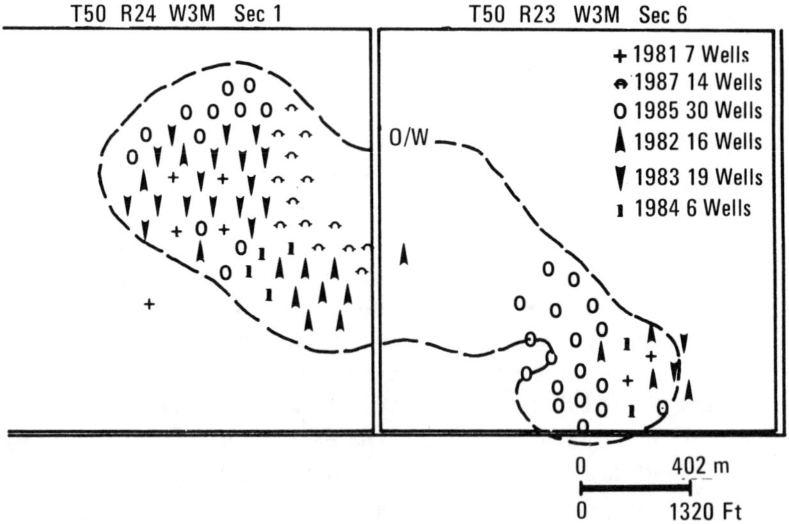


**Fig. 4-17** Pikes peak development through 1987 (Adapted from Miller et al., 1989)

The wells were completed as open hole, or with a perforated cased hole, close to the base of the zone to diminish the steam override effect (Figure 4-18). Cased hole completions proved to provide better control of the completion interval at wells near the water-oil contact. No thermal packers or insulated tubing have been used at Pikes Peak.

**Performance.** The steam stimulation performances have been good and 20 to 30 percent of the OOIP has been recovered. The decision to initiate a steam drive pilot was taken after carefully considering that poor future results could be expected if cyclic steam injection was continued. The selection criteria were based also on the interwell communication response indicated by an increase in fluid production or temperature or a decrease in produced water salinity (Miller et al., 1989). The conversion to steam drive of initial 1A11 pattern was followed by six additional patterns (Figure 4-19).

An observation well was drilled 110 ft away from the 1A11 pattern after eight months of steam drive. It was cored and logged to provide information regarding vertical conformance and residual oil saturation. The core analyses indicated that

- The top 30 ft were swept by steam leaving 12 percent oil saturation in the "A" interbeds and 9 percent oil saturation in the homogeneous sand.
- a transition zone was 5 ft with an average oil saturation of 42 percent.
- the bottom 30 ft of the homogeneous sand had an average oil saturation of 72 percent (initial saturation 85 to 90 percent).

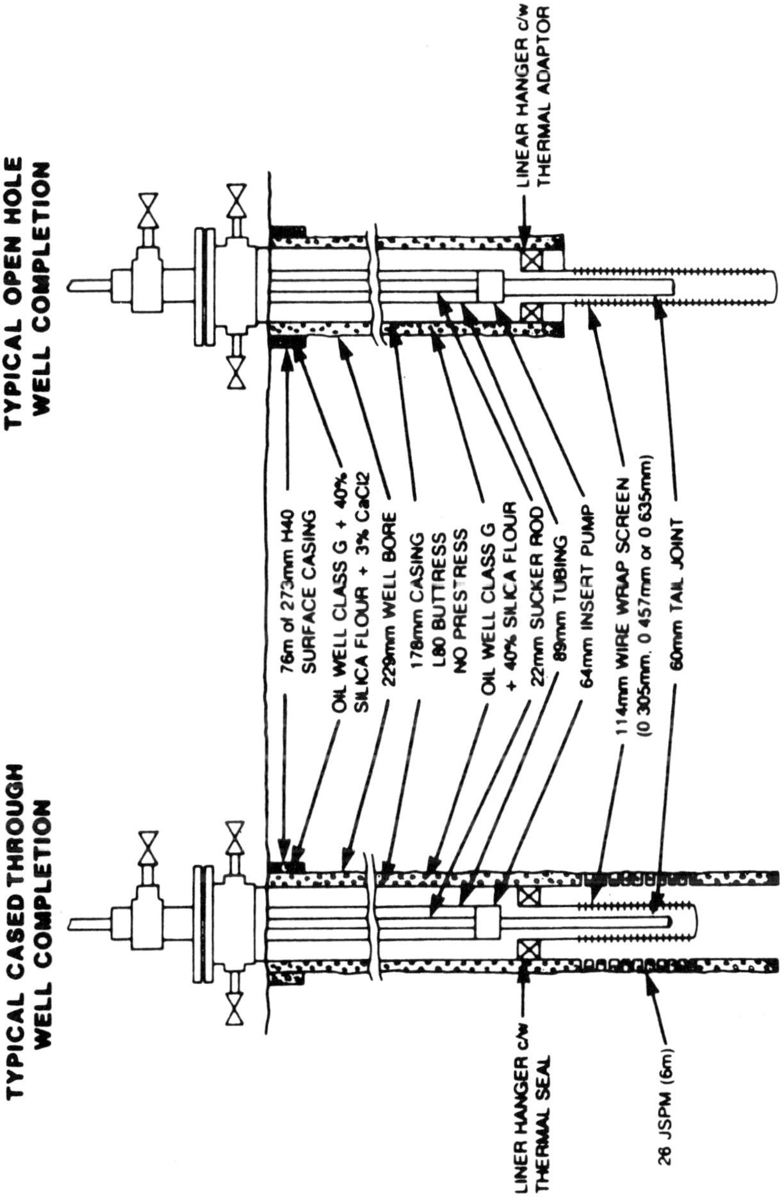


**Fig. 4-18** Pikes peak steam pilot well completions (From Miller et al., 1989)

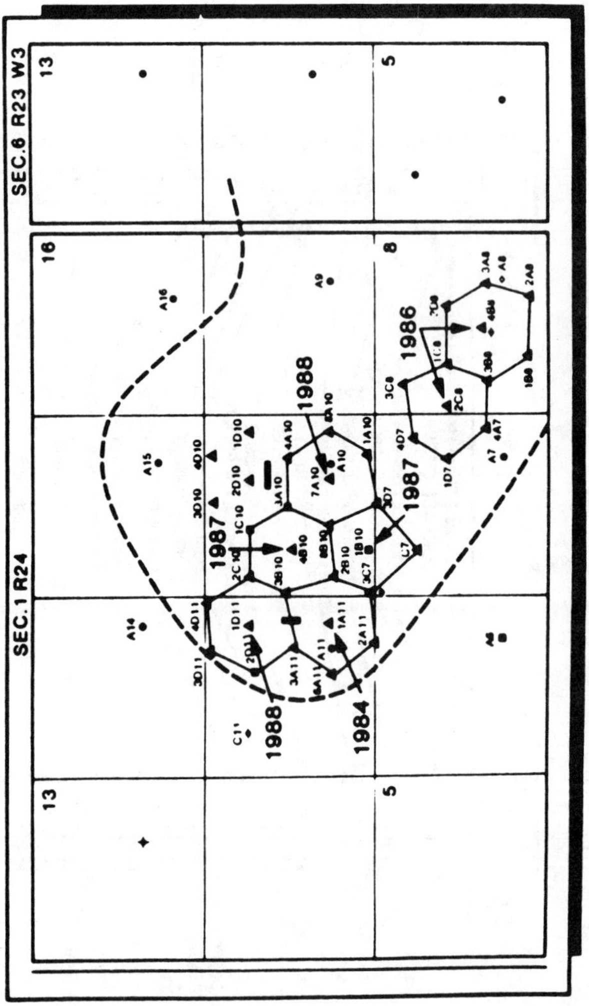


**Fig. 4-19** Steam drive conversion sequence (From Miller et al., 1989)

To reduce the steam override effect and to improve vertical conformance, a foam test was conducted in the 1A11 pattern. The results were not those anticipated; only a slight vertical sweep improvement was observed, since foam diverted the steam outside the pattern boundary (Miller et al., 1989). The use of a larger quantity of foam with improved qualities is still being considered.

The oil production by steam flood was 50 to 60 percent of the total area oil production (4400 bbl/day), and the cumulative steam-oil ratio achieved was 3.9.

**Comments.** The experience gained from the Pikes Peak oil reservoir steam flood can be considered interesting and useful because

- Steamflood has proven to be successful when applied to reservoirs with high initial oil viscosity if the reservoir is preheated through cyclic steam injection and interwell communication encouraged by small well spacing is achieved.
- The initial well's response is characterized by a drop in the salinity of produced water, a sharp increase of fluid temperature, an increase in water production, and, after two to three weeks, an increase in oil production. This indicates that steam entrains the oil rather than forming an oil bank.
- The steam override effect is still present with lower vertical sweep results and the use of foam-surfactant injection or other agents (polymers) has to be improved to recover more of the remaining bottom oil.

### Tar Sand Steam Injection

Tar sands are reservoirs containing crude bitumen, that is, oils with gravity less than 10 °API at 60 °F and with high viscosity. The world bitumen resources are very large—more than 4 trillion bbl—and are located principally in Canada, 60 percent; Venezuela, 25 percent, and USSR, 14 percent. There are still deposits yet to be discovered. In the United States the amount of 34.4 billion bbl of bitumen exist in Utah, Texas, California, and Missouri.

In order to produce oil from tar sands through wells, a large amount of heat is needed to reduce the bitumen's viscosity. The current status and development of bitumen recovery using thermal methods and new approaches to its inherent problems are presented by Carrigy (1988). Because only a small fraction of crude bitumen reserves is located "close enough to the surface to be exploited by mining methods," the remaining reserves can be reached through wells drilled from the surface or using a combination of mining and well drilling. The heat carrier agent is introduced through the well into the reservoir by cyclic steam injection or is continuously injected in a steam drive process. Heat can also be introduced into the reservoir, igniting the oil around the wellbore and sustaining by air injection the combustion front that results. When continuous injection of steam in steam drive processes (i.e., air in in situ

**TABLE 4-3.** Properties of Tar Sand Reservoirs at Locations Tested by Steam Soak

| | 1 | 2 | 3 | 4 |
|---|---|---|---|---|
| Depth, ft | 3,800 | 1870 | 1400 | 935 |
| Net thickness, ft | 210 | 110 | 115 | 110 |
| Gravity, °API | 8.9 | 5 | 9.3 | 7 |
| Reservoir temperature, °F | 140 | 100 | 55 | 52 |
| Oil viscosity at reservoir temperature, cp | $2.15 \times 10^3$ | $>4.0 \times 10^4$ | $100 \times 10^3$ | $1.6 \times 10^6$ |

1. JOBO II, Venezuela (Officina formation)
2. Oxnard, California (Vaca sand)
3. Cold Lake, Alberta, Canada (Clearwater formation)
4. Carbonate trend, Alberta, Canada (Grosmont II formation)

From Carrigy (1988).

combustion) is used, communication between wells by cyclic steam injection (steam soak) has to be established first.

The main properties of some tar sand reservoirs and locations tested by steam soak and steam drive are given in Tables 4-3 and 4-4, respectively (Carrigy, 1988).

More than 50 steam injection field tests have been conducted in tar sand reservoirs worldwide and have demonstrated that steam is an important heat carrier agent in the development of bitumen resources. The principal constraints regarding steam injection in tar sand reservoirs involve the possibility of establishing communication between wells at the commercial well spacing used, the accentuated steam override tendency, and sand invasion.

The new approaches to tar sand oil recovery involve drilling horizontal wells along the lower half of the formation to affect a larger rock volume with steam. By also using steam plus additives (for instance $CO_2$ or surfactant), a

**TABLE 4-4.** Properties of Tar Sand Reservoirs at Locations Tested by Steam Drive

| | 1 | 2 | 3 | 4 |
|---|---|---|---|---|
| Depth, ft | 2500 | 1500 | 260 | 1800 |
| Net thickness, ft | 80 | 52 | 110 | 90 |
| Gravity, °API | 9 | −0.5 to −2.0 | 7 | 9 |
| Reservoir temperature, °F | 110 | 95 | 55 | 62 |
| Oil viscosity at reservoir temperature, cp | $25 \times 10^3$ | $20 \times 10^6$ | $>1.0 \times 10^6$ | $2.2 \times 10^5$ |

1. Cat Canyon, California (S-1 B sand)
2. Street Ranch, Texas (San Miguel-4 sand)
3. Athabasca, Alberta, Canada (McMurray formation)
4. Peace River, Alberta, Canada (Bullhead group)

From Carrigy (1988).

miscible effect is developed and the override tendency is reduced. Downhole steam generator equipment developed for tar sand reservoirs would also be an appropriate choice to reduce heat losses before steam is injected into the formation. Based on the same principles of oil displacement and entrainment by steam, new approaches combining mining and petroleum drilling methods for reaching a reservoir will create greater possibilities for recovering more oil from tar sands.

## 4-8 SCREENING CRITERIA

Based on the results obtained in the field, screening criteria for finding out whether an oil reservoir is a good candidate for steam injection were established. New field results helped widen the applicability ranges of steam injection. For instance, limiting his consideration to steamflood processes only, Chu (1985) developed a valuable screening guide based on 28 steamflood project results. A screening guide is helpful in preliminary considerations for selecting possible applications for reservoirs in a large geological basin. However, detailed studies, tests, and predictions have to be carried out before a decision is made for a specific reservoir. The key reservoir characteristics and crude properties which influence the success of a steamflood project are as follows:

| | |
|---|---|
| *Depth* | Above a depth of 300–400 ft to avoid parting pressure of adjacent formations. Limited to 5000 ft due to heat losses. Higher limit possible using downhole steam generators. |
| *Reservoir Pressure and Temperature* | Not critical for steamflood. |
| *Formation gross thickness* | Between 15 and 300–400 ft, with thinner pay zones resulting in higher heat losses to adjacent formations. |
| *Porosity* | Higher than 18 to 20 percent. |
| *Permeability (absolute)* | Formations with very good permeability (between 250 and 1000 md) or higher. |
| *Oil saturation at start of project* | Higher than 40 to 50 percent, steamflood is not successful after waterflood. |
| *Oil density (°API)* | Less than 36 °API (as steam is applied also to light oils). |
| *Oil viscosity* | The upper limit can be decreased to 6000 cp or less by cyclic steam injec- |

| | |
|---|---|
| | tion. In general between 200 cp and 2000–3000 cp. |
| *Factors favoring steam injection* | Shallow and dip oil reservoirs, thick pay zones with very good to excellent permeability, cheap and high-quality water source. |
| *Adverse factors* | Strong reservoir nonuniformity, highly water-sensitive clay content, low interwell communication. |

As soon as screening criteria are met by a reservoir's given characteristics, a quick estimate of the steam-oil ratio (SOR) can be made using the two equations developed by regression analysis (Chu, 1985). The SOR (steam-oil ratio), or the reciprocal (OSR), correlates to reservoir depth $D$, ft; thickness $h$, ft; porosity $\phi$; permeability $k$, md; oil saturation $S_o$; and oil viscosity, $\mu$, cp.

$$\text{SOR} = 18.744 + (1.453 \times 10^{-3} D) - (50.88 - 10^{-3} h) - (0.8864 \times 10^{-3} k) - (0.591 \times 10^{-3} \mu) - 14.79 S_o - (0.2938 \times 10^{-3} kh/\mu) \tag{4-15}$$

Taking into consideration that the calorific power of 1 barrel of oil can produce 13 to 14 barrels of steam and also the costs other than fuel alone, an efficient steamflood process must end when the instantaneous SOR reaches a value at or near 8 (Chu, 1985). When the SOR value calculated using Eq. 4-15 is higher than 5, then SOR must be taken as the reciprocal of OSR obtained using Eq. 4-16.

$$\text{OSR} = (-112.53 \times 10^{-4}) + (0.2779 \times 10^{-4} D) + (1.579 \times 10^{-4} h) - (13.57 \times 10^{-4} \theta) + (7.232 \times 10^{6} \mu) + (1.043 \times 10^{-5} kh/\mu) + 0.5120 \phi S_o \tag{4-16}$$

which takes into consideration the formation dip, $\theta$ (rad).

**Example 4-6.** Make a quick prediction and estimate the SOR of a reservoir prospect having $D = 2800$ ft, $h = 70$ ft, $k = 600$ md; $S_o = 0.55$, $\phi = 0.25$, $\mu = 400$ cp, and $\theta = 15°$ (0.26 rad)

$$\text{SOR} = 18.744 + 4.0684 - 3.5616 - 0.5318 - 0.2364 - 8.1345 - 0.0308 = 10.31$$

$$\text{OSR} = (-112.53 \times 10^{-4}) + 0.077812 + (110.53 \times 10^{-4}) - (3.5282 \times 10^{-4}) + (28.998 \times 10^{-4}) + (15.015 \times 10^{-4}) + 0.0704 = 0.152$$

$$\text{the reciprocal} = \frac{1}{0.152} = 6.58 = \text{SOR}$$

As we observe, the formation dip improves the steam drive process efficiency and reduces the steam override tendency. More oil is produced per barrel of steam injected.

Steam injection is the most effective enhanced oil recovery process based on the amount of oil produced. Although cyclic steam injection has contributed much to oil production, its future use will be for stimulating and preparing wells for steamflood and in situ combustion.

## QUESTIONS AND PROBLEMS

**4-1** Describe cyclic steam injection and explain when and why the process results in higher oil rates and also high recoveries.

**4-2** Make a schematic representation of the steam drive process and explain the various zones formed in the reservoir.

**4-3** Steam generators are used to inject 1400 bbl/day wet steam having 75 percent quality at 1000 psia sand-face pressure into a formation 56 ft thick. Assuming radial and uniform propagation of heat, calculate the total heat injected and the reservoir heated radius after 1 year of injection. The heat capacity per cubic foot of steam-saturated rock is 34 Btu/ft$^3$ $\times$ °F, the reservoir temperature is 150 °F, and the heat lost to overburden is 20 percent of the available heat above the reservoir temperature.

**4-4** Wet steam of 75 percent quality is injected in a well at a rate of 828 bbl/day. Find the heat injection rate if the steam injection pressure is 420 psia and the initial formation temperature is 116 °F.

**4-5** What is the cumulative heated area after 2 years of steam injection with $11.192 \times 10^6$ Btu/hr at 420 psia? The productive formation has 42 ft thickness and 32 Btu/ft$^3$ $\times$ °F heat capacity. The thermal conductivity and thermal diffusivity of the cap rock is 0.96 Btu/ft $\times$ hr $\times$ °F and 0.68 ft$^2$/day, respectively, and the formation temperature is 118 °F.

**4-6** Using the same reservoir characteristics as in 4-5, find the cumulative oil displaced by steam for each 100-day time increment. The formation has a porosity of 21 percent and the oil saturation of 57 percent was reduced by steamflooding to 9 percent.

**4-7** Steam of 65 percent quality was injected for 5 years at a rate of 10,000 bbl/day with 640 psia sand-face injection pressure. The effect of the steam drive process was the reduction of the oil saturation to 50 percent (in absolute value). Calculate the oil recovery by steam if the initial reservoir temperature was 115 °F, the production area 250 acres, and the gross thickness 50 ft.

## REFERENCES

AYDELOTTE, S. R. and G. A. POPE, "A Simplified Predictive Model for Steamdrive Performance," *Journal of Petroleum Technology* (May 1983), pp. 991–1002.

CARRIGY, M. A., "Thermal Recovery from Tar Sands," Paper 12556 SPE presented at the 1988 California Regional Meeting, Long Beach, California, March 23–25, 1988.

CHU, C., "State-of-the-Art Review of Steamflood Field Projects," *Journal of Petroleum Technology* (October 1985).

FAROUQ ALI, S. M., "A Comprehensive Wellbore Steam-Water Flow Model for Steam Injection and Geothermal Applications," *Society of Petroleum Engineers Journal* (October 1981).

FAROUQ ALI, S. M., and R. F. MELDAU, "Current Steamflood Technology," *Journal of Petroleum Technology* (October 1979), pp. 1332–342.

FAROUQ ALI, S. M., and R. F. MELDAU, "Improved Oil Recovery," *Steam Injection* (Oklahoma City, OK: Interstate Oil Compact Commission, 1983), Chapter VII, p. 339.

GAGNER, M. J., *The "200" Steamflood Demonstration Project* (Washington, D.C.: U.S. Department of Energy DOE/ET/12059-9, October 1986).

GATES, C. F., and H. J. RAMEY, JR., "Better Technology Opens Way for More Thermal Projects," *Oil and Gas Journal* (July 13, 1964).

GOMAA, E. E., "Correlations for Predicting Oil Recovery of Steamflood," *Journal of Petroleum Technology* (February 1980).

LAKE, LARRY W., *Enhanced Oil Recovery* (Englewood Cliffs, NJ: Prentice-Hall, 1989), p. 4.

LAUWERIER, H. A., "The Transport of Heat in an Oil Layer Caused by the Injection of Hot Fluid," *Applied Science Research*, No. 5, Sec. A (1955), p. 145.

MARX, J. W., and R. H. LANGENHEIM, "Reservoir Heating by Hot Fluid Injection," *Petroleum Transactions* (AIME), 216 (1959), p. 312.

MATTHEWS, C. S., "Steamflooding," *Journal of Petroleum Technology* (March 1983).

MILLER, K. A., L. G. STEVENS, and B. J. WATT, "Successful Conversion of the Pikes Peak Viscous Oil Cyclic Steam Project to Steamdrive," SPE 18774, 1989 California Regional Meeting, Bakersfield, California, April 5–7, 1989.

MYHILL, N. A., and G. L. STEGEMEIER, "Steam-Drive Correlation and Prediction," *Journal of Petroleum Technology* (February 1978).

PATZEK, T. W., and M. T. KOINIS, "Kern River Steam Foam Pilots," SPE DOE 17380, *Sixth Symposium on Enhanced Oil Recovery*, Tulsa, Oklahoma, April 17–20, 1988.

PEAKE, W. R., "Steamflood Material Balance Applications," *SPE Reservoir Engineering* (August 1989).

PRATS, M., *Thermal Recovery*, Henry L. Doherty Series, Monograph 7 (Dallas, TX: Society of Petroleum Engineers of American Institute of Mining and Metallurgical Engineers, 1982).

RAMEY, H. J., JR., "How to Calculate Heat Transmission in Hot Fluid Injection," in *Fundamentals of Thermal Oil Recovery* (Dallas, TX: Petroleum Engineer, 1965), p. 165.

RUBINSTEIN, L. I., "The Total Heat Losses in Injection of a Hot Liquid into a Stratum," *Neft I Gas, Moscova* Vol. 2 (1959), p. 41.

SEBA, R. D., JR., and G. E. PERRY, "A Mathematical Model of Repeated Steam Soaks of Thick Gravity Drainage Reservoirs," *Journal of Petroleum Technology* (January 1969).

SMITH, CH. R., "Mechanics of Secondary Oil Recovery," reprint the 1985 edition published in 1966 (Malabar, Florida: Robert E. Krieger Publishing Company, 1985), p. 473.

STOVALL, S. L., "Recovery of Oil from Depleted Sands by Means of Dry Steam," *Oil Weekly*, Vol. 74, no. 9 (August 13, 1934), p. 17.

VAN POOLLEN, H. K., and associates, *Fundamentals of Enhanced Oil Recovery* (Tulsa, Oklahoma: PennWell, 1980), pp. 19–30.

WHITE, P. D., and J. T. MOSS, *Thermal Recovery Methods* (Tulsa, Oklahoma: PennWell, 1983), Chapter 6.

WILLIAMS, R. L., et al., "An Engineering Economic Model for Thermal Recovery Methods," Paper 8906, 1980 SPE Annual California Regional Meeting, Los Angeles, California, April 9–11, 1980.

WILLMAN, B. T., V. V. VALLEROY, et al., "Laboratory Studies of Oil Recovery by Steam Injection," *Petroleum Transactions* (AIME), 222 (1961), p. 681.

# Chapter 5

# In Situ Combustion

## 5-1 GENERAL

In the porous rock of an oil reservoir, the oil can be ignited around the wellbore by means of an igniter or by a spontaneous reaction of the oil to the air injected into the formation.

A burning front is built up, and the combustion is sustained by continuous injection of air or oxygen enriched air.

A small portion of the oil in place is burned furnishing heat to the rock and its fluids. The heat generated

- reduces the viscosity of the oil, increasing its mobility.
- increases sweep efficiency and reduces oil saturation.
- vaporizes some of the liquids in the formation generating steam and hot gases.
- produces miscible fluids by condensation of the light components of the vaporized oil.

The continuous injection of air or oxygen-enriched air develops efficient pressure maintenance and/or gas drive mechanisms. As a result the oil is more easily displaced to the well producers by the slow movement of the burning front through the reservoir rock.

The idea of in situ combustion was patented in 1923 by Wolcott and Howard. The first field test attempts to ignite oil in a reservoir were conducted after 1930 in the Soviet Union and in the United States. Negative results were reported because of the well's reduced injectivity. The field tests were started again in the United States during 1952 and extended after the first laboratory results were published in 1953 and 1954 by Kuhn and Grant.

## 5-2 LABORATORY STUDIES

In situ combustion laboratory experiments are conducted using oxidation cells and long or short combustion tubes of various sizes.

### Oxidation Cells

The oxidation cells are used to obtain information regarding the reactivity of different oils in a porous medium and the mechanism of the reactions.

The oxidation cell, into which air is injected continuously, is a core sample saturated with oil and heated to 940 °F. The temperature in the core sample, which is increased linearly with time, is recorded by thermocouples, the oxygen is consumed, and the effluent gases are analyzed and measured.

In general, a reactive oil exhibits two successive oxidation reactions as shown in Figure 5-1.

**1.** The low-temperature oxidation reaction ($T < 600$ °F) affects the lighter components of the crude. Small amounts of $CO_2$ and CO are formed even though $O_2$ consumption is high. The oxygen remains incorporated in the hydrocarbon chain which is not destroyed by combustion (Burger and Sourieau, 1985).

In a reservoir, the oxygen consumed in the first reaction is reduced because the oil is heated by the combustion gases that flow downstream.

**2.** The high-temperature oxidation reaction takes place in the narrow zone of the reservoir closest to the combustion front and affects the heavier components of the crude. It is more important than the low-temperature reaction because the oil is cracked, resulting in volatile and gaseous fractions and a residue of cokelike deposits on the sand grains.

The residue of cokelike deposits, which constitute the principal fuel source, is burned to maintain the combustion front.

The results obtained by the oxidation cell experiments also showed the reactivity of different oils. For example, a paraffin-base and high-API-gravity

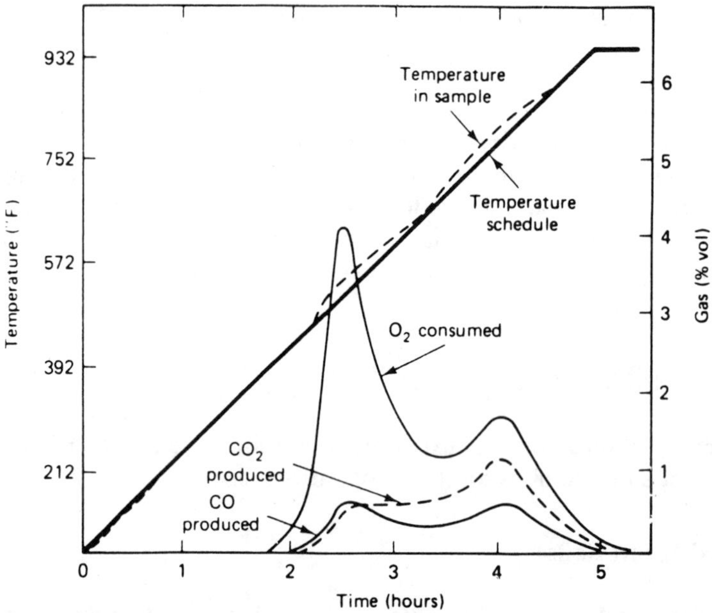


**Fig. 5-1** Differential thermal analysis of a crude oil (From Burger and Sahuquet, 1972)

oil could be completely flushed out from the path of the combustion zone by the hot fluids, leaving no cokelike deposit behind (Van Poollen, 1980). The asphaltic and naphthemic base crude oils have aromatic compounds and therefore form the greatest amount of coke. They are good candidates for the in situ combustion process because it is this cokelike residue which sustains the combustion front.

It also has been pointed out that the formation of coke is improved by the catalytic effect of different metals such as nickel, vanadium, chromium, and so on, and by a larger specific area of the porous rock (matrix), as in a clay sand. These conditions may give a light oil sufficient fuel availability to ensure the propagation of a combustion front.

### Combustion Tubes

The combustion experiments were also conducted using short or long stainless steel combustion tubes operated to provide adiabatic conditions. The assembly consists of a high-pressure jacket (1000 to 2000 psi), 3 to 7 ft long and 6 to 10 in. in diameter. The combustion tube, filled with a mixture of reservoir sand, oil, and water, is fitted inside the jacket and surrounded by heater collars or bands. An insulating material protects the combustion tube against heat losses in the annulus, which is gas pressurized. The orientation of the tube can be horizontal as in Figure 5-2 or vertical as in Figure 5-3.

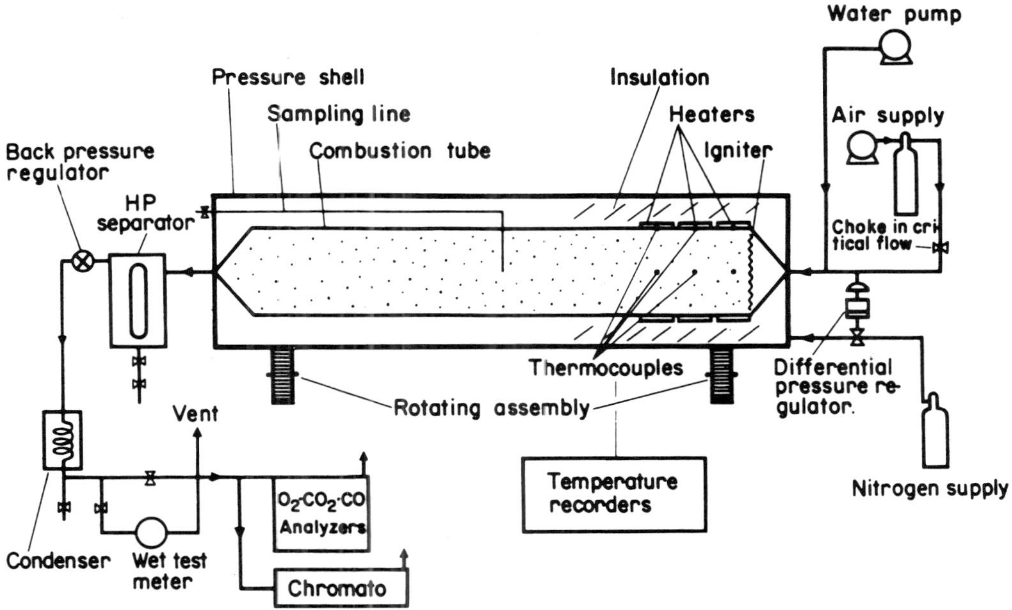


Fig. 5-2 Laboratory equipment for in situ combustion (From Burger and Sahuquet, 1973)

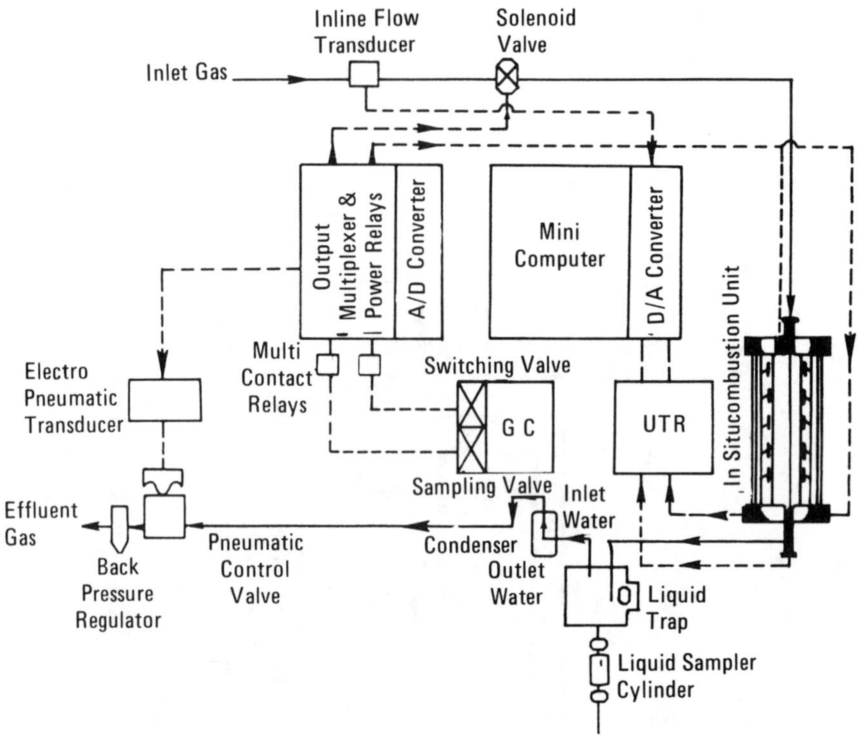


**Fig. 5-3** Flow diagram for the automated in situ combustion assembly (From Vossoughi and Willhite, 1982)

In the horizontal combustion tube system, to avoid gravitational segregation effects, the pressure jacket can be supported on steel rollers that enable the cell to rotate. After the ignition of the crude oil, air is injected to sustain the combustion front. The exhaust gases are measured and analyzed continuously, the oil and water produced are separated, and their production history is determined. In the vertical combustion tube orientation, air is injected from the top and all the effluents exit from the bottom of the tube.

The laboratory experiments, field tests, and commercial developments of in situ combustion contributed to a better qualitative description of the process.

## 5-3 QUALITATIVE DESCRIPTION OF IN SITU COMBUSTION

The mechanism of oil displacement by a combustion front is complex. On a cross section made between an injection well and a producer in a reservoir formation with uniform permeability, the combustion front has practically an elongated S shape. The density differences between injected air and reservoir liquids make evident the gravitational segregation and the tendency of gas to override (Figure 5-4).

This tendency is accentuated when the reservoir thickness is larger and is attenuated when the formation is inclined. Under steady-state conditions the various zones formed in the reservoir during the in situ combustion process are as follows:

*Zone 1* is the combustion front, where the oxygen is consumed in burning the coke deposited on rock grain surfaces and steam is one of the products formed. Zone 1 has the highest temperature, from 600 to 1200 °F.

*Zone 2* is left behind by the combustion front as a hot and clean sand that heats the injected air before the air reaches the front.

*Zone 3* is the vaporization zone ahead of the combustion front, where the lighter hydrocarbons and the interstitial water are vaporized and the heavier hydrocarbons are thermally cracked, leaving the coke-like deposit on the sand grains.

*Zone 4* is the condensation zone where the steam and the hydrocarbon gases move forward into the cooler reservoir, condense, and a large amount of heat is released. The oil displacement is increased by the oil's lower viscosity, higher mobility, and the miscible effect of the mixture between the condensed gas and the oil bank.

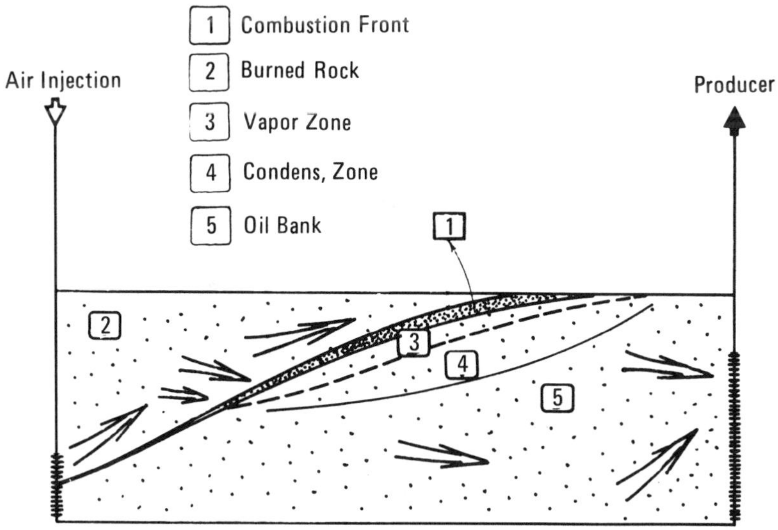


**Fig. 5-4** Schematic representation of in situ combustion process and the various zones as formed in the oil reservoir

*Zone 5* is characterized by a water saturation higher than the interstitial water saturation (water bank) which pushes the oil bank ahead to the producers (Latil, 1980).

The knowledge of the displacement mechanism of the in situ combustion process makes it possible to improve its efficiency. For instance, considering the tendency of gas to override, the injectors are perforated only in the lower half of the pay zone (limited entry) and the producers avoid the top of the formation (Figure 5-4).

Since the specific heat of the injected air is too low to carry the heat accumulated in the burned zone, more than half of the heat generated remains behind the combustion front (zone 2). The heat is lost by conduction in the adjacent formations. To achieve a better thermal efficiency, after starting off the process as a dry combustion, water is injected in combination with the injection of air. This process is called wet combustion.

## 5-4 WET COMBUSTION

The wet combustion process, also known as COFCAW (combination of forward combustion and waterflood) (Parish and Craig, 1969), transfers the accumulated thermal energy forward (upstream) from the combustion front. Water is the transferring agent because of its high heat capacity and its latent heat of vaporization.

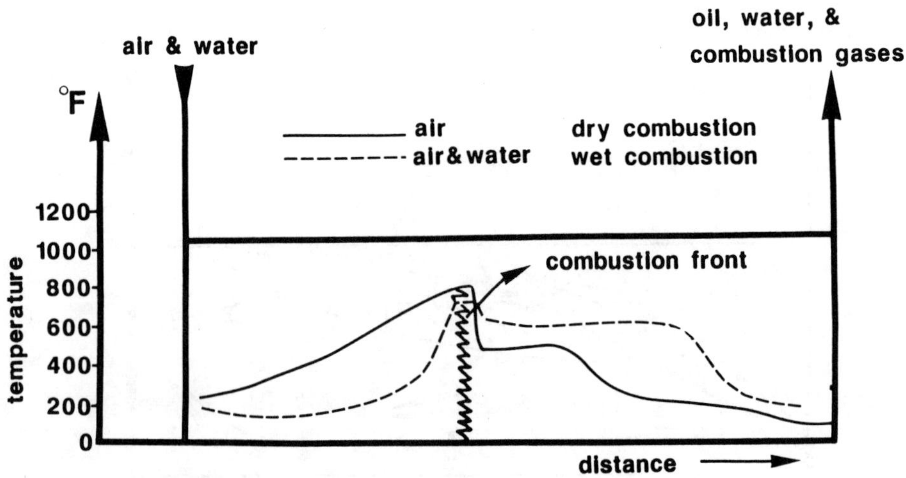


**Fig. 5-5** Temperature profiles in dry and wet combustion

Air and water are injected concurrently or alternately into the injection well. The injected water flashes into superheated steam, passes through the combustion front, and transfers heat to the area ahead of the front.

A comparison of the temperature profiles shows that the wet combustion process operates at lower temperatures behind and at the combustion front. Ahead of the combustion front the hot zone (the steam plateau) is increased in size (Figure 5-5).

The advantages of the process are evident: A much larger area of oil saturated rock is affected by higher temperatures, oil mobility and sweep efficiency is increased, less fuel at the combustion front is necessary, and less air is required to sweep the reservoir.

The superiority of wet combustion over dry burning was substantiated with field data from the Bodcau in situ combustion project in Bossier Parish, Louisiana (Joseph and Push, 1980). However, in oil reservoirs with low permeability and higher content of dirty sand (swelling clay), the introduction of water into formation may reduce injectivity and increase air injection pressure.

## 5-5 REVERSE COMBUSTION

When oil is too viscous to flow under reservoir conditions but the reservoir has an adequate air permeability, it is possible to produce oil by reverse combustion. In this case, the combustion front moves counter to the air flow (Figure 5-6). After ignition, the well is put into production and another well is used for injection. The front moves in the same way in which a cigar is consumed—by expelling the air instead of inhaling it (Crawford, 1971). The process has limited use and was tested in a tar sand reservoir in Bellamy Field (Trantham and Marx, 1965).

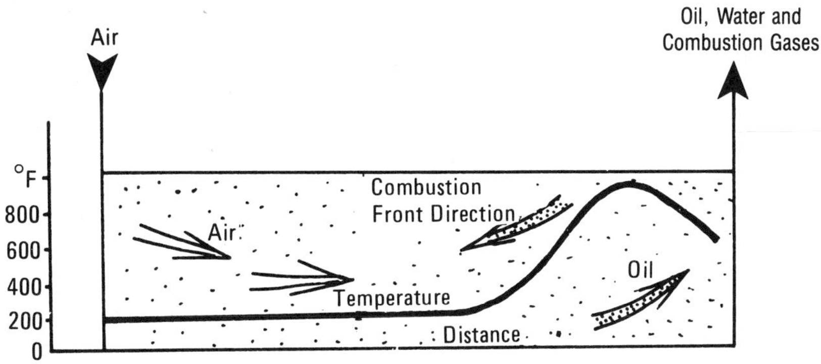


**Fig. 5-6** Temperature profile in reverse combustion

## 5-6 COMBUSTION PARAMETERS

### Description

The main parameters obtained by running laboratory experiments or performing pilot tests and field developments are as follows:

1. *The self-sustained combustion temperature*, $T_c > 600$ °F is the high-temperature oxidation reaction when the oil is cracked and a sufficient energy level of the crude oil–oxygen reaction is assured to sustain the combustion.
2. *The atomic H/C ratio* of fuel burned, $n$ = from about 0.5 to 2, is the average number of hydrogen atoms per carbon atom and is a characteristic of the various crude oils, explaining the different values obtained for the fuel content.
3. *The fuel content*, $C_u$ = from 0.7 to 2.5 $lb_m/ft^3$, is the amount of coke available for combustion that is deposited on the rock as a result of distillation and thermal cracking.
4. *The minimum air flux density*, $U_{min}$ = from 1.2 to 4 $scf/ft^2 \times hr$, is that which is necessary to obtain the self-sustained combustion temperature. Each square foot of a porous medium saturated with crude oil must receive in one hour from 1.2 to 4 scf of air. The air flux density value increases when the combustion of the crude oil results in a higher amount of coke deposited and consumed as fuel. The minimum air flux density is given by the equation

$$U_{min} = V_b \times C_a, \text{ scf/ft}^2 \times \text{hr} \tag{5-1}$$

   where *the rate of the burning front advance*, $V_b$ = from 0.125 to 0.5 ft/day (when the formation thickness is 20 to 30 ft), is necessary to maintain the minimum air flux; and *the air required (consumed)*, $C_a$ = from 160 to 400 $scf/ft^3$, is the volume of air required to burn through a cubic foot of reservoir rock.
5. *The air oil ratio*, *AOR* = from 3000 to 20,000 scf/bbl, is the ratio of the amount of air injected (in standard conditions: 14.7 psia, 60°F) to the amount of oil produced; *AOR* is the parameter used for characterizing the performance of an in situ combustion process.

## 5-7 CALCULATIONS

*The atomic H/C ratio*, $n$, is calculated from the combustion equation for 1 lb-mol of combustion gas produced or using Eq. 5-2

$$n = \frac{1.06 + 2Y_{CO} - 5.06(Y_{O_2} + Y_{CO_2} + Y_{CO})}{Y_{CO_2} + Y_{CO}} \cdots \tag{5-2}$$

where $Y$ is the mole fraction or volume fraction in combustion gases (Benham and Poettmann, 1958).

**Example 5-1 (adapted from White and Moss, 1983).** In an oxidation cell the effluent gas composition recorded was

$$14.9\%\ CO_2,\quad 1.15\%\ CO,\quad 0.19\%\ O_2,\quad \text{and}\quad 83.76\%\ N_2$$

Assuming the injected air composition is 21 percent $O_2$ and 79 percent $N_2$, calculate the atomic H/C ratio of the coke deposited.

SOLUTION

**(a)** The combustion equation is determined for 100 lb-mol of combustion gas produced (White and Moss, 1983):

$$\overbrace{83.76\ N_2 + \text{mol}\ O_2}^{\text{Air Injected}} + \overbrace{\text{mol C} + \text{mol H}}^{\text{Fuel Burned}} =$$

$$\overbrace{83.76\ N_2 + 14.9\ CO_2 + 1.15\ CO + 0.19\ O_2}^{\text{Gas Produced}} + \overbrace{\text{mol}\ H_2O}^{\text{Water}}$$

The number of moles of oxygen corresponding to 83.76 mol of $N_2$ in the air injected is

$$\text{mol}\ O_2 = \left(\frac{83.76}{79}\right)100\ (0.21) = 22.26 \text{ mol of oxygen}$$

The number of moles of carbon burned is obtained knowing the amount of gas $CO_2$ and CO produced:

$$\text{mol C} = 14.9 + 1.15 = 16.05 \text{ mol of carbon}$$

The number of moles of hydrogen consumed represents the hydrogen required to form the water produced. The water is produced with the remaining oxygen from the oxygen balance:

$$\begin{aligned}
\text{oxygen in} &= O_2 \rightarrow 22.26 \times 2 = 44.52 \text{ mol atoms}\\
\text{oxygen out} &= CO_2 \rightarrow 14.9 \times 2 = 29.80 \text{ mol atoms}\\
&\quad CO \rightarrow 1.15 \times 1 = 1.15 \text{ mol atoms}\\
&\quad O_2 \rightarrow 0.19 \times 2 = 0.38 \text{ mol atoms}\\
&\quad \text{Total} = 31.85 \text{ mol atoms}
\end{aligned}$$

The remaining oxygen is $44.52 - 31.85 = 12.67$ mol atoms, which produces 12.67 mol of water. The 12.67 mol of water produced requires $12.67 \times 2 = 25.34$ mol of hydrogen. The atomic H/C ratio of the fuel is

$$n = \frac{25.34}{16.05} = 1.58$$

**(b)** Using Eq. 5-2,

$$n = \frac{1.06 + 2(0.0115) - 5.06(0.0019 + 0.149 + 0.0115)}{0.149 - 0.0115} = 1.621$$

The results obtained are nearly the same, and the value of the atomic H/C ratio can be expressed as $n = 1.6$ for this particular case.

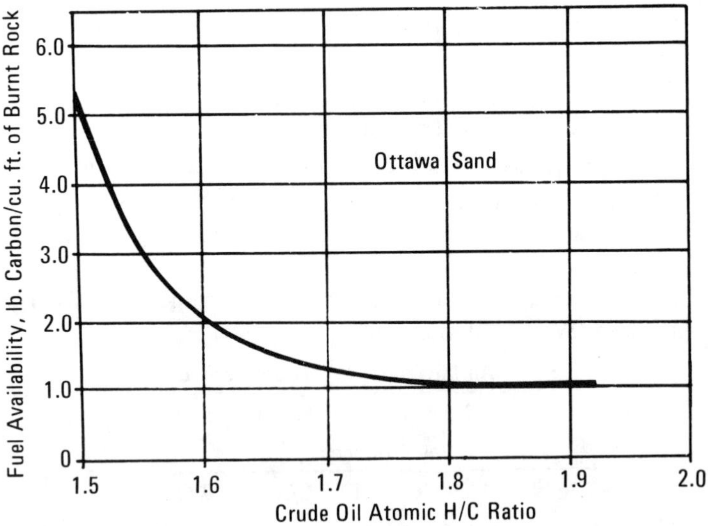


**Fig. 5-7** Correlation of fuel availability (content) with atomic H/C ratio (Alexander et al., 1962)

*The fuel content*, $C_u$, is calculated from the area under the CO and the $CO_2$ concentration curves (Fig. 5-1) and from the airflow rate. Figure 5-7, where the fuel content *Cu* has been correlated to the atomic H/C ratio (Alexander and Martin, 1962), can be also used for an initial estimate of fuel content.

*The air required*, $C_a$, needed to burn through a cubic foot of reservoir rock depends on the amount of coke deposited, which varies for different crude oils. Also, there is a minimum amount of air necessary to burn 1 pound of fuel (coke). The minimum amount of air needed for different types of crude oil can be calculated using the atomic H/C value.

**Example 5-2.** Calculate the minimum amount of air needed to burn 1 pound of coke knowing $n$ = H/C = 1.6.

SOLUTION The molecular weight of the fuel is $CH_{1.6}$ = 12 + (1.6) = 13.6. The mol content for 1 pound of fuel is

$$\frac{12}{13.6} = 0.882 \text{ lb-mol of carbon}$$

and

$$\frac{1 \times (1.6)}{13.6} = 0.118 \text{ lb-mol of hydrogen}$$

The minimum oxygen necessary is

$$\left(\frac{0.882 \text{ lb-mol of carbon}}{12} + \frac{0.118 \text{ lb-mol of hydrogen}}{4}\right) \times 379 \text{ scf/lb-mol} = 39 \text{ scf/lb}_m$$

and the minimum amount of air necessary is

$$\frac{39 \text{ scf/lb}}{0.21} = 185.7 \text{ scf/lb}_m \text{ (for 100\% oxygen utilization)}$$

Usually a minimum of 160 to 92 scf of air is necessary to burn 1 $lb_m$ of fuel. The minimum air necessary can also be calculated by knowing that 100 Btu of heat is released from each scf of air, and each pound of fuel can generate an average of 18,000 Btu/$lb_m$ (Smith, 1985).

**Example 5-3.** Calculate the air required to burn through a cubic foot of reservoir rock using the data of Example 5-2 and 90 percent oxygen utilization.

SOLUTION

$$C_a = C_u \times \text{minimum air necessary} \tag{5-3}$$

For $n = 1.6$ from Figure 5-7

$C_u = 1.85$ $lb_m$ carbon/ft$^3$ of burned rock

$C_a = 1.85 \text{ lb/ft}^3 \times 185.7 \text{ scf/lb}_m \times 0.9 = 309.2 \text{ scf/ft}^3$

## 5-8 AREA OF APPLICATION AND PILOT TESTS

The in situ combustion process has been proved as a very useful method to recover more oil, especially heavy oil. A state-of-the-art review of in situ combustion projects has been given by Farouq Ali in 1972, by Chu in 1977 and 1982, and by White in 1985, among others.

### Reservoir and Fluid Characteristics

The in situ combustion process has been successfully applied to a variety of reservoirs having

- depth, between 160 ft (Suplacu de Barcau, Romania) and 11,400 ft (West Heidelberg, Mississippi, United States)
- formations, sand and sandstones
- average net pay thickness between 4 ft (Gloriana, Texas) and 120 ft (Brea Olinda, California, and West Newport, California)
- dip between 0 and 5° (Bellevue, Louisiana, and Glen Hummel, Texas) or less than 45° (Midway Sunset, California)
- porosity between 16 and 37 percent
- absolute permeability between 40 and 8000 md

The majority of in situ combustion field projects have been conducted in reservoirs with heavy crude oil, between 15 and 22 API, and viscosities between 10 and 6000 cp at reservoir temperature. There are also successful projects in light oil reservoirs with viscosities less than 10 cp such as May-Libby, Louisiana, with 3 cp, or Sloss, Nebraska, with 0.8 cp (Parrish, Pollock, and Craig, 1974).

The in situ combustion method can also be used for very viscous oils or for tar sand reservoirs. In these formations, to obtain an acceptable pressure level and oil mobility, the process must be combined with fracturing techniques and steam stimulation.

Regarding the reservoir oil saturation at the start of the process, the values vary between 30 percent at Sloss, Nebraska, and 94 percent at East Tia Juana, Venezuela. An oil saturation of 30 to 40 percent at the start of the process means that the reservoir has been waterflooded previously or has been produced by cyclic steam and/or steam drive (Counihan, 1977). A value of oil saturation higher than 85 percent (for example, 94 percent—East Tia Juana, Venezuela) shows an oil–wet rock reservoir.

In the life of a reservoir, the in situ combustion process can be applied:

- as the primary method for developing production of a heavy oil reservoir without natural oil mobility characteristics.
- as a secondary method, after the natural depletion of reservoir.
- as a tertiary method, after waterflooding or cyclic steam and/or steam drive operations.

### Pilot Tests

Prior to field development of in situ combustion, pilot tests involving a small portion of the reservoir are conducted in order to check the laboratory results and to obtain

- data to design surface injection facilities (injection pressure, injection rate).
- information regarding formation damage, sand invasion, emulsions, corrosions, and so on.
- data about the existence of preferential fluid flow, directional permeabilities, gas fingering, impermeable barriers, and so on.
- production rates, temperatures, and effluent gas measurements; burning front configuration and velocity, and so on.
- information regarding the oil recovery (only if the field pilot is considered a confined pattern).

The conventional pattern for most pilots is the inverted five-spot array with one injector in the middle and four producers in the corners of a square (Figure 5-8a).

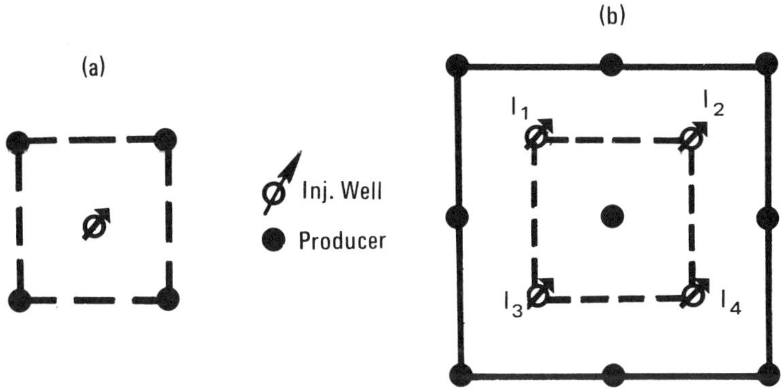


**Fig. 5-8** Open inverted (a) and confined direct (b) five-spot

If the combustion is performed in four adjacent inverted five-spot patterns with injection wells $I_1$, $I_2$, $I_3$, and $I_4$ (Figure 5-8b), the four injection wells then encompass a direct five-spot pattern which may be considered a confined pattern with a central production well P.

The cumulative oil production of the well P divided to the initial oil in place, corresponding to the direct five-spot pattern area, gives an estimate of the oil recovery.

**Example 5-4.** A combustion test in a confined pattern was conducted on a depleted oil reservoir with a current oil recovery of 10 percent. Estimate the final oil recovery expected after the commercial development of the in situ combustion method, given the following:

| | |
|---|---|
| Confined area | 1.25 acres |
| Net thickness | 20 ft |
| Effective porosity | 24% |
| Irreductible water saturation | 25% |
| Oil formation volume factor | |
| Initial | 1.12 |
| Current | 1.05 |
| Cumulative oil production of the central well P, as the effect of combustion | $\Delta N_c = 12{,}470$ bbl |

SOLUTION The initial oil in place is given by

$$N = 7758 \frac{\phi(1 - S_{wi})}{B_{oi}} A \times h \tag{5-4}$$

$$N = 7758 \frac{\text{bbl}}{\text{acre-ft}} \times \frac{(0.24)(0.75)}{1.12} 1.25 \text{ acres} \times 20 \text{ ft}$$

$$N = 31{,}170 \text{ bbl of oil}$$

The increase in oil recovery as the result of in situ combustion is

$$E_{R_c} = \frac{\Delta N_c}{N} = \frac{12{,}470}{31{,}170} = 0.40 \text{ or } 40\%$$

The final oil recovery expected is

$$0.10 + 0.40 = 0.50 \qquad \text{or} \qquad E_R = 50\%$$

After the evaluation of the pilot test results, implementation of in situ combustion is expanded using different flood patterns.

## 5-9 FIELD DEVELOPMENT

The expansion of the in situ combustion process to full scale may use well patterns based on the five-spot or staggered line drive when directional permeabilities are known to exist. Existing wells are utilized if the old wells are in good working order. New injection and production wells have to be drilled as well.

Basically, the in situ combustion process is similar to gas injection. One of the most important factors to be considered is the influence of the gravitational effect on the flow of fluids.

In Figure 5-9a, Well 1 will produce more oil at higher rates because of a slightly isobatic difference. In Figure 5-9b, the combustion gas segregation upward is accentuated as the formation dip increases and air is injected in the higher regions. Also, the air's tendency to override the top of the formation is diminished and the oil flows downdip where it can be produced at a lower air-oil ratio and at higher rates.

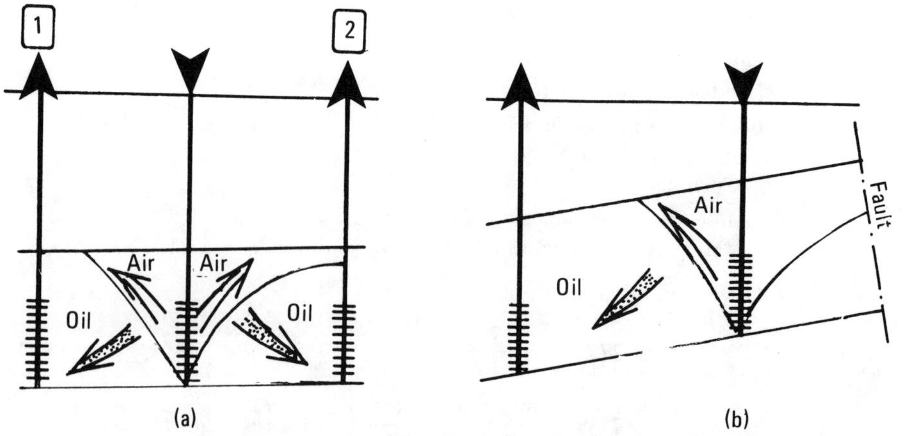


**Fig. 5-9** Gravitational effect on the fluids flow in an in situ combustion process

For this reason combustion operations in dip reservoirs should start upstructure, at the uppermost part of the reservoir, and progress downward. Where large gas caps exist, combustion operations have to be avoided or started down the structure. For flat reservoirs, where the injected air will not move updip, uniform well patterns with central injection wells are used.

The main characteristics and results of in situ combustion projects are presented through three examples of field applications. The Moco Zone project, in California, was one of the first commercial in situ combustion processes initiated. The Suplacu de Barcau Field, Romania, is the largest developed process known. The Heidelberg Field, Mississippi, is the deepest applied process.

### The Moco Zone, California, United States

The Moco Zone project is one of the Midway Sunset fields in California (Gates and Sklar, 1985).

**Characteristics.** The reservoir is an anticline at a depth of 2100–2700 feet. The six major sands of the pay zone have 129 feet total net thickness and are inclined up to 45° to the north and 20° to the south (Figure 5-10).

The heavy crude oil in this reservoir has 14.5 °API gravity and 110 cp viscosity at reservoir temperature. The ultimate oil recovery under natural depletion was estimated at 17 percent. Considering the gravitational effect and the pay zone's high specific permeability of 1575 md, the above recovery value is possible and could even be exceeded if the reservoir is produced for an extended period.

**Performances.** Combustion was initiated in 1960 by spontaneous ignition after about 18 days of air injection in a well located high in the structure. As the process was continued the combustion front formed and moved downward on the flanks of the anticline. After 10 years the total air injection rate into five injectors was about 6000 Mcf/day and the oil rate from 30 producers was 1600 bbl/day. The cumulative air-oil ratio was 2890 scf/bbl and the oil recovery reached at the same time was 24 percent of the initial oil in place.

**Experience.** As one of the first in situ combustion processes successfully operated, this project provided useful experience regarding results and operating conditions.

- After the compressor capacity reaches the projected injection rate, any decrease of this rate corresponds with an oil rate decline.
- The pressure that forms in the secondary gas cap high in the structure as a result of the presence of combustion gas and air has to be kept constant, with no leakage to the adjacent formations or through old wells.

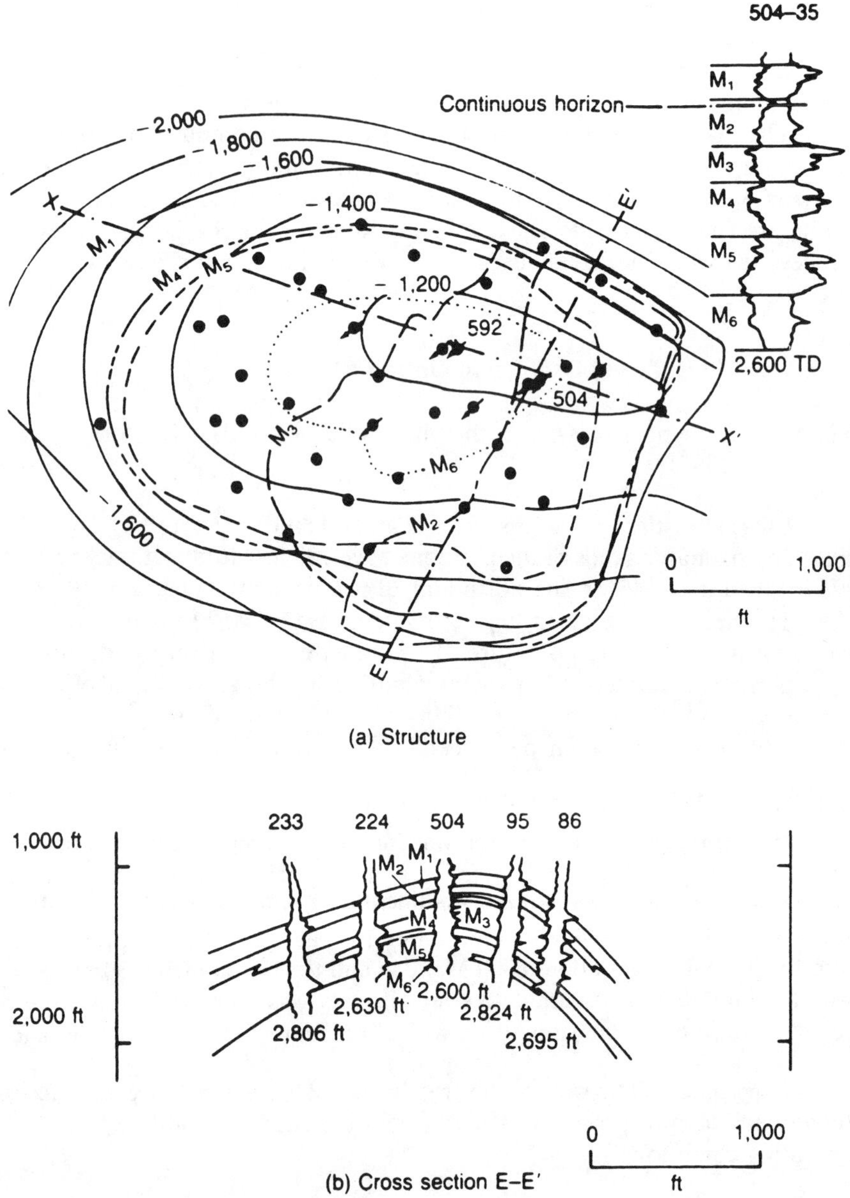


**Fig. 5-10** Moco Zone structure and cross section showing centrally located injection wells (From Gates and Sklar, 1985)

- Any increase in the air injection rate means more fluid displacement downdip, which must be sustained with improved well capacity rates or with more production wells.
- Low air-oil ratio and high oxygen efficiency for the wells located downstructure is due to the gravitational effect.
- Spontaneous ignition may take place not at the formation face near the wellbore but in the formation a short distance from the well. In this case the liner or the well casing can be damaged by the back movement of the burning front. For this reason cemented and perforated liner installation is recommended.
- Sand invasion is prevented by the classic method of gravel packing.
- The bottom hole temperatures ought to be measured frequently at the producers. Hot temperatures have to be prevented by injecting cooling water into the tubing-casing annulus.

### Suplacu de Barcau, Romania

Suplacu de Barcau field, situated in northwestern Romania, is the world's largest in situ combustion process (Carcoana et al., 1976, 1983, 1990; Gadelle et al., 1981).

**Characteristics.** The shallow oil reservoir of Suplacu de Barcau (164- to 656-ft depth) consists of slightly shaley unconsolidated average to coarse sands with 0.32 porosity and 1700 to 2000 md permeability. The reservoir, an east-west monocline with 2 to 7° dip north has a net pay thickness of 33 feet and an original oil in place reserve of 295 MM bbl. A general view of the field is shown in Figure 5-11. The central part of the zone subjected to combustion

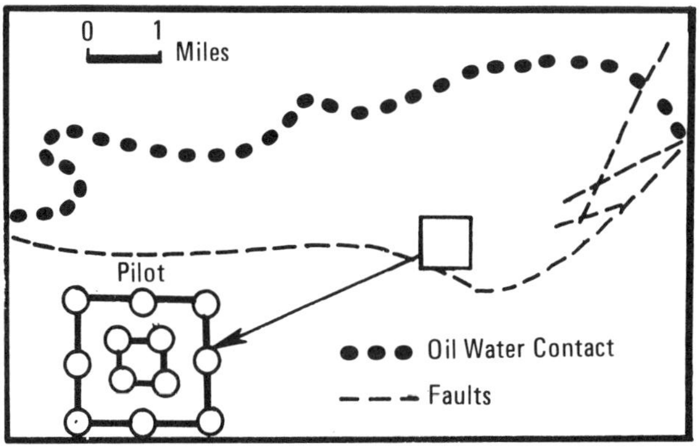


**Fig. 5-11** Suplacu de Barcau field (From Carcoana, 1990)

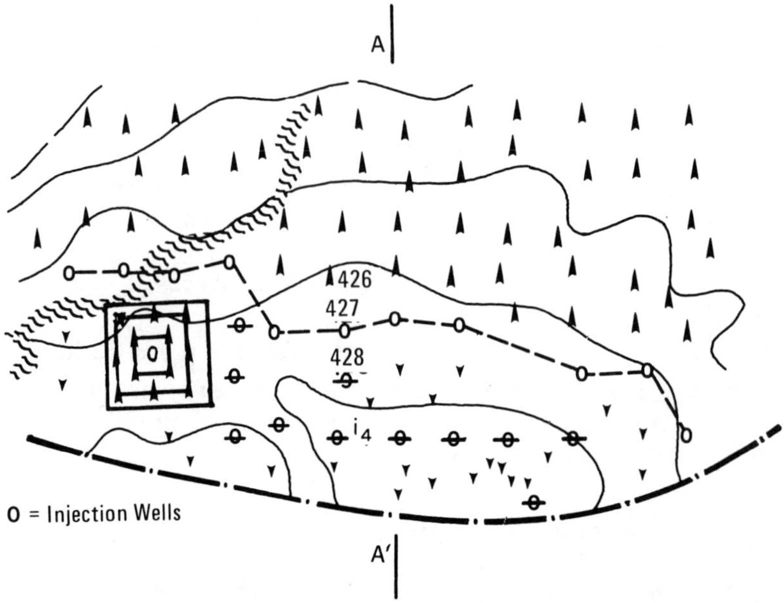


**Fig. 5-12** Line drive development of combustion front

is presented in Figure 5-12, and a cross section of the reservoir is shown in Figure 5-13.

**Process feasibility.** The reservoir was discovered in 1959, and the first well's oil production rate was up to 10 bbl/day per well. Despite the formation's shallow depth, the production rate was not encouraging, and the estimated final recovery was 9 percent of original oil in place. It was felt that the only way to increase the oil recovery and production rate would be to use thermal methods. The lack of large-capacity steam generators led to the decision favoring in situ combustion.

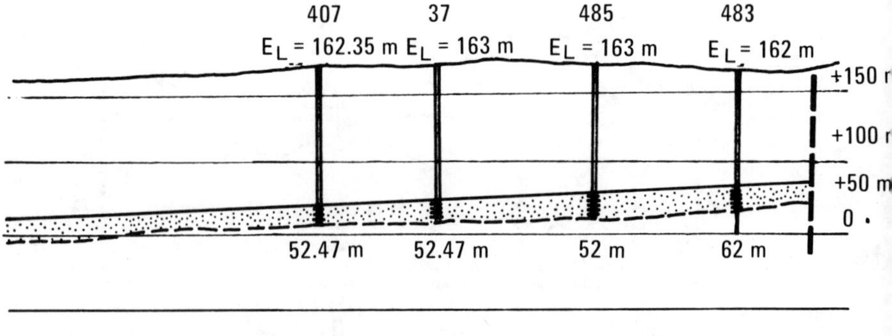


**Fig. 5-13** Cross section A-A′

**Ignition.** The field test began in 1964 on a 1.25-acre inverted five-spot pilot located up structure. The combustion chamber of a gas-fueled burner device (Aldea and Petcovici, 1968; Burger and Sourieau, 1985) was placed at the top of the formation. An electric heating element is used to ignite a mixture of combustible gas and primary air (Figure 5-14). Secondary air is injected through the annulus. The air injected combines with the hot burned gas at the outlet of the combustion chamber and heats the formation around the well. The ignition of the oil is achieved and is sustained by the continuous air injection. Other ignition methods were also used, such as electrical ignition with a 30-KW electric heater or chemical ignition with linseed oil. In this last case a slug of 6–12 bbl of oil with a high oxidation rate (linseed oil) is placed at the bottom of the well in front of the formation and is followed by air injection. The formation temperature increases and the oil around the wellbore is ignited.

**First results and expansion.** The test results were very encouraging. The oil rate increased up to 30 times at the wells located down on the north side of the pilot. Immediately the pilot was expanded to a 5-acre nine-spot pattern using the same injection well, and between 1967 and 1971 six other 8–10-acre nine-spot patterns were subjected to combustion. The total amount of air injected into the zone reached 15.9 MM scf/day ($445 \times 10^3$ $m^3$/day) and the average air-oil ratio was 8.4 M scf/bbl. Other production wells situated downdip, outside the north side of the patterns, were influenced by combustion and showed increased oil rates and combustion gas in the effluent. The combined behavior of the producers made evident the contact made between the combustion fronts of adjacent patterns and later, the development of a line drive. As a consequence, the nine-spot patterns were replaced starting in 1970 by line drive exploitation. In this way, the combustion front moves downdip from the line of injectors.

In 1976 a new compression plant started up with 10 1600-KW compressors, each capable of supplying 5.65 MM scf/day of air at 220 psia pressure. In 1980 the air injection rate was 64 MM scf/day ($1.81 \times 10^6$ $m^3$/day) through 45–50 injection wells separated by a normal well spacing of 330 feet and supplying a combustion front 3 miles long. Wet combustion was tested in nine wells in 1976 and expanded to 20 wells in 1978–1979. The injection cycle was 10 days of air injection and 2 days of water injection with an injected water-air ratio between 0.09 to 0.18 bbl/M scf. Also, water injection in the burned zone was performed in wells surpassed by the combustion front. The injection capacity increased to 100 MM scf/day ($2.8 \times 10^6$ $m^3$/day) in 1983 and to 120 MM scf/day ($3.4 \times 10^6$ $m^3$/day) in 1988, with 100 combustion wells and 600 producers influenced by a combustion front 5 miles (8 km) long.

**Performance.** Because of the good results of combustion, the reservoir was gradually developed for production. The producers were drilled all over the productive area in a regular pattern except in the area where Suplacu de

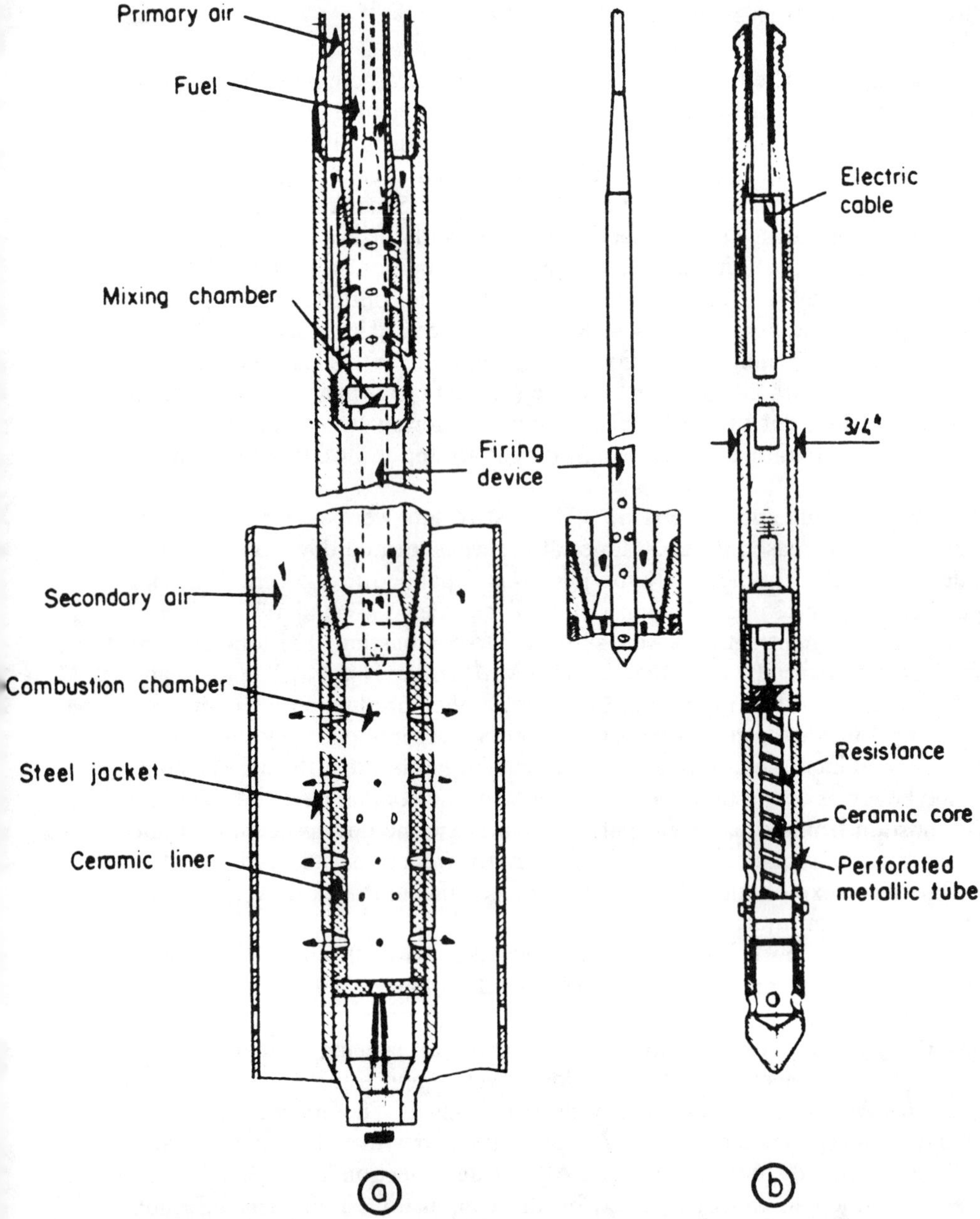


**Fig. 5-14** Diagram of a burner (Document ICPPG), Romania. (a) General diagram. (b) Detail of the ignition device

Barcau village overlays the reservoir. The combustion process starting up the structure and moving down influenced the first two rows of wells, which produced with significant oil rate increases. The next three rows of wells were not as influenced by combustion and had low rate increases and/or combustion gases in the effluent. All these wells are considered to be wells influenced by combustion. The total oil production of the reservoir is the oil from the influenced zone plus the oil produced by wells located in areas still unaffected by combustion.

As shown in Figure 5-15 the increase in oil production corresponded to the increase of the air injection capacity. Thus, oil production from the influenced zone averaged 2200 bbl/day (350 $m^3$/day) in 1974, with 84 producers influenced by combustion. It rose to 6300 bbl/day (1000 $m^3$/day) in 1978 and to 10,400 bbl/day (1650 $m^3$/day) in 1987 with approximately 600 wells affected by the total process.

The average air-oil ratio was maintained between 9.5 M scf/bbl and 11.3 M scf/bbl (1600–2000 $sm^3/m^3$) within the 1973–1979 period and increased to 14.2 M scf/bbl (2500 $sm^3/m^3$) after 1985.

**Oil recovery.** An initial evaluation of the ultimate oil recovery factor was made in July 1970 for the zone affected by combustion at that time. The wells were separated into five groups, each having similar characteristics with regard to oil and gas flow rate histories, downhole temperatures, and the amount of CO, $CO_2$, and $O_2$ present in the gas produced. Production forecasts were made for each group of wells using the production rate decline curves, and the ultimate recovery rate factor was estimated at 35 percent of OOIP. By 1975 the recovery rate had already reached 38.2 percent and in 1980 47.5 percent for the same zone initially affected by combustion. Evaluation of the ultimate oil recovery made after 1980 for the entire reservoir showed a value of 52 percent of OOIP.

It was not possible to determine precisely the location of the *combustion front*. Initially a combustion front is cylindrical in shape (1) until the influence of the producing well causes cusping (2) (Figure 5-16). The oil rate increases, the temperature of the liquids produced increases slowly and their characteristics are changed. The water produced becomes yellow, the pH and salinity decrease, and the sulfate and iron content increase. The oil, less viscous, has a characteristic odor. The produced gases consist of 11 to 15 percent $CO_2$, nitrogen, traces of CO, sometimes $H_2S$, and possibly unreacted $O_2$. The presence of the oxygen may indicate air channeling, a low oxygen efficiency, or a quenched combustion. The amount and composition of the combustion gas produced makes it possible to calculate the amount of carbon burned and the reservoir volume swept by the combustion front. Also, fluid temperatures measured periodically at the producers were plotted as isothermal contour maps. The isothermal curves obtained at different times were compared to determine the prevailing directions and velocities of the combustion front. Figure 5-17 shows the combustion front location at the end of 1970 before the replacement of the nine-spot well pattern with line drive.

The *transfer of air injection* to producers already passed by the combustion front was made with the intention of improving the vertical sweep efficiency and causing the line of injectors to approach the combustion front.

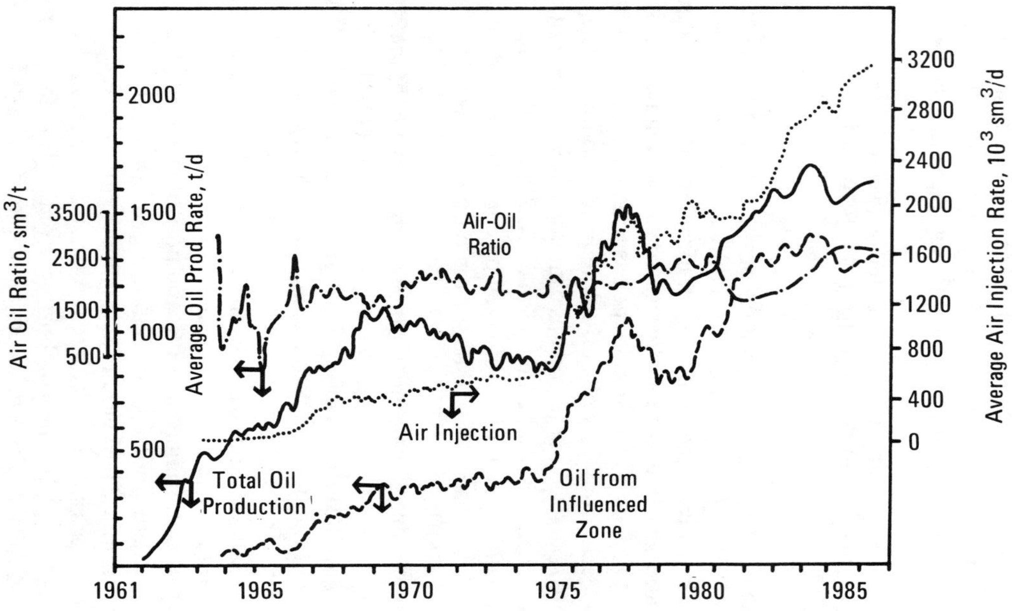


**Fig. 5-15** Field performance (From Turta et al., 1988)

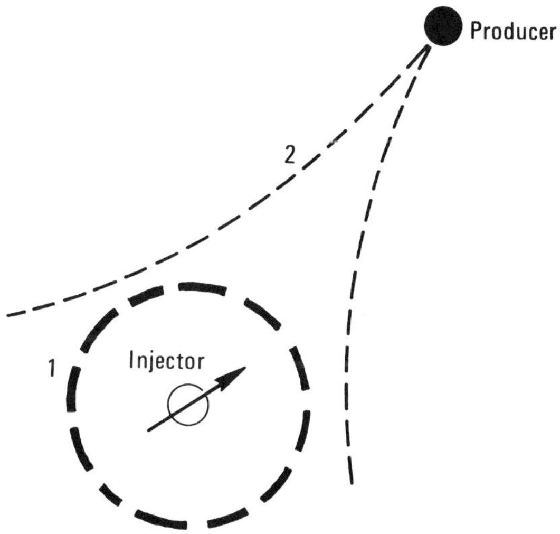


**Fig. 5-16** Initial shape of combustion front

Observation wells drilled in the burned zone showed that approximately the top half of the pay zone was burned, followed by a coke layer and below, the unburned oil saturated zone. For example, the cross section (Fig. 5-18a) illustrates the moment when the combustion front approaches producer "428." The temperature measured at well "428" is rising, the oil rate has a sharp exponential decline and hot combustion gases are produced. To avoid damages, well "428" has to be shut down, and wells "427" and "426" should be ready to produce the oil bank. As the combustion front advances, the distance between the line of injectors and the wells influenced by combustion increases. The approach of the injectors line was made by transferring air injection from well "$I_4$" to well "428" and so on (Figure 5-18b). Air efficiency and vertical sweep efficiency were increased. The air injection was transferred directly, since the formation temperature was appropriate for spontaneous ignition to take place.

*The oil recovery* was checked systematically using detailed contour maps for the top and base of the formation (Machedon et al., 1975). The original oil in place was calculated for each section of the reservoir between two isobaths from 10 to 10 m (33 ft). The cumulative oil produced by the wells in each section gives the recovery factor at any particular time (Figure 5-19). The flow of oil between sections takes place only downward, and the amount that leaves one section is produced through the wells of the section below.

**Difficulties.** One of the major difficulties observed at the beginning was the leakage of combustion gases to the surface. Because formerly producing wells were unplugged or improperly sealed, leaking casings were acting as conduits for gas and/or steam to move to the surface. Mud and steam volcanoes

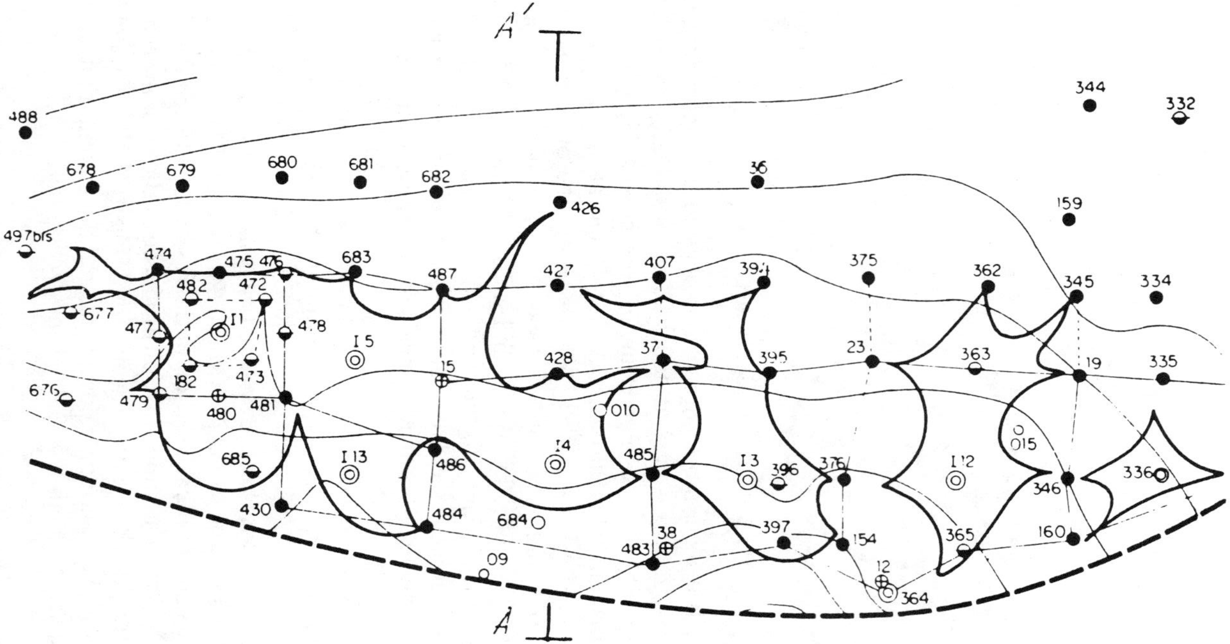


Fig. 5-17 Suplacu de Barcau field (Romania). Combustion zone at the end of 1970 (From Petcovici, 1970)

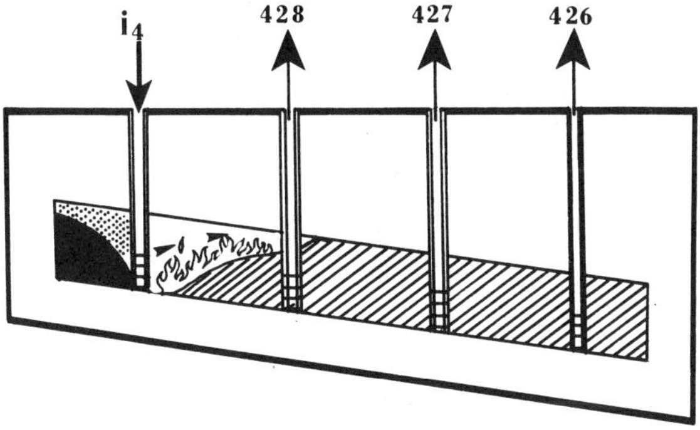


**Fig. 5-18a** First producer row reached by combustion front (From Carcoana, 1990)

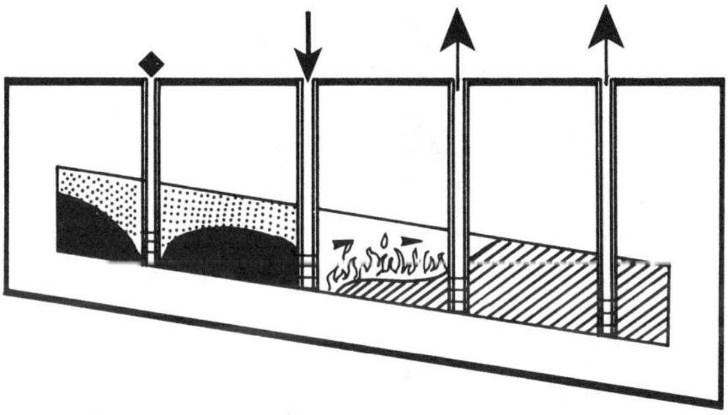

**Fig. 5-18b** Transfer of air injection from well I4 to well 428 (From Carcoana, 1990)

appeared through or between wells adjacent to air injectors located high on the structure. Also, the presence of hazardous gases was detected in the basements of some houses in the area where a 700-family village overlapped the combustion zone. Damaged old wells or noneffective well completions had to be sealed by repeatedly injecting high-viscosity mud below the fracture pressure. Even after the combustion front had advanced down the structure and the injection line was far from the area, gas leakage still appeared. A huge "valve" to the secondary gas cap formed, indicating the need to maintain the balance between air injection and the fluid produced at a pressure below the formation parting pressure. Any pressure increase was corrected either by more fluid produced through infill drilling ahead of the front and steam stimulation of the low productivity wells or by reducing the air injection rate. As a matter of fact, by using these means the control of a uniform and isobathic advancement of the burning front down structure was achieved.

• Despite the fact that 130-ft (40-m) high stacks were designed to vent the large volume of effluent gas, their efficiency diminished with atmospheric

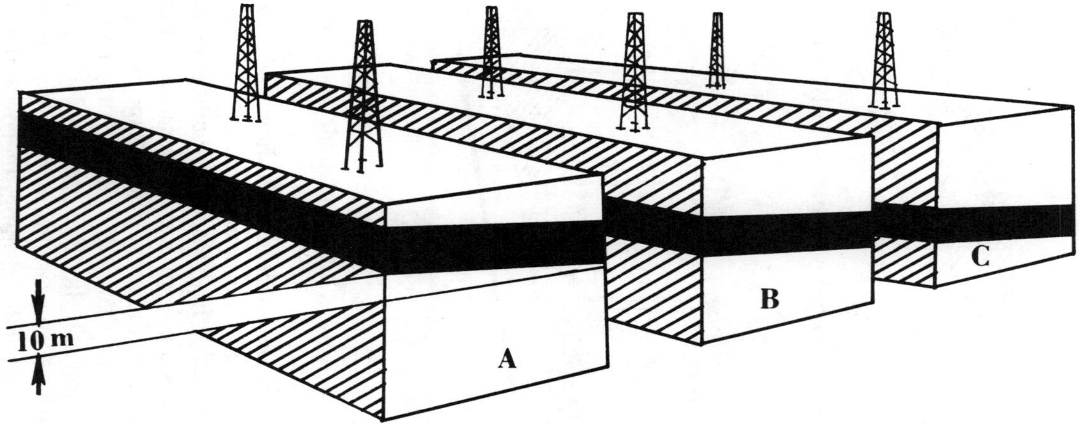


Fig. 5-19 Idealized reservoir sections between two contours (From Carcoana, 1990)

pressure increases. This caused the air around the area to "smell" and sometimes the presence of $CO_2$ and CO was detected.

• From the need to operate the producers with practically no back pressure on the formation (open casings and pipes to oil tanks) arose the threat of combustion gases in the surrounding area. Relocating the village situated in this oil field was considered as a last resort because of the difficulty of controlling this threat.

• Other difficulties encountered were injectivity reduction, air and/or combustion channeling, stable emulsions, excessive heating of some producers, and failure to reduce the back pressure on the formation. Remedial operations to prevent invading sand and coke from plugging the borehole or downhole equipment were also greatly needed. It is important to emphasize that the cyclic steam injection process was used to efficiently heat cold zones ahead of the combustion front or to clear the producer's perforated interval. Figure 5-20 shows the area affected by combustion at the end of 1987.

### Heidelberg Field, Mississippi, United States

The Heidelberg Field is represented by the Cotton Valley sands 4 and 5, which are the major oil reservoirs where the deepest in situ combustion project (average depth 11,400 ft) was commercially developed (Huffman and Benton, 1983).

**Characteristics.** The structure is a 10° dip monocline lying on the west flank of a salt intrusion dome. The reservoirs with a total effective thickness of 65 feet and a productive area of 350 acres are confined updip by

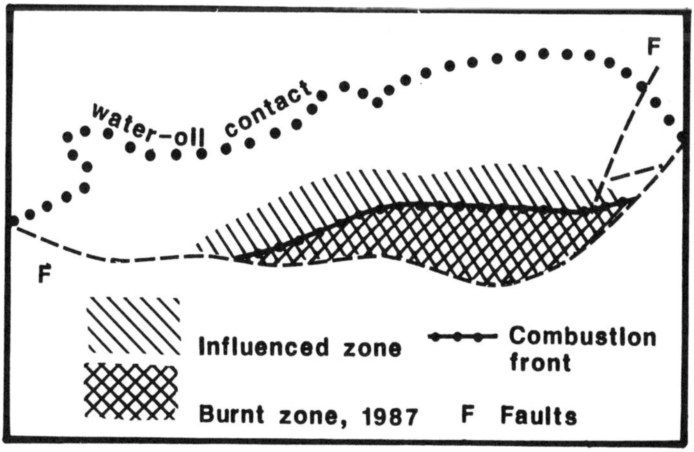


**Fig. 5-20** Area affected by combustion (From Carcoana, 1990)

the salt and downdip by an immobile asphalt tar deposit. The 27 ° API gravity oil (15° downdip, near the tar deposit) has a 6-cp viscosity at reservoir temperature (220 °F). The initial pressure was 5100 psig and the saturation pressure 930 psig. The high degree of undersaturation corresponds to a constant volume type reservoir, characterized by expansion of reservoir rock and fluid content recovery mechanism.

**Performance.** The primary recovery mechanism explains the sharp decline of primary production, the low oil recovery value (6.1 percent of original oil in place), and a pressure decline from the initial value of 5100 psig to 1500 psig.

**In situ combustion performance.** To recover more of the remaining oil, a pressure maintenance project was initiated using air injection for in situ combustion.

Beginning in 1971, air was injected into sand 5 through well 5-6-2 located upstructure. This well was converted to air injection (spontaneous ignition). The presence of combustion gas after three months and an increase of the oil rate after six months was observed in wells 5-12-1 and 6-1-1 located downstructure. Air injection was extended to the other two wells, and additional producers were drilled downdip. Sand 4 was also involved in the process, and because production and injection data indicated communication between sand 4 and 5 in different wells, behind the pipe, both sands were considered a single hydrodynamic unit (Figure 5-21).

The performance history of the project is shown in Figure 5-22. The same direct relationship between air injection and oil production is observed.

At the $1 \times 10^6$ scf/day ($28 \times 10^3$ m$^3$/day) air injection rate, oil production increased until the end of 1976 to an average of 400 bbl/day compared with 47 bbl/day before the process started. By adding new producers drilled downstructure, treating producers with acid stimulation, and expanding air injection into two other wells, oil production was increased to 2000 bbl/day in 1980. The injection rate and pressure at this time were about $5 \times 10^6$ scf/day and 2500 to 3000 psig, respectively. The cumulative air-oil ratio was 4 M scf/bbl, and combustion efficiency was high. The result of experimental flue gas injection demonstrated that flue gas could be injected without problems.

In the final design, one injector was for flue gas injection in sand 4, one was for air injection in sand 5, and the third air injector was a dual well in sands 4 and 5. Air for the project was supplied by five compressors with a total capacity of about $6.2 \times 10^6$ scf/day at a discharge pressure of 5000 psig.

From a total of 14 producers, 7 wells were producing from sand 4 and 7 from sand 5. In August 1982 the oil recovery reached 16 percent, and the ultimate oil recovery from both sands was estimated to be 41 percent of the original oil in place (Huffman and Benton, 1983).

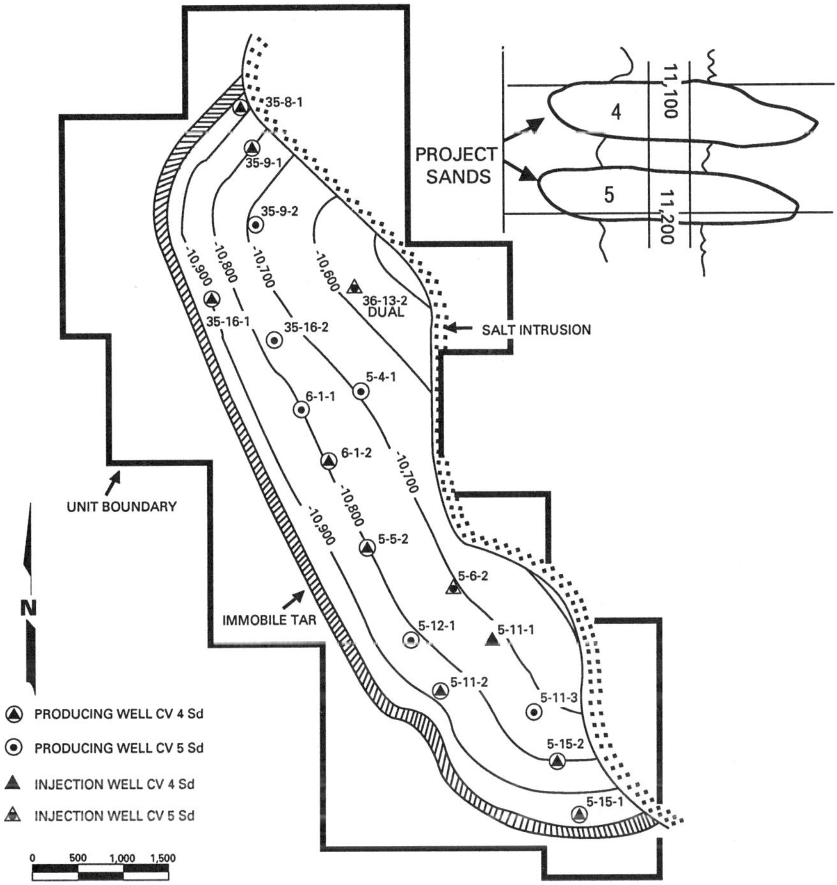


**Fig. 5-21** Top of sand 4 (From Huffmann et al., 1983)

**Experience.** The very successful West Heidelberg project is the deepest in situ combustion-assisted pressure maintenance operation. It has been in operation for 19 years and has provided the following valuable experience.

- In situ combustion is not limited by depth, especially when the reservoir is depleted by primary production.
- Air injection is effective, assuring a pressure maintenance mechanism for oil displacement to the downstructure wells in addition to the support of a combustion front.
- Even for deep reservoirs, additional new producers drilled downstructure ahead of the combustion front are justified by the existing wells response.

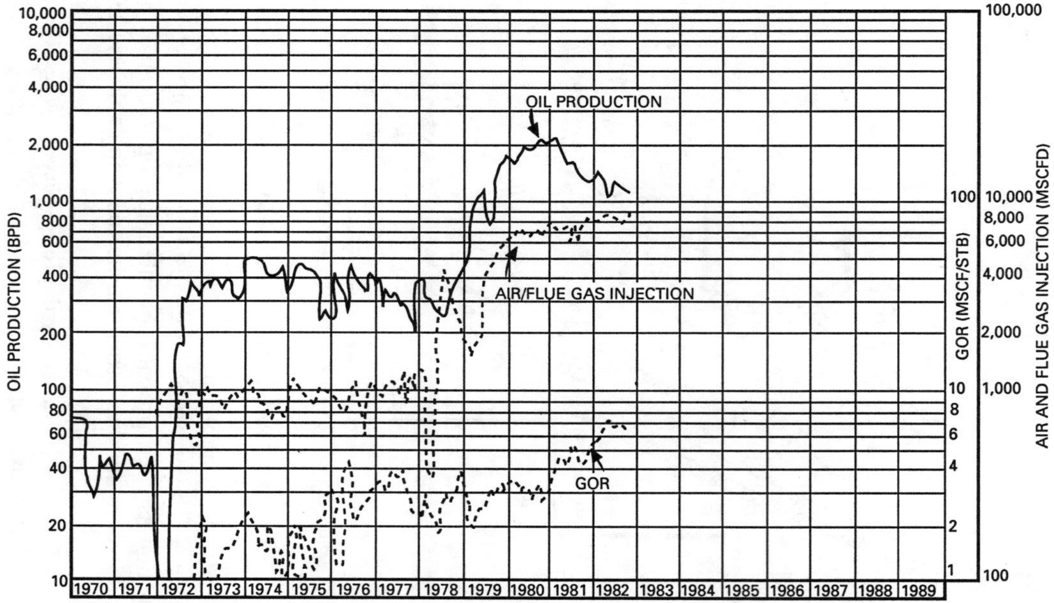


**Fig. 5-22** Project performance sands 4 and 5 (From Huffman et al., 1983)

- A program of well workover, stimulation by acidization, infill drilling, and expanded air injection assured the success of the West Heidelberg combustion process.
- High-pressure injection of clean, dry air has been continued by maintaining the compressor units and lines with alternate lubrication, cooling, and washing operations.

## 5-10 PATTERN SWEEP, INVASION, AND DISPLACEMENT EFFICIENCIES

Two of the most important facts to know in order to control and evaluate an in situ combustion process is the volume of the burned zone and also the reservoir volume affected by heat and its carrying agents.

The specific notations defined by SPE (1986) are

$E_A$ = areal efficiency (used in describing results of model studies only), is the area swept in a model divided by total model reservoir area (see $E_p$).

$E_p$ = pattern sweep efficiency (developed from areal efficiency by proper weighting for variations in net pay thickness, porosity, and hydrocarbon saturation), is the hydrocarbon pore space enclosed behind the injected fluid or heat front divided by the total hydrocarbon pore space of the reservoir or project.

$E_I$ = invasion (vertical) efficiency, is the hydrocarbon pore space invaded (affected, contacted) by the injection fluid or heat front divided by the hydrocarbon pore space enclosed in all layers behind the injected fluid or heat front.

$E_D$ = displacement efficiency, is the volume of hydrocarbons displaced from individual pores or small groups of pores divided by the volume of hydrocarbon in the same pores just prior to displacement.

$E_V$ = volumetric efficiency, is the product of pattern sweep and vertical efficiencies.

$$E_V = E_p E_I \tag{5-5}$$

$E_{Db}$ = displacement efficiency from the burned portion of the in situ combustion pattern (considered 100 percent).

$E_{Du}$ = displacement efficiency from the unburned portion of in situ combustion pattern.

$E_R$ = overall reservoir recovery efficiency, is the volume of oil recovered divided by the volume of hydrocarbon in place at the start of the project.

$$E_R = E_p E_I E_D = E_V E_D \tag{5-6}$$

The scheme of a combustion front advancing in an oil reservoir is given in Figure 5-23 with an explicit presentation of the notions just defined (Burger and Carcoana, 1975).

The burned and unburned zones, the pattern sweep efficiency and the vertical efficiency appear clearly enough. It is important to note that pilot tests in the field and in commercial scale development of in situ combustion pointed to the existence of two unburned zones:

- *Unburned zone I*: inside the horizontal projection of the area covered by the combustion front; this zone was not observed in the lab experiments.
- *Unburned zone II*: strictly speaking, the zone in front of the combustion zone.

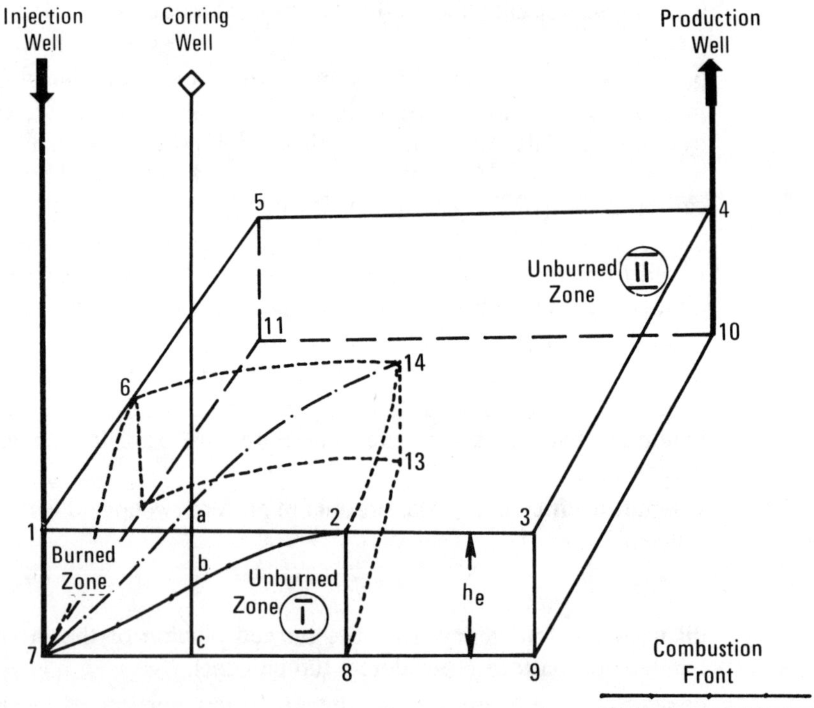


**Fig. 5-23** Unburned zones, pattern sweep, and invasion efficiencies

**TABLE 5-1.** Pattern Sweep Efficiency

| Field | Pattern Sweep Efficiency, % |
|---|---|
| Delaware Childers, Oklahoma | ~100 |
| South Belridge, California | ~100 |
| S. Oklahoma, Oklahoma | 85 |
| S.E. Kansas (Humbolt Chanute) | 70 |
| Shannon, Wyoming | 43 |
| Delhi, Louisiana | 50 |
| Suplacu de Barcau, Romania | 85 |
| A13 Sloss, Nebraska | 50 |
| Niitsu, Japan | ~100 |

## Sweep Efficiencies

The pattern sweep efficiency $E_p$ is determined approximately in the field by controlling the location of the combustion front at a given time. For an enclosed five-spot well pattern the flow of injected air and combustion gases relative to the movement of the oil corresponds to an infinite mobility ratio for choosing the areal sweep efficiency. Different modeling techniques show in this case an areal sweep efficiency of 62.6 percent, limited by practical considerations to 55 percent (Nelson and McNiel, 1961).

Some of the sweep efficiency values obtained in different five-spot pattern projects are presented in Table 5-1.

The higher values occurred because the inverted five-spot patterns were not confined.

The vertical sweep efficiency $E_I$ is determined by taking postcombustion core samples. These indicate thickness of the burned zone and the amount of coke formed in the zone swept by the combustion front. It can also be determined taking into consideration the fact that the resistivity of the burned zone is 2 to 10 times higher than before combustion. The vertical sweep efficiency $E_I$ is largely dependent on the thickness of the formation and has higher values for thinner pay zones.

The Delhi (Louisiana) project, where the thickness of the pay zone was only 2.5 ft, was characterized by a vertical sweep efficiency $E_I$ value close to unity. In the North Tisdale project, core samples taken from a well drilled in the area swept by combustion showed vertical efficiency $E_I = 27\%$. This time the burned portion of the formation thickness was located in the middle of the pay zone and not in the upper half as usual. Other $E_I$ values reported ranged between 26 percent (South Oklahoma) to 50 percent (Trix Liz, Texas) and 60 percent (Emma Fry III, Illinois, and South Belridge, California), with an average value of 35 to 40 percent obtained at Suplacu de Barcau (Romania).

It is important to note that the burned zone is a clean sand with improved porosity and permeability.

The displacement efficiency $E_{Du}$ for the *unburned zone I* is known by the oil saturation values taken from core samples. For instance, the North Tisdale project had a displacement efficiency $E_{Du} = 50\%$ of the initial oil content, and the South Belridge (California) oil saturation value within the *unburned zone I* was less than half of the initial oil content. In the Suplacu de Barcau project the bottom coring of three wells drilled in the area swept by the combustion front shows an average value of $E_{Du} = 43\%$ (Turta, 1972, 1974).

The displacement efficiency $E_{Du}$ for the *unburned zone II* is very difficult to estimate. The oil from the *unburned zone II* is influenced by heat conduction. This increases its mobility and thereby makes it more easily displaced by the uncontrolled movement of the combustion front fluids. A bottom coring well drilled on a water-based mud in the *unburned area II* in the North Tisdale combustion project showed the following oil saturations (Martin and Alexander, 1971):

| | |
|---|---|
| 40% | at the top of the production formation |
| 32% | in the median zone |
| 50% | at the bottom of the pay zone |

compared with $S_o = 60\%$ at the start of the project. Assuming an average value of 40 percent, the displacement efficiency in *zone II* is $E_{Du} = 33\%$.

## 5-11 OIL CONSUMED IN SITU

As a result of distillation and thermal cracking, the oil consumed in situ deposits on the rock the coke available for combustion. The amount of oil (fuel) consumed is a function of the reservoir characteristics and of the volumetric sweep efficiency, $E_V$ of the process. For instance, assuming reservoir rock porosity is 26 percent and oil saturation at the start of the project is 70 percent, the oil consumed for a fuel content $C_u$ = from 0.8 to 2.6 $lb_m/ft^3$ varies between 5.8 and 19 percent for $E_V = 100\%$ and between 1.7 and 5.7 percent for $E_V = 30\%$.

**Example 5-5.** Calculate the oil consumed in an in situ combustion process developed on a reservoir with 23 percent porosity and 63 percent oil saturation. The fuel content is 1.7 $lb_m/ft^3$ and the volumetric efficiency is 26 percent.

SOLUTION Oil in place (per unit volume)

$$N = V \times \phi \times S_o = 1\ ft^3 \times 0.23 \times 0.63 = 0.1449\ ft^3$$

Oil consumed per 1 $ft^3$ of burned rock

$$C_u = 1.7 \text{ lb}_m$$

Given the specific gravity coke $SG = 1.2$, the volume of the oil consumed is

$$N_c' = \frac{C_u}{SG \times \rho_{water}} \tag{5-7}$$

$$= \frac{1.7 \text{ lb}_m}{1.2 \times 62.5 \text{ lb}_m/\text{ft}^3} = 0.0226 \text{ ft}^3 \text{ per 1 ft}^3 \text{ of burned rock}$$

and the oil consumed as a percentage of the oil in place, when $E_V = 0.26$, is given by

$$S_{o_{cons}} = 100 \frac{N'c}{N} \times E_V \tag{5-8}$$

$$= 100 \times \frac{0.0226}{0.1449} \times 0.26 = 0.04 \times (100) \text{ or } 4\% \text{ of the oil in place}$$

The value of the parameter $S_{o_{cons}}$ can also be known every moment by solving the ratio

$$S_{o_{cons}} = \frac{\text{cumulative air injected}}{\begin{matrix}\text{minimum air necessary} \\ \text{to burn 1 lb}_m \text{ of fuel}\end{matrix} \times \begin{matrix}\text{oil in place at the start} \\ \text{of the process}\end{matrix}} \tag{5-9}$$

**Example 5-6.** Calculate the oil consumed after 5 years of in situ combustion developed as a primary recovery method. The oil reservoir ($SG = 0.950$) has $157 \times 10^6$ bbl OOIP reserve, and the combustion process is sustained by the injection of $700 \times 10^3$ $ft^3$ air/day through each of the 12 injection wells.

SOLUTION Cumulative air injected is

$$12 \text{ wells} \times 700 \times 10^3 \text{ scf/day} \times 365 \text{ days} \times 5 \text{ yr} = 15.33 \times 10^9 \text{ scf}$$

Minimum air necessary to burn 1 pound of fuel is usually between 160 to 192 scf/$lb_m$. Assuming a value of 186 scf/$lb_m$ and coke's specific gravity $SG = 1.2$, the minimum air necessary to burn 1 bbl of fuel is given by

$$186 \text{ scf/lb}_m \times SG \times \rho_w \text{ lb/ft}^3 = 186 \times 1.2 \times 62.55 = 13{,}961 \text{ scf/ft}^3$$

or

$$\frac{13{,}961 \text{ scf}}{5.614 \text{ bbl}} = 2487 \text{ scf/bbl}$$

For a value of 0.9 oxygen utilization, the oil consumed is

$$S_{o_{cons}} = \frac{15{,}330 \times 10^6 \text{ scf} \times 0.90}{2487 \text{ scf/bbl} \times 157 \times 10^6 \text{ bbls}} \times 100 = 3.5\% \text{ of OOIP}$$

## 5-12 OIL RECOVERY

The value of the oil recovery reported from most of the in situ combustion experiments and projects ranges between 40 and 60 percent of the oil in place at the start of the process. Higher recovery values are suspected to be the result of oil coming from outside the pattern or zone taken into consideration.

The pattern and vertical sweep efficiencies $E_p$ and $E_I$ are estimated using the procedures just described. The displacement efficiency $E_{Du}$ for the *unburned zone I* inside of the area swept by the combustion front is also determined using these procedures. The displacement efficiency in the *zone II* outside of that area is unknown and may be expected to be less than in the underlying zone.

A range of possible oil recovery can be obtained by considering

1. The minimum assumption, when the displacement efficiency is applied only to the *unburned zone I* (no oil produced from the *unburned zone II*).
2. The maximum assumption, when the displacement efficiency is applied to both zones *I* and *II*.

For the unit volume of rock, the oil produced by in situ combustion in reservoir conditions (Burger and Carcoana, 1975) is

$$N_{p_1} = S_o \times \phi \times E_{Du} \times E_p(1 - E_I) + (S_o - S_{o_{\text{cons}}})\phi E_V \tag{5-10}$$

$$N_{p_2} = S_o \times \phi \times E_{Du}(1 - E_V) + (S_o - S_{o_{\text{cons}}})\phi E_V \tag{5-11}$$

and the corresponding recovery factor is

$$E_{R_1} = \frac{N_{p1}}{S_o \phi} \quad \text{and} \quad E_{R_2} = \frac{N_{p2}}{S_o \phi} \tag{5-12}$$

The recovery factor $E_R$ is higher than $E_{R_1}$ and less than $E_{R_2}$ and refers to the oil in place at the start of the project.

**Example 5-7.** Given

| | |
|---|---|
| Oil saturation at the start of the project | $S_o = 0.70$ |
| Effective rock porosity | $\phi = 0.32$ |
| Pattern sweep efficiency | $E_p = 0.55$ |
| Vertical sweep efficiency | $E_I = 0.35$ |
| Displacement efficiency in zone I | $E_{Du} = 0.43$ |
| Oil consumed | $S_{o_{\text{cons}}} = 0.065$ |

calculate the oil recovery.

SOLUTION

**(a)** Calculate oil recovery when displacement efficiency is applied only to the *unburned zone I* and no oil is produced from the zone outside of the area swept by the combustion.

$$\begin{array}{c}\text{oil produced from the } \textit{unburned zone I} \\ \text{inside the area swept by combustion}\end{array} + \begin{array}{c}\text{oil produced from} \\ \text{the } \textit{burned} \text{ zone}\end{array}$$

$$N_{p1} = S_o \times \phi \times E_{Du} \times E_p(1 - E_{\mathrm{I}}) + (S_o - S_{o_{\mathrm{cons}}})\phi E_V$$

$$= 0.70 \times 0.32 \times 0.43 \times 0.55 \times (1 - 0.35)$$

$$+ (0.70 - 0.065) \times 0.32 \times 0.55 \times 0.35$$

$$= 0.0344 + 0.0391 = 0.0735$$

$$E_{\mathrm{R}_1} = \frac{N_{p1}}{S_o \phi} = \frac{0.0735}{0.70 \times 0.32} = 0.328 \text{ or } 32.8\%$$

**(b)** Calculate oil recovery when displacement efficiency is applied to both *zones I and II*.

$$\begin{array}{c}\text{oil produced from the } \textit{unburned} \\ \textit{zones I and II}\end{array} + \begin{array}{c}\text{oil produced from} \\ \text{the } \textit{burned} \text{ zone}\end{array}$$

$$N_{p2} = S_o \times \phi \times E_{Du}(1 - E_V) + (S_o - S_{o_{\mathrm{cons}}})\phi E_V$$

$$= 0.70 \times 0.32 \times 0.43 \times (1 - 0.35 \times 0.55) + 0.0391$$

$$= 0.0777 + 0.0391 = 0.1168$$

$$E_{\mathrm{R}_2} = \frac{N_{p2}}{S_o \phi} = \frac{0.1168}{0.70 \times 0.32} = 0.521 \text{ or } 52.1\%$$

So, the oil recovery based on the data of the problem must be higher than 32.8 percent and less than 52.1 percent of the oil in place at the start of the project.

Figure 5-24 presents a correlation between oil recovery and volume burned at different initial gas saturations. The correlation based on field data was developed by Gates and Ramey (1980) and can be used to estimate the value of the oil recovery.

For instance, the rock volume burned (Example 5-7) $E_V = 0.55 \times 0.35 = 0.1925$ corresponds to a recovery of 40 percent, assuming no initial gas saturation.

Ultimate or final oil recovery of a reservoir is the ratio of the cumulative oil produced from the reservoir (a hydrodynamic unit) under different recovery mechanisms to the original oil in place

$$E_{\mathrm{R_{final}}} = \frac{\Delta N}{N} 100\% \tag{5-13}$$

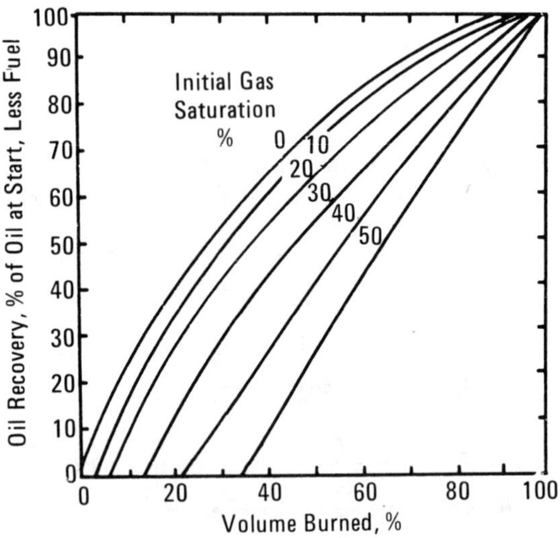


**Fig. 5-24** Estimated oil recovery versus volume burned (From Gates and Ramey, 1980)

**Example 5-8.** Calculate the ultimate oil recovery obtained from an oil reservoir which has been produced under different recovery mechanisms according to the following information:

| | |
|---|---|
| Original oil in place | 10,000,000 bbl |
| Oil produced by primary methods | 600,000 bbl |
| Oil produced by water flooding | 1,100,000 bbl |
| Oil produced by in-situ combustion | 3,200,000 bbl |

SOLUTION Looking at the history of the reservoir, we can assume that oil viscosity and an unfavorable water-oil mobility ratio explains the low amount of oil produced by primary and waterflood methods.

Because of the reservoir's depth and higher water saturation, it was decided to use in situ combustion as the next recovery mechanism so more oil could be obtained. The results presented in Table 5-2 show the following:

- The primary method recovered 6 percent of the oil in place at the start of the project, which represents the OOIP, and recovery also means final oil recovery attained by primary method.
- The waterflood method recovered 11.7 percent of the oil in place at the start of the project; the final oil recovery after waterflooding is 17 percent, expressed as a ratio between the total amount of oil produced (primary + waterflood) and the OOIP.
- The in situ combustion method recovered 38.5 percent of the oil in place at start of the project; the final oil recovery after in situ combustion was 49 percent—

**TABLE 5-2.** Oil Recovery Factor

| Oil in Place (MM bbl) | | Oil Produced (MM bbl) | | Oil Recovery | |
|---|---|---|---|---|---|
| | | | | of Oil in Place at Start of Project | $E_{R\,final}$ |
| Original | 10 | By primary method | 0.6 | $\frac{0.6}{10} \times 100 = 6\%$ | 6% |
| After primary (start waterflood) | 9.4 | By waterflood | 1.1 | $\frac{1.1}{9.4} \times 100 = 11.7\%$ | $\frac{1.7}{10} \times 100 = 17\%$ |
| After waterflood (start combustion) | 8.3 | By in situ combustion | 3.2 | $\frac{3.2}{8.3} \times 100 = 38.5\%$ | $\frac{4.9}{10} \times 100 = 49\%$ |
| Total oil produced | | | 4.9 | $E_{R\,final} = \frac{4.9}{10} \times 100 = 49\%$ | |

the ratio between the total amount of oil produced (primary + waterflood + in situ combustion) and the OOIP.

The amount of oil produced under one or more recovery mechanisms can also be referred to as the original oil in place.

In the example, the effect of waterflood increased the final oil recovery with 11 percent absolute value from 6 to 17 percent, and the effect of in situ combustion increased the final oil recovery with 32 percent absolute value ($3.2/10 \times 100 = 32\%$), from 17 to 49 percent.

The final (ultimate) oil recovery of a reservoir can only be improved by changing its recovery mechanism. Reaching the final oil recovery, within the limits of the same recovery mechanism, depends on the reservoir characteristics and operating conditions.

## 5-13 SCREENING CRITERIA

The dry or wet in situ combustion process has been used at a wide range of oil reservoirs to enhance oil recovery. It is possible to use a screening guide to assess the technical feasibility of the process on a specific oil reservoir (Carcoana, 1976; Chu, 1982; White, 1985).

### Reservoir Characteristics

| | |
|---|---|
| Depth and pressure | Depth higher than 300 ft, the upper limit is conditioned by pressure which increases the cost of air compression. Air injection in shallow reservoirs should avoid reaching the parting pressure of the formation to prevent environmental pollution. |
| Temperature | Not critical. The higher the temperature, the better the chances for spontaneous ignition. |
| Thickness | Greater than 10 ft. |
| Porosity | Higher than 16 to 18 percent. |
| Permeability (absolute) | Higher than 100 md for viscous oils or shallow reservoirs with limited air injection pressures; higher than 30 md for less viscous oils and deeper formations. |
| Oil viscosity | Less than 5000 cp (a moderate high viscosity). |
| Reservoir oil saturation at start of project | Higher than 25 to 30 percent. (In situ combustion may be applied after waterflooding.) |

| | |
|---|---|
| Formation type | Sands and sandstones; also carbonates with primary porosity. |
| Oil type | With asphalt and naphtha compounds; also rocks with metals content, dirty sands. |

In addition to these screening criteria, the recovery mechanism and the reservoir geometry must be considered to assure project success, whether or not other factors are favorable or adverse. Factors favoring the in situ combustion development include

- high values for $\phi k$, $S_o$, effective/gross thickness, and temperature.
- low vertical permeability.
- dependable well cementing job.

Adverse factors are

- high rock heterogeneity for gas and airflow.
- Large gas-cap area.
- emulsifying tendency of the produced oil, sanding, corrosion, and heat damage to casing, cement, tubing, and pumps.
- large thickness and low temperature of the horizontal productive formations.
- the decrease of the vertical efficiency by the gravity override to the upper portion of the reservoir.

Research and more experience are needed to prevent and/or eliminate some of these adverse factors.

## 5-14 INJECTION OF OXYGEN-ENRICHED AIR OR PURE OXYGEN

The performance of in situ combustion can be improved by injecting, instead of air, oxygen-enriched air or pure oxygen. Moore received a U.S. patent on this method in 1965, but only recently has interest in its use increased and several companies now have lab experiments and projects in operation (Cady and Moss, 1981; White and Fairfield, 1982; Moore and Bennion, 1987; Petit, 1987).

Evident advantages of the use of oxygen instead of air are

- The decrease in the amount of gas to be compressed and injected because large volumes of nitrogen are now eliminated. This increases the possibility of using in situ combustion in low permeability reservoirs and reduces the volume of vent gases.

- Much higher solubility of $CO_2$ in the oil and water phases with oil viscosity reduction and significant swelling of the oil. This increases the oil mobility and recovery and reduces the time to complete the project.
- The produced gas, rich in carbon dioxide, can be used as a source for other EOR projects.

However, when using oxygen as injected agent, strict safety precautions are required to prevent the formation of a flammable mixture of oxygen-hydrocarbon gas (Burger and Sourieau, 1985).

The oxygen supply and injection system must be free of hydrocarbon material and not be oxygen reactive. In the reservoir any unreacted oxygen breakthrough should be prevented by shutting in production wells. The downhole and surface equipment is exposed to highly corrosive conditions due to the high partial pressure of $CO_2$ produced together with a hot water phase.

## 5-15 DESIGN OF AN IN SITU COMBUSTION FIELD PILOT

To illustrate the way in which an in situ combustion pilot test can be designed, the method given by Nelson and McNiel (1961) and by C. R. Smith (1985) is presented in Example 5-9.

The schedule of the total cubic feet of air, $Q_a$, injected into a five-spot inverted well pattern proposed by Nelson and McNiel refers to a three-phase air injection flow rate (Figure 5-25).

The first phase is the initial period where the flow rate increases linearly with time up to time $t_1$ when it reaches the constant airflow rate designed.

The burning front moves at a constant velocity $V_b$ to the point where the air compressor capacity is reached (constant airflow rate designed). It remains essentially radial for about 10 percent of the pattern area.

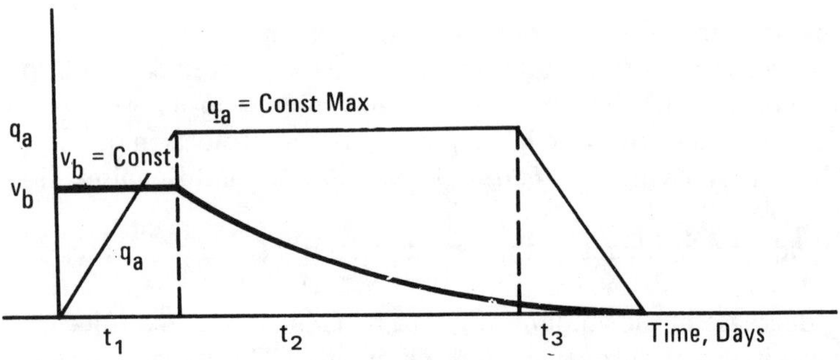


**Fig. 5-25** Air injection schedule (From Nelson and McNiel, 1961)

During this period the cumulative air injected is

$$Q_{a_1} = \frac{r}{V_b} \times \frac{q_{a_{max}}}{2} \text{ ft}^3 \tag{5-14}$$

where

$$\frac{r}{V_b} = t_1 \text{ days} \tag{5-15}$$

In the second phase the air injection continues with a maximum constant rate $q_{a_{max}}$ and with decreasing velocity.

Assuming for the third phase (at the end of combustion) the same time period $t_1 = t_3$ and the same amount of air injected $Q_{a_1} = Q_{a_3}$, the air injected in the second step is given by

$$Q_{a_2} = Q_a - Q_{a_1} - Q_{a_3} \tag{5-16}$$

and

$$t_2 = \frac{Q_{a_2}}{q_{a_{max}}} \text{ days} \tag{5-17}$$

Therefore, the total time required for the project is

$$T = 2t_1 + t_2 \tag{5-18}$$

In designing the compression facilities for the five-spot pilot project, it is necessary to calculate the approximate injection pressure required.

The air injection pressure can be calculated as follows:

$$p_{iw}^2 = p_w^2 + \frac{q_a \mu_a T}{0.703 k_g h} \left( \ln \frac{a^2}{r_w V_b t_1} - 1.238 \right) \tag{5-19}$$

where

$p_{iw}$ = injection well bottom hole pressure, psia
$p_w$ = production well bottom hole pressure, psia
$q_a$ = maximum air rate, scf/day
$\mu_a$ = air viscosity, cp
$T$ = reservoir temperature, ° Rankine
$k$ = effective permeability to air*, md
$h$ = effective thickness, ft
$a$ = well-to-well spacing, ft
$r_w$ = production well radius, ft
$V_b$ = burning front velocity, ft/day
$t_1$ = time to reach maximum air rate, days

---

* It is assumed to be 5 percent of the absolute permeability.

In the field as air injection is begun, a peak pressure value is observed during the first weeks. Treatment operations of the productive formation surrounding the wellbore are required to clean away the material deposited and to restore permeability.

Also, it is recommended that the air injection pressure calculated using the relationship given should be verified by prior reservoir air injectivity tests made with a mobil compressor unit. The injection pressure must not exceed the fracturing pressure of the formation in order to avoid channeling and gas losses to the adjacent formations.

Knowing the air injection pressure required and the maximum air injection rate, the size of the compressor needed and the number of compression stages can be determined.

**Example 5-9.** An oil reservoir has been produced in primary by dissolved gas drive as the recovery mechanism. The reservoir has the following characteristics:

| | |
|---|---|
| Depth | $D = 2700$ ft |
| Productive area | $A = 300$ acres |
| Effective thickness | $h_e = 30$ ft |
| Absolute permeability | $k = 600$ md |
| Effective porosity | $\phi = 30\%$ |
| Irreducible water saturation | $S_{wi} = 25\%$ |
| Oil gravity | $\gamma_o = 20$ °API |
| Oil viscosity | $\mu_o = 80$ cp |
| Oil formation volume factor | |
| Initial | $B_{oi} = 1.07$ |
| Actual | $B_{oa} = 1.04$ |
| Primary oil recovery | $E_{Ra} = 7\%$ |
| Reservoir pressure | |
| Initial | $p_i = 1250$ psig |
| Actual | $p_a = 450$ psig |
| Reservoir temperature | $T = 130$ °F |
| Well radius | $r_w = 3$ in. |

The screening criteria and the first lab tests indicated that the reservoir is a good candidate for in situ combustion and that a pilot test in the field should be designed.

The oil has naphtenic components and the oxidation cell shows the amount of coke deposited

$$C_u = 1.5 \text{ lb}_m/\text{ft}^3$$

For an inverted five-spot well pattern encompassing 10 acres (Figure 5-26), calculate

1. Total air required, $Q_{air}$
2. The maximum air injection rate, $q_a$, ft³/day

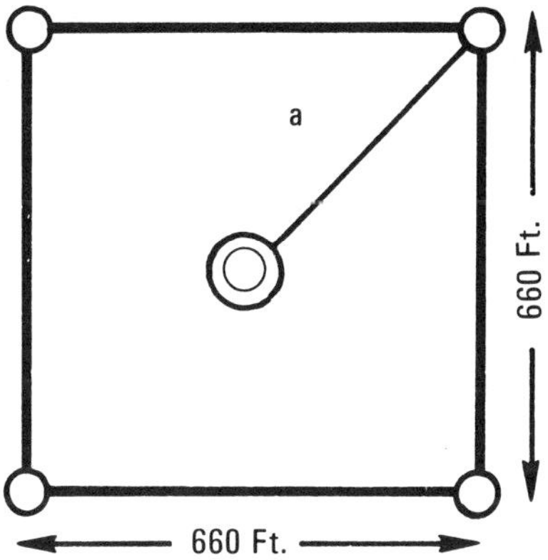


**Fig. 5-26** Ten acres inverted five-spot

3. The injection schedule and project time required
4. Air injection pressure, $p_a$
5. Oil consumed as fuel
6. Final oil recovery

SOLUTION

**(a)** *The total air required*, $Q_{\text{air}}$, is given by

$$Q_{\text{air}} = V_{\text{rock}} \times C_a \times E_p \times E_I \tag{5-20}$$

where $C_a$ is the air required to burn through a cubic foot of reservoir rock (Eq. 5-3).

$$C_a = C_u \times \text{minimum air necessary}$$

The minimum amount of air needed to burn 1 pound of fuel is usually between 160 and 192 scf/lb$_m$. Assuming

| | |
|---|---|
| Minimum air needed | 180 scf/lb$_m$ |
| Sweep pattern efficiency, $E_p$ | 0.55 |
| Invasion (vertical) efficiency, $E_I$ | 0.50 |
| Oxygen utilization | 90% |

the air required

$$C_a = 1.5 \text{ lb}_m/\text{ft}^3 \times 0.9 \times 180 \text{ scf/lb}_m = 243 \text{ ft}^3 \text{ air per cubic foot of rock}$$

and total air required

$$Q_a = 10 \text{ acres} \times 43{,}560 \text{ ft}^2/\text{acre} \times 30 \text{ ft} \times 243 \text{ ft}^3/\text{ft}^3 \times 0.55 \times 0.50$$
$$= 873 \times 10^6 \text{ scf}$$

**(b)** *The maximum air injection rate*, $q_{a_{\max}}$, is given by

$$q_{a_{\max}} = q_D \times U_{\min} \times a \times h \text{ scf/day} \tag{5-21}$$

where $q_D$ is the dimensionless flow rate calculated as the limiting value at which a combustion front can exist for the minimum air flux assumed to apply to the system. Practically speaking, $q_D$ represents the air-oil mobility ratio. The results based upon potentiometric studies on a five-spot well system (Smith, 1985) show the areal sweep efficiency at breakthrough and the $q_D$ values.

| $E_p$ (%) | $q_D$ |
|---|---|
| 50 | 3.39 |
| 55 | 4.77 |
| 57.5 | 6.06 |
| 62.6 | $\infty$ |

- $U_{min}$ is the minimum air flux density at which the rate of heat loss equals the rate of heat generation and the burning front reaches zero velocity.
- $a$ is the shortest flow line between injector and producer along which air flows with maximum rate, $q_a$.

From Eq. 5-1

$$U_{min} = C_a \times V_b$$

where $V_b$ is the velocity of the burning front as it advances (between 0.125 and 0.5 ft/day); in our example assumes $V_b$ = 0.125 ft/day. The maximum injection rate is

$$q_{a_{max}} = 4.77 \times 0.125 \text{ ft/day} \times 243 \text{ scf/ft}^3 \times 466.7 \text{ ft} \times 30 \text{ ft}$$

$$= 2.02 \times 10^6 \text{ scf/day}$$

Therefore, an air compressor with a discharge capacity of $q_{a_{max}} = 2.02 \times 10^6$ scf/day is required.

(c) *The injection schedule* has three successive air injection phases:

- *Phase 1*. The burning front moves at a constant velocity $V_b$ = 0.5 ft per day (for efficient displacement) and remains radial until $r$ = 117.7 ft (10 percent of the pattern area), when the designed airflow rate, $q_{a_{max}}$, is reached (Figure 5-27). Beyond this point, the burning front advance will drop below 0.5 ft/day.
- *Phase 2*. The air injection continues with $q_{a_{max}} = 2.02 \times 10^6$ scf/day.
- *Phase 3*. Has the same duration as phase 1 with the same amount of air injected but decreasing velocity of the burning front, $Q_{a_3} = Q_{a_1}$.

Hence,

$$Q_{a_1} = \frac{117.7 \text{ ft}}{0.5 \text{ ft/day}} \times \frac{2.02}{2} \times 10^6 \text{ scf/day} = 237.7 \times 10^6 \text{ scf}$$

$$t_1 = \frac{117.7}{0.5} = 235 \text{ days}$$

$$Q_{a_2} = 873 \times 10^6 \text{ scf} - 2 \times 237.7 \times 10^6 \text{ scf} = 397.5 \times 10^6 \text{ scf}$$

$$t_2 = \frac{397.5 \times 10^6 \text{ scf}}{2.02 \times 10^6 \text{ scf/day}} = 197 \text{ days}$$

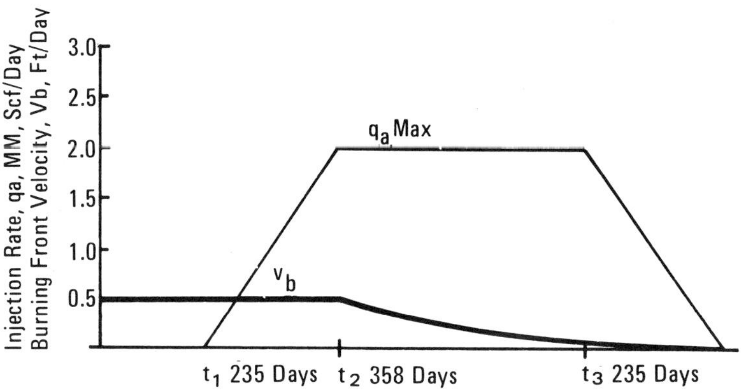


**Fig. 5-27** Air injection schedule

Therefore, total time required for the project is

$$T = 2 \times 235 \text{ days} + 197 \text{ days} = 667 \text{ days or } 1.83 \text{ years}$$

**(d)** *The air injection pressure* is calculated using Eq. 5-19, assuming production well bottom hole pressure $p_w = 35$ psia and air viscosity $\mu_a = 0.02$ cp.

$$p_{iw} = \left[35^2 + \frac{2.02 \times 10^6 \times 0.02(130 + 460)}{0.703 \times 30 \times 30}\left(\ln\frac{660^2}{0.125 \times 235 \times 3/12} - 1.238\right)\right]^{1/2}$$

$$= 607 \text{ psia}$$

**(e)** *The oil consumed as fuel* is expressed as a percentage of the oil in place at start of combustion. The oil in place at start of combustion is calculated knowing the oil saturation, $S_o$. For the reservoir being produced by a dissolved gas drive mechanism, a material balance equation can be written.

$$\begin{matrix}\text{original oil} \\ \text{in place, } N\end{matrix} = \begin{matrix}\text{oil in place at} \\ \text{start of combustion}\end{matrix} + \text{oil produced}$$

or

$$N = \frac{V_p(1 - S_{wi})}{B_{oi}} = \frac{V_p S_o}{B_o} + \Delta N$$

and the oil saturation $S_o$ is given

$$S_o = (1 - S_{wi})\frac{B_o}{B_{oi}}\left(1 - \frac{\Delta N}{N}\right) \tag{5-22}$$

or

$$S_o = (1 - 0.25)\frac{1.04}{1.07}(1 - 0.07) = 0.6779$$

The oil in place at start of combustion

$$N_{\text{actual}} = 7758 \times 10 \times 30 \times 0.30 \times 0.6779$$

$$= 473{,}354 \text{ bbl in reservoir conditions}$$

The percentage of oil consumed using Eq. 5-8 is

$$S_{o_{cons}} = 100\frac{1.5 \text{ lb}_m/\text{ft}^3}{1.2 \times 62.5 \text{ lb}_m/\text{ft}^3}$$

$$\times \frac{7758 \text{ bbl/acres-ft}(10 \text{ acres})(30 \text{ ft})}{473{,}354 \text{ bbl}} 0.55 \times 0.50$$

$$= 2.70\% \text{ of the oil in place at start of combustion}$$

or the percentage of oil consumed using Eq. 5-9 is

$$S_{o_{cons}} = \frac{873 \times 10^6 \text{ scf} \times 100}{180 \text{ scf/lb}_m \times 1.2 \times 62.5 \text{ lb}_m/\text{ft}^3 \times 473{,}354 \text{ bbl} \times 5.614 \text{ ft}^3/\text{bbl}}$$

$$= 2.43\% \text{ of the oil in place at start of combustion}$$

the amount of oil consumed is

$$473{,}354 \text{ bbl}\left(\frac{2.70 + 2.43}{100 \times 2}\right) = 12{,}141 \text{ bbl}$$

and

$$\frac{12{,}141 \text{ bbl}}{7758 \text{ bbl/acres-ft } 10 \text{ acres} \times 30 \text{ ft} \times 0.3(1 - 0.25)} = 0.023$$

or 2.3% of the original oil in place

**(f)** *Final oil recovery* is the total oil produced, by a primary mechanism and by combustion, from the original oil in place.

Oil produced by combustion is given by Eqs. 5-10 and 5-11, assuming the displacement efficiency $E_{Du} = 0.40$.

$$N_{p_1} = 0.6779 \times 0.30 \times 0.40 \times 0.55 \times (1 - 0.5)$$

$$+\left(0.6779 - \frac{0.027 + 0.0243}{2}\right) \times 0.3 \times 0.55 \times 0.5 = 0.07618$$

$$N_{p_2} = 0.6779 \times 0.30 \times 0.40(1 - 0.55 \times 0.50) + 0.05381 = 0.11278$$

and the recovery factor by combustion ranges between

$$E_{R_1} = \frac{0.07618}{0.6779 \times 0.3} = 0.374 \quad \text{and} \quad E_{R_2} = \frac{0.11278}{0.6779 \times 0.3} = 0.554$$

and can be taken as (0.374 + 0.554)/2 = 0.464 or 46.4 percent of the oil in place at start of combustion.

The final oil recovery of the reservoir is

$$E_{R_{final}} = 0.07 + \frac{0.464(473{,}354)}{523{,}665}$$

$$= 0.4894 \text{ or } 49\% \text{ from the original oil in place}$$

If a pilot successfully passes evaluation, plans are made to develop in situ combustion as a full-scale process. A number of five-spot well patterns have to be operated simultaneously and successively until the entire reservoir pro-

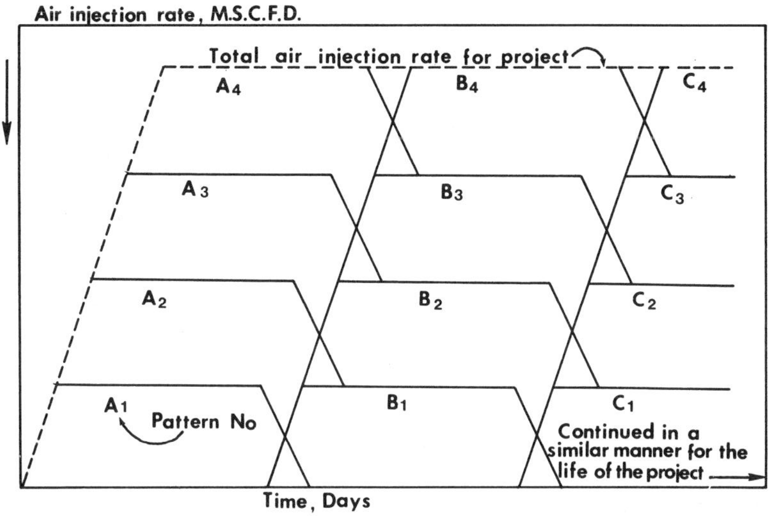


**Fig. 5-28** Programming air injection, where a number of five-spot patterns are to be operated (From Nelson and McNiel, 1961)

ductive area is influenced by combustion. Nelson and McNiel provided a program for the simultaneous operation of a number of patterns (Figure 5-28).

To develop an in situ combustion process, the specific reservoir geometry and characteristics must be considered and a total air injection rate determined for the project such that a compressor's loading approaching 100 percent could be maintained.

For instance, in a five-spot well pattern development (Figure 5-28), pattern $A_1$ is first ignited. When this pattern reaches its maximum injection rate, pattern $A_2$ is ignited, and so on. The four patterns are fully operated after $4 \times t_1$ days, and in this way the duration of the project is unnecessarily extended. The delay can be avoided if the four patterns are simultaneously ignited (in a two- to three-week period). As it was pointed out, each oil reservoir has its own geometry and characteristics. The reservoir engineer must apply the most appropriate development program based on these characteristics to facilitate the flow of reservoir liquids to producers, to increase combustion efficiency, and to assure the most effective utilization of the compressors' capacity.

## QUESTIONS AND PROBLEMS

**5-1** An oxidation cell experiment reveals two successive oxidation reactions. Which is the more important reaction and why?

**5-2** How does the reactivity of different oils indicate whether or not they are good candidates for in situ combustion?

**5-3** What are the conditions that give a light oil sufficient fuel availability to ensure the propagation of a combustion front?

**5-4** Make a schematic representation of the in situ combustion process and explain the various zones as formed in the oil reservoir.

**5-5** Describe wet combustion and enumerate the advantages of the process.

**5-6** Calculate the atomic H/C ratio of the coke deposited in an oxidation cell experiment where the following data were obtained.

| Injected Air Composition | Produced Gas Composition |
|---|---|
| 21% $O_2$ | 14.5% $CO_2$ |
| 79% $N_2$ | 0.8% CO |
| | 0.2% $O_2$ |
| | 84.5% $N_2$ |

**5-7** Assuming oxygen utilization is 85 percent and the atomic H/C ratio of the oil is 1.8, calculate the amount of air required to burn through a cubic foot of reservoir rock.

**5-8** Describe the in situ combustion operating experience gained from
**(a)** the Moco Zone project (United States)
**(b)** the Suplacu de Barcau Field (Romania)
**(c)** the Heidelberg Field (United States)

**5-9** Identify on Figure 5-23 both unburned zones I and II, and calculate the invasion efficiency and pattern sweep efficiency when the average burned zone thickness is 40 percent and volumetric efficiency is 22 percent.

**5-10** An oil reservoir 600 acres in extent with a 40-ft productive formation thickness, 26 percent porosity, and 0.65 oil saturation is subjected to an in situ combustion process developed on a commercial scale. Calculate the oil consumed as fuel, knowing that 1.9 $lb_m$ of fuel is needed per 1 $ft^3$ of burned rock and that 0.24 is the volumetric efficiency.

**5-11** Assuming oxygen utilization is 90 percent and 180 scf air is necessary to burn 1 pound of fuel, calculate the cumulative amount of injected air necessary to consume 3.59 percent of the oil in place at start of combustion (see 5-10).

**5-12** Using the Ramey and Gates correlation, calculate the oil recovery of a reservoir under in situ combustion when the volumetric efficiency is 40 percent and the initial gas saturation is 10 percent.

**5-13** The following data were obtained from an oil reservoir which in primary produced 16 percent of the OOIP under dissolved gas drive recovery mechanism.

| | |
|---|---|
| Productive area | 300 acres |
| Formation thickness | 25 ft |
| Effective porosity | 28% |
| Irreducible water saturation | 0.24 |
| Primary recovery | 16% (dissolved gas drive) |
| Oil formation volume factor | |
| Initial | 1.1 |
| Actual | 1.04 |
| The fuel content | 1.4 $lb_m/ft^3$ |
| Coke specific gravity | 1.15 |

| | |
|---|---|
| Minimum air necessary to burn 1 pound of fuel | 183 scf/$lb_m$ |
| Pattern sweep efficiency | 0.6 (estimated) |
| Vertical efficiency | 0.4 (estimated) |
| Displacement efficiency in Zone I | 0.36 |

Calculate the final oil recovery attained after in situ combustion.

**5-14** Explain how low vertical permeability favors in situ combustion development and enumerate the advantages and disadvantages of the use of oxygen instead of air as the injected agent.

**5-15** The following data indicate the reservoir "Z" is a good candidate for in situ combustion, and a pilot test in the field must be designed.

| | |
|---|---|
| Depth | 3600 ft |
| Productive area | 500 acres |
| Formation gross thickness | 56 ft |
| Effective thickness | 28 ft |
| Absolute permeability | 320 md |
| Effective porosity | 23% |
| Irreducible water saturation | 30% |
| Oil formation volume factor | |
| Initial | 1.15 |
| Actual | 1.06 |
| Actual oil recovery | 18% |
| Reservoir pressure | |
| Initial | 1600 psig |
| Actual | 650 psig |
| Reservoir temperature | 150 °F |
| Well radius | 3 in. |
| Amount of coke deposited | 2 $lb_m/ft^3$ |
| Minimum air required | 190 scf/$lb_m$ |
| Well pattern | 5 acres inverted five-spot |

Assuming appropriate values for volumetric and displacement efficiencies, calculate

**(a)** total air required
**(b)** maximum air injection rate
**(c)** injection schedule
**(d)** injection pressure
**(e)** oil consumed
**(f)** final oil recovery

## REFERENCES

ALEXANDER, J. D., W. L. MARTIN, and J. M. DEW, "Factors Affecting Fuel Availability and Composition During In-Situ Combustion," *Journal of Petroleum Technology* (October 1962).

ALDEA, G., V. PETCOVICI, et al., *In Situ Combustion Field Test Pilot in Romania* (original title: *Aplicarea Experimentala a Metodei de Exploatare Prin Combustie Subterana in Romania*), Petrol, Si Gaze, Vol. 19, No. 1, pp. 26–36 (1968).

BENHAM, A. L., and F. H. POETTMANN, "The Thermal Recovery Process—An Analysis of Laboratory Combustion Data," *Transactions of the AIME*, Vol. 213 (1958), pp. 83–85.

BURGER, J., A. CARCOANA, et al., "Basic Research on In-Situ Combustion and Recent Field Results," Paper no. PD-14-4 presented at the Ninth World Petroleum Congress, Tokyo, May 1975.

BURGER, J., and B. SAHUQUET, "Chemical Aspect of In-Situ Combustion, Heat of Combustion and Kinetics," *Society of Petroleum Engineers Journal*, Vol. 12 (1972), pp. 410–412.

BURGER, J., and B. SAHUQUET, "Laboratory Research on Wet Combustion," *Journal of Petroleum Technology*, Vol. 25 (1973).

BURGER, J., P. SOURIEAU, et al., *Thermal Methods of Oil Recovery* (Houston, TX: Gulf, 1985; Paris: Editions Technip), p. 284.

CADY, G. V., and J. T. MOSS, "The Design and Installation of an Oxygen Supported Combustion Project in Lindbergh Field, Alberta, Canada," Paper presented at the 1981 World Oil and Gas Show, Dallas, Texas, December 14–17, 1981.

CARCOANA, A., "Results and Difficulties of the World's Largest In-Situ Combustion Process: Suplacu de Barcau Field, Romania," SPE/DOE paper 20248 presented at the SPE/DOE Seventh Symposium on Enhanced Oil Recovery, Tulsa, Oklahoma, April 22–25, 1990.

CARCOANA, A., and GH. ALDEA, *Increasing Ultimate Oil Recovery* (original title: *Marirea Factorulni Final de Recuperare la Zacamintele de Hidrocarburi*) (Bucarest, RO: Ed. Technica, 1976).

CARCOANA, A., V. C. MACHEDON, et al., "In-Situ Combustion—An Effective Method to Enhance Oil Recovery in Romania," Special paper SP8, presented at the Eleventh World Petroleum Congress, London, 1983.

CHU, C., "A Study of Fireflood Field Projects," *Journal of Petroleum Technology* (February 1977).

CHU, C., "State-of-the-Art Review of Fireflood Field Projects," *Journal of Petroleum Technology* (January 1982).

COUNIHAN, T. M., "A Successful In-Situ Combustion Pilot in the Midway Sunset Field, California," Paper presented at the 1977 SPE California Regional Meeting, Bakersfield, California, April 13–15, 1977.

CRAWFORD, P. D., "Thermal Recovery Guide Helps Select Projects," *World Oil* (August 1, 1971).

FAROUQ ALI, S. M., "A Current Appraisal of In-Situ Combustion Field Test," *Journal of Petroleum Technology* (April 1972).

GADELLE, C. P., et al., "Heavy Oil Recovery by In-Situ Combustion," SPE paper 8905 presented at SPE 50th California Regional Meeting, Los Angeles, California, April 9–11, 1980.

GADELLE, C. P., J. C. BURGER, et al., "Heavy Oil Recovery by In-Situ Combustion—Two Field Cases in Romania," *Journal of Petroleum Technology* (November 1981).

GATES, C. F., and H. J. RAMEY, JR., "A Method for Engineering In-Situ Combustion Oil Recovery Projects," *Journal of Petroleum Technology* (February 1980).

GATES, C. F., and I. SKLAR, *Combustion as a Primary Recovery Process—Midway Sunset Field*, 1985 ed., SPE Reprint Series, No. 7 "Thermal Recovery Processes" (Society of Petroleum Engineers, 1985), p. 359.

GRANT, B. F., and S. E. SASZ, "Development of an Underground Heat Wave for Oil Recovery," *Transactions of the AIME* (1954).

HOWARD, F. A., "Method of Operating Oil Well," U.S. Patent #1,473,348 (XI-1923).

HUFFMAN, G. A., J. P. BENTON, et al., "Pressure Maintenance by In-Situ Combustion, West Heidelberg Unit, Jasper County, Mississippi," *Journal of Petroleum Technology* (October 1983).

JOSEPH, C., and W. H. PUSH, "A Field Comparison of Wet and Dry Combustion," *Journal of Petroleum Technology* (September 1980).

KHUN, C. S., and R. L. KOCH, "In-Situ Combustion—Newest Method of Increasing Oil Recovery," *Oil and Gas Journal* (August 1953).

LATIL, M., *Enhanced Oil Recovery* (Houston, TX: Gulf, 1980).

MACHEDON, V., et al., "Evolution et Controle du Processus de Combustion Souterraine," *Proceedings*, *International Symposium on Hydrocarbons*, *Exploration*, *Drilling and Production Techniques*, Paris, 1975.

MARTIN, W. L., J. D. ALEXANDER, et al., "Thermal Recovery at North Tisdale Field, Wyoming," SPE paper no. 3595 (1971).

MOORE, R. G., D. W. BENNION, et al., "New Insights into Enriched Air In-Situ Combustion," SPE paper 16740, 1987 SPE Annual Technical Conference and Exhibition, Dallas, Texas, September 27–30, 1987.

MOORE, T. V., U.S. Patent #3,208,519 (1965).

NELSON, T. W., and J. S. MCNIEL, "How to Engineer an In-Situ Combustion Project," *Oil and Gas Journal* (June 5, 1961).

PARRISH, D. R., and F. F. CRAIG, JR., "Laboratory Study of a Combination of Forward Combustion and Waterflooding—The COFCAW Process," *Journal of Petroleum Technical* (June 1969).

PARRISH, D. R., C. B. POLLOCK, and F. F. CRAIG, JR., "Evaluation of COFCAW as a Tertiary Recovery Method, Sloss Field, Nebraska," *Journal of Petroleum Technology* (June 1974).

PETCOVICI, V., "Considerations sur les Possibilités du Controle du Front de Combustion In Situ sur Champ," *Revue de l'Institut Français du Petrole*, Vol. 25 (1970), pp. 1355–1374.

PETIT, H. J. M., "In-Situ Combustion with Oxygen-Enriched Air," SPE paper 16741, 1987 SPE Annual Technical Conference and Exhibition, Dallas, Texas, September 27–30.

SMITH, CH. R., *Mechanism of Secondary Oil Recovery* (Malabar, FL: Robert E. Krieger, 1985), p. 406.

*SPE Letter and Computer Symbols Standard* (Richardson, TX: Society of Petroleum Engineers, 1986).

TRANTHAM, J. C., and J. W. MARX, "Bellamy Field Test Oil from Tar by Counterflow Underground Burning," SPE of AIME Paper 1269 (October 1965).

TURTA, A., "Results of Continuous Bottom Coring of $K_3$, $K_1$, and $K_2$ Wells Drilled in the Pattern Area of the In-Situ Combustion Pilot in Suplacu de Barcau Oil Field," technical paper (Cimpina, RO: ICPPG, 1972, 1974).

TURTA, A., GH. ALDEA, and M. ZAMFIR, "The In Situ Combustion Industrial Exploitation of Suplacu de Barcau, Panonian Field, Romania," Paper presented at Fourth Unitar/UNDP International Conference on Heavy Crude and Tar Sands, Edmonton, Canada, August 7–12, 1988.

VAN POOLLEN, H. C., and associates, *Enhanced Oil Recovery* (Tulsa, OK: PennWell, 1980).

VOSSOUGHI, S., G. P. WILLHITE, et al., "Automation of an In-Situ Combustion Tube and Study of the Effect of Clay on the In-Situ Combustion Process," *Society of Petroleum Engineers Journal* (August 1982).

WHITE, P. D., "In-Situ Combustion Appraisal and Status," *Journal of Petroleum Technology* (November 1985).

WHITE, P. D., and W. H. FAIRFIELD, "Oxygen Fireflood Is Source of Carbon Dioxide," *Oil and Gas Journal* (July 19, 1982).

WHITE, P. D., and J. T. MOSS, *Thermal Recovery Methods* (Tulsa, OK: PennWell, 1983), pp. 72–73.

WOLCOTT, E. R., "Method of Increasing the Yield of Oil Wells," U.S. Patent #1,457,479 (June 1923).

# Chapter 6

# Polymer Flooding

## 6-1 GENERAL

The waterflood field operations to increase the oil recovery, despite the successful development, resulted also in poor and incomplete sweeps of the reservoir volume. The first attempts to improve sweep efficiency in waterfloods were made by Detling (1944). He patented a number of additives, including water-soluble polymers, to increase the viscosity of injected water and the volume of the reservoir affected. In the next two decades over 25 patents for additives to be used under different reservoir conditions were issued and the first results of laboratory experiments dealing with their use were published (Sandiford and Pye, 1964). Because of their lower cost, the water-soluble polymers prevailed over other additives (molasses, glycerin, glycols, etc.) tested in the field. After 1964, field test results and other significant laboratory studies (Mungan et al., 1966; Gogarty, 1967; Burcik, 1968; Maerker, 1972) made possible the development of polymer flooding as a method to enhance oil recovery.

## 6-2 PRINCIPLE AND METHOD DESCRIPTION

### Water and Oil Mobilities

When water displaces oil through the reservoir's porous volume, the velocity of the displacement is proportional to the water mobility

$$\lambda_w = \frac{k_w}{\mu_w}$$

where $k_w$ is the effective permeability of the rock to water in the water swept zone of the reservoir and $\mu_w$ is the water viscosity. Water is the displacing phase and oil is the displaced phase. The oil also has a mobility

$$\lambda_o = \frac{k_o}{\mu_o}$$

where $k_o$ is the effective permeability of the rock to oil in the oil bank and $\mu_o$ is the oil viscosity. Since the flow takes place under the same pressure gradient, the flow velocity is the same in the water swept area and in the oil bank zone only when $\lambda_w = \lambda_o$ and the ratio $\lambda_w/\lambda_o = 1$.

### Mobility Ratio Concept

The mobility ratio concept described by Craig (1980) is defined as the mobility of the displacing phase to the mobility of the displaced phase. In waterflooding,

$$M_{w-o} = \frac{\lambda_w}{\lambda_o} = \frac{k_w\,\mu_o}{\mu_w\,k_o}$$

or, dividing by the absolute permeability, the water-oil mobility ratio

$$M_{w-o} = \frac{k_{rw}\,\mu_o}{\mu_w\,k_{ro}} \tag{6-1}$$

where $k_{rw}$ is the relative permeability to water at the average water saturation behind the waterfront and $k_{ro}$ is the relative permeability to oil ahead of the flood front at irreducible water saturation.

It is obvious that when the mobility ratio is greater than unity, since $\mu_w$ is less than $\mu_o$, water flows at a higher velocity through the path of least resistance and breaks through into producing wells prematurely. The mobility ratio in this case is deemed to be unfavorable.

### Polymers Reduce Water-Oil Mobility Ratio

The role of water-soluble polymers is to increase the water viscosity and also to reduce the permeability of the rock to water, in other words, to reduce the water-oil mobility ratio close to unity or less. Then, the volumetric sweep efficiency (areal × vertical) will be improved and a higher oil recovery will be achieved at breakthrough with polymer flooding than with waterflooding.

After water breakthrough into the producers, the flow of the two phases (water and oil) in the swept area of the reservoir is controlled by the fractional flow equation of Buckley and Leverett (1942),

$$f_w = \frac{1}{1 + (k_o/\mu_o)(\mu_w/k_w)} \tag{6-2}$$

where

$f_w$ = fraction of water in the flowing stream passing any point in the swept area (i.e., the water cut)

$k_o, k_w$ = effective rock permeabilities to oil and to water, respectively, at one given water saturation at one point in the reservoir (Craig, 1980)

$\mu_o, \mu_w$ = oil and water viscosities

Equation 6-2, simplified by ignoring the capillary pressure and gravitational effects, can also be expressed using relative rock's permeabilities to oil and water.

Again, it is easy to observe then when water viscosity $\mu_w$ increases and the permeability of the rock to water $k_w$ decreases, the fractional flow of oil

$$f_o = 1 - f_w = 1 - \frac{1}{1 + (k_{ro}/\mu_o)(\mu_w/k_{rw})} \tag{6-3}$$

will increase, improving the rate of oil recovery.

Permeability reduction and a higher water viscosity will increase the resistance to flow of the polymer solution diverting it toward areas unswept by water.

**Example 6-1.** Given the water-oil relative permeability curves of an oil reservoir (Figure 6-1) and the ratio of viscosities $\mu_o/\mu_w = 6$, calculate the water-oil mobility ratio $M_{w-o}$ and the water cut $f_w$ at $S_w = 50\%$.

SOLUTION Assuming the residual oil saturation in the waterflooded area of the reservoir, $S_{or} = 0.20$, and the irreducible water saturation in the oil bank ahead of the waterflood front, $S_{wi} = 0.20$, we read

$$k_{rw} = 0.57 \text{ at } S_{or} \qquad \text{and} \qquad k_{ro} = 1.0 \text{ at } S_{wi}$$

the water-oil mobility ratio

$$M_{w-o} = \frac{k_{rw}}{\mu_w}\frac{\mu_o}{k_{ro}} = \frac{0.57}{1.00} \times 6 = 3.4$$

At $S_w = 50\%$

$$k_{rw} = 0.13 \qquad \text{and} \qquad k_{ro} = 0.24$$

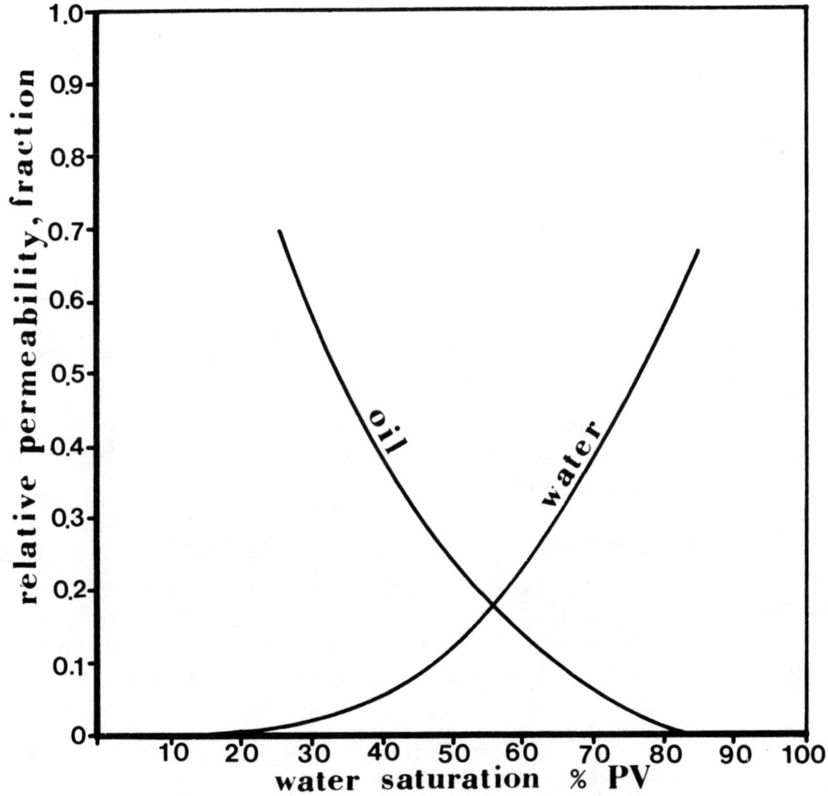


**Fig. 6-1** Water-oil relative permeability characteristics

the water cut

$$f_w = \frac{1}{1 + (k_{ro}/\mu_o)(\mu_w/k_{rw})} = \frac{1}{1 + (0.24 \times 1)/6 \times 0.13)} = 0.76$$

### Method Description

In polymer flooding, a slug of 0.3 or higher PV of polymer solution is injected into the reservoir with a prior injection of a low-salinity brine (freshwater) slug. The polymer slug is followed by another freshwater slug and by continuous drive water injection. The schematic cross-section view of a polymer injection is given in Figure 6-2.

The polymer solution slug is injected between two freshwater buffers in order to attenuate the direct contact with the saline reservoir water. The saline water reduces the polymer solution's viscosity.

Many studies of the characteristics of polymers as water mobility control agents have been reported. A state-of-the-art review on polymer flooding

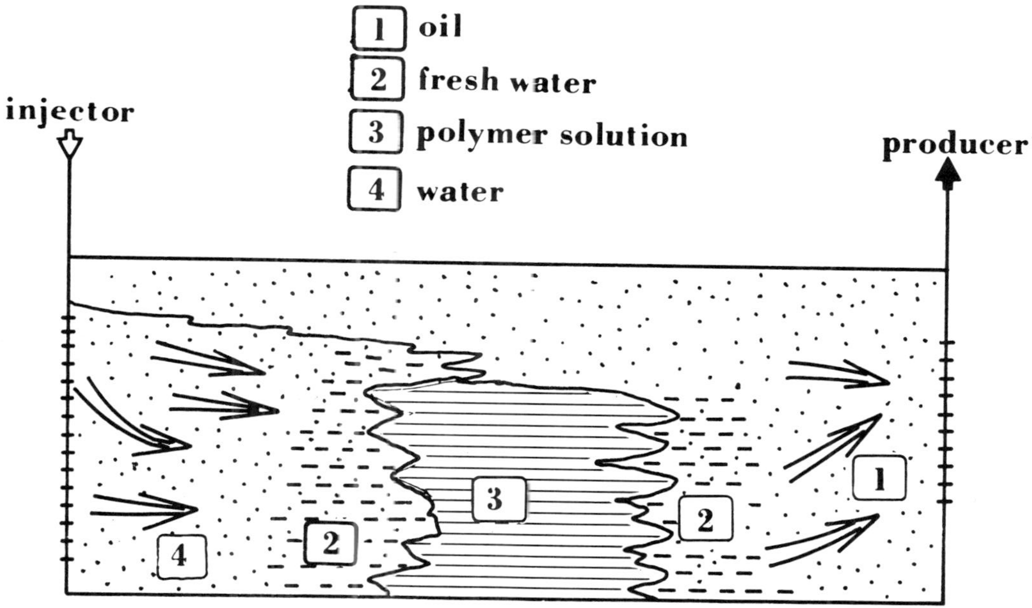


Fig. 6-2 Schematic view of polymer flood

and on the application of polymers for mobility control is given by Needham and Doe (1987) and Gao (1987), respectively.

Knowing the principle and method description it is now understood that polymer flooding does not reduce the residual oil saturation. It improves oil recovery over waterflooding by increasing the reservoir volume contacted. Also, in comparison to waterflooding, polymer flooding accelerates oil production, and a higher recovery is obtained at breakthrough. Polymer flooding is successful when applied in the early stages of a waterflood process or when applied at the beginning when the mobile oil saturation is high. It has little or no effect in uniform waterflooded reservoirs containing low-viscosity oil or having high water saturation at the start of the process. Reservoirs with high permeability variations, rapid water breakthrough into the producers, and low recoveries can also be flooded with polymer solutions, but the risk factor is higher.

## 6-3 POLYMER TYPES

There are two principal types of polymers being used in field applications: hydrolyzed polyacrylamide (HPAM) and polysaccharides biopolymer or xanthan gum.

### Polyacrylamides

The polyacrylamide is obtained by the polymerization of the acrylamide monomer. Through hydrolysis some of the acrylamide monomers are converted to carboxylate groups with a negative charge. The hydrolyzed polyacrylamide has a 20 to 40 percent degree of hydrolisis, a molecular weight higher than $3 \times 10^6$ and a linear chain molecular structure. The long molecular chain of the partially hydrolyzed polyacrylamide, in solution with freshwater, eases the flow through the tortuous porous space of the rock reservoir. In saline water, the electrolytes in solution cause the molecules to coil. This obstructs the flow through the porous space and reduces the viscosity of the solution. The hydrolyzed polyacrylamide solution is sensitive to salts and must be prepared with freshwater. Other susceptibilities of HPAM solutions are caused by the presence of oxygen, which is a source of instability and chemical degradation, and by temperature and mechanical degradation. The HPAM molecules' long chains may be broken, especially at high velocity and temperature conditions, when the injected solution passes through the well's perforated interval and flows through the porous space of the formation near the wellbore.

Being less expensive and providing higher residual resistance to drive water injection, polyacrylamide is more widely used in the field than polysaccharide as a water mobility control agent. However, recent tests have shown that microbial attack of polyacrylamide can be a potentially serious problem (Grula and Sewell, 1982) and biocides, such as formaldehyde, need to be used to prevent the viscosity loss caused by microbes.

### Polysaccharides

The polysaccharide biopolymer is obtained from sugar in a fermentation process caused by the bacterium *Xanthomonas campestris*. The polysaccharide's molecular structure gives the molecules a great stiffness (Needham and Doe, 1987), their behavior being like rigid-rod molecules (Gao, 1987). As a consequence, in contrast to polyacrylamide polymer, the viscosity of a polysaccharide biopolymer solution is not affected by salinity, and shearing effects can be tolerated. Despite these advantages, the polysaccharide biopolymer is expensive, and its stability degrades at temperatures above 200 °F. Biodegradation by enzymes of polysaccharide polymers is also common and usually results in a decrease in the solution's viscosity. Formaldehyde is the most effective biocide added to biopolymers to prevent biodegradation.

## 6-4 APPARENT VISCOSITY AND RESISTANCE FACTOR

### Apparent Viscosity

Polymers are used in aqueous solutions at low concentrations of 300 to less than 2000 parts per million (ppm) or 0.03 to less than 0.20 percent. The viscosity values of polymer solutions measured with an Ostwald viscometer, for different concentrations, are plotted practically as a straight line on a semilog paper. The measured viscosity of the low-concentration polymer solutions used in the field is only 1 to 1.5 cp. It is interesting to observe that the behavior in a porous medium of these low-concentration polymer solutions suggests a much higher viscosity. Indeed, the viscosity of the polymer solution flowing in the reservoir is from 5 to 25 times higher when calculated indirectly using Darcy's equation (apparent viscosity), assuming the same effective permeability (Figure 6-3).

In reality, the effective permeability of the formation to a polymer solution is lower than it is to water without polymer. It is difficult to separate the effect of the permeability reduction from that of the viscosity increase. As we pointed out, what is important is that the total effect can be expressed as mobility reduction and this total effect can be measured.

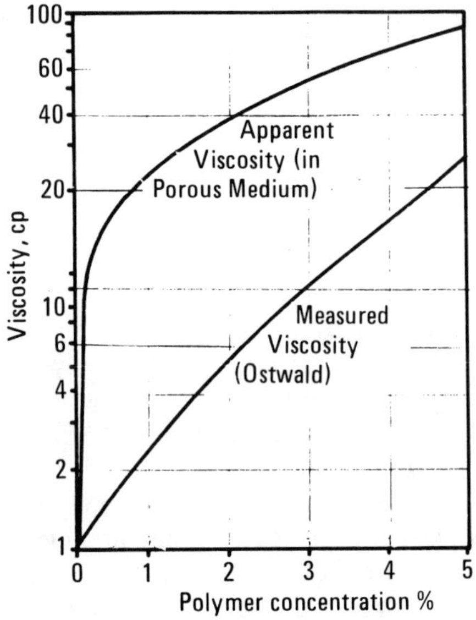


**Fig. 6-3** Resistance effect of polymer solution in porous media (From Pye, 1964)

### Resistance Factor

The measure of the mobility reduction is known as the *resistance factor*, $R$

$$R = \frac{\lambda_w}{\lambda_p} = \frac{k_{rw}/\mu_w}{k_{rp}/\mu_p} = \frac{k_{rw}\,\mu_p}{\mu_w\,k_{rp}} = \frac{M_{w-o}}{M_{p-o}} \tag{6-4}$$

where

$\lambda_p$ = water-soluble polymer mobility

$k_{rw}, k_{rp}$ = relative permeabilities to water and to polymer solution, respectively

$\mu_p$ = viscosity of the polymer solution (apparent)

$M_{w-o}, M_{p-o}$ = water-oil and polymer solution-oil mobility ratios, respectively

A plot of the resistance factor $R$ as function of the ratio $V_{inj}/V_p$ (cumulative injected volume per porous volume) is given in Figure 6-4.

The data were obtained by flowing a 300-ppm polymer solution through the porous volume of a core sample. A rapid increase of the resistance factor from 1 to 8 was observed for the first 20 pore volumes injected. Continuing the polymer solution injection, the value of the resistance factor remained practically constant. This tendency of the resistance factor to stabilize must be observed in the laboratory tests for $R$ values less than 10 or 12 to avoid in the field high injection pressures or blockages.

As a matter of fact, polymers with high resistance factors can be used in profile improvement to plug the more permeable streaks near injectors and to

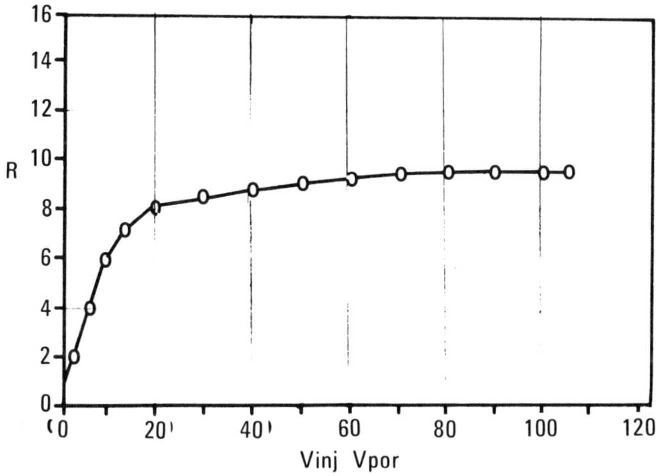


**Fig. 6-4** The resistance factor "R" function of cumulative volume injected

reduce the variation in permeability. Also, polymers gelled with crosslinkers can be used in an effort to plug reservoir high-permeability zones far from injcctors.

The advantages of polymers as water mobility control agents in porous mediums are indicated by the large resistance factor values obtained using water-containing low-polymer concentrations and by their ability to stabilize flow resistance.

### Residual Resistance Factor

The measure of the reduction of rock's permeability to water after polymer flow is known as the *residual resistance factor*, $R_R$.

$$R_R = \frac{(k_{rw}/\mu_w) \text{ before polymer flow}}{(k_{rw}/\mu_w) \text{ after polymer flow}} \tag{6-5}$$

Permeability reduction is observed after flushing with brine, following injection of a polymer solution, a Berea sandstone core sample. The original permeability of the core, having been reduced by adsorption on the rock surface and by mechanical entrapment of polymer molecules, cannot be recovered.

The existence of residual resistance effects has economic importance. Expenditures for polymer occur only during the injection period. Long afterward, the residual resistance factor effect continues at no added expense.

Biopolymer polysaccharides are not retained on rock surfaces. This is the reason they do not exhibit the residual resistance effect.

## 6-5 POLYMER RETENTION

Polymer retention, expressed by adsorption of the polyacrylamide on rock surfaces and by entrapment of polymer molecules in small pore spaces, explains the permeability reduction.

### Adsorption and Entrapment

The polyacrylamide polymer adsorbs on the surface of most rock reservoirs. For instance, calcium carbonate has a greater affinity for polymer than does silica. The adsorbed polymer layers represent both an additional resistance to flow and a loss of polymer. Indeed, when adsorption takes place, polymer solutions leaving the porous medium have a lower concentration than before. The reduced polymer concentration is used as a measure of adsorption. The higher the polymer concentration before flowing through the porous space, the higher will be the adsorption on the rock surface.

The porous space in a rock reservoir offers a variety of opening sizes. The long chain of the polymer molecule can easily flow into a large pore opening but cannot leave it if the other end has a smaller opening. Then the polymer molecule is trapped. Entrapment can also take place when the flow is restricted or stopped. Then the molecule loses its elongated shape and coils up. When the flow of polymer molecules through the porous medium is restricted in pores with small openings, only the passage of brine is permitted. The small openings not contacted by flowing polymer molecules form the so-called "inaccessible pore volume" (Dawson and Lang, 1972). Up to 30 percent of the total pore volume may not be accessible to polymer molecules. This allows polymer solutions to advance and displace oil at a rate faster than predicted. In other words, the effective porosity for a polymer solution is less than the effective porosity for brine.

### Molecular Weight and Screen Factor

Laboratory tests on partially hydrolyzed polyacrylamide solutions of 500 ppm with molecular weights ranging from about $3 \times 10^6$ to $10 \times 10^6$ showed that the mobility reduction, resistance factor, and permeability reduction increase when the molecular weight increases. The resistance factor and the permeability reduction, respective to the molecular weight of the polymer, have been correlated by the so-called "screen factor" (Jennings et al., 1971).

The measurement is made by comparing the time required for a given polymer solution volume to run out of the device to the time required to drain the same volume of solvent (water). The time required, read between the two guide marks, corresponds to 2.44 in.$^3$ (40 cm$^3$) volume and is 8 to 10 seconds for water (Knight, 1973).

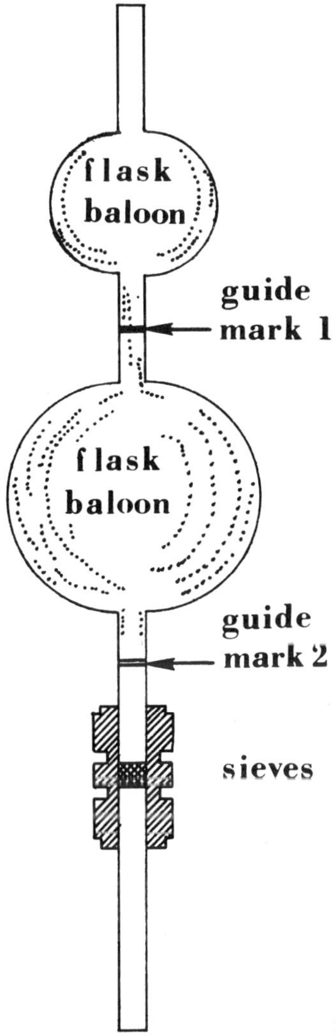


**Fig. 6-5** Screen viscometer (From Knight, 1973)

A typical correlation of the resistance factor to the screen factor is given in Figure 6-6.

As we observe, the screen viscometer does not measure viscosity since it is more sensitive to changes in polymer quality than solution viscosity. The screen viscometer measures flow effects due to the viscoelastic properties of the polymer solution. It can tell us in the field if a specific polymer solution charge has a resistance factor that is too low or too high and has to be discarded instead of being injected into the reservoir.

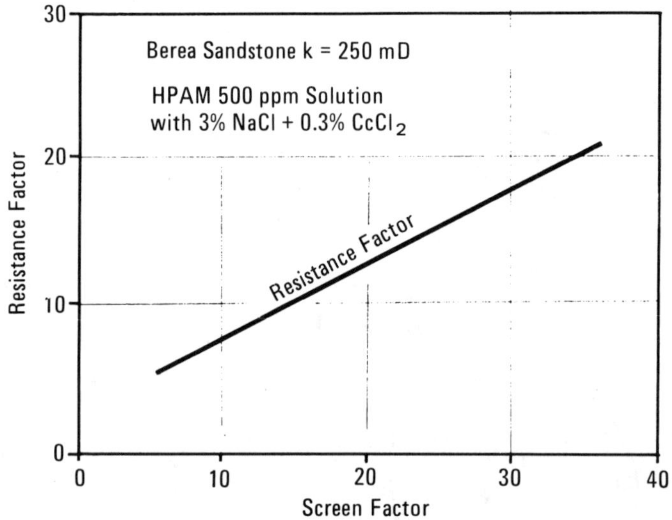


**Fig. 6-6** Typical correlation of the resistance factor with the screen factor (From Jennings et al., 1971)

## 6-6 FIELD PROJECTS AND RESULTS

After the first field experiments and project evaluations made by Jewett and Schurz in 1970, the application of water-soluble polymers to field projects has increased. Currently polymer flooding is being used on a commercial scale as a method of recovering more oil.

### Field Projects

Comprehensive surveys of the published results of full-scale polymer flood case histories have been conducted by Manning et al. (1983) and by Needham and Doe (1987), among others. An adapted statistical analysis of the fieldwide polymer flood projects is shown in Table 6-1.

The fieldwide projects were performed principally in sandstone reservoirs. The carbonate lithologies made a relatively small contribution to polymer flooding. The pay zones were characterized by depths between 400 ft and 11,000 ft, with a 4000 ft average, by 117 °F mean temperature with 229 °F maximum value, and by a good average porosity of 19 percent. The rock permeabilities ranged from the very low value of 1.5 md to a maximum of 7400 md with 453 md the mean value. The permeability variation (Lorenz coefficient) ranged from 0.07 (practically a uniform reservoir) to 0.96 (for a highly heterogeneous reservoir) with the mean value being 0.69. The viscosities of the reservoir crude oils where polymer flooding was applied ranged from very low

**TABLE 6-1.** Polymer Flood Statistics

| | Fieldwide Projects | | | | |
|---|---|---|---|---|---|
| Parameter | Number of Projects | Min | Max | Mean | Standard Deviation |
| Depth, ft | 87 | 400 | 10,800 | 4005 | 2201 |
| Temperature, °F | 88 | 46 | 229 | 117 | 34 |
| Porosity | 87 | | | 0.19 | |
| Average permeability, md | 80 | 1.5 | 7,400 | 453 | 1107 |
| Permeability variation | 71 | 0.07 | 0.96 | 0.69 | 0.19 |
| Oil viscosity, cp | 82 | 0.072(?) | 435 | 21.45 | 51.48 |
| Water-oil ratio at start | 42 | 0 | 75 | 5.3 | 11.77 |
| Average polymer concentration, ppm | 48 | 51 | 600 | 279 | 149 |
| Water-to-oil mobility ratio | 49 | 0.1 | 40 | 7.86 | 9.71 |
| Oil recovery, % OOIP | 20 | 0 | 14 | 3.85 | 3.62 |
| Oil recovery, STB/acre-ft | 23 | 0 | 143.4 | 34.4 | 40.07 |
| Oil recovery, STB/lb | 18 | 0 | 11.20 | 3.74 | 3.50 |

From Manning et al. (1983).

values to a maximum 435 cp with an average value of 21.4 cp. The water-to-oil mobility ratio ranged from 0.1 to 40, the mean value being 7.86. The water-oil ratio at start of the polymer flood had values corresponding to water-free oil production rates (WOR = 0) up to a maximum of WOR = 75 (water cut 98.6 percent) in postwaterflood applications.

The polymer used in the majority of these applications was polyacrylamide, and only a few projects utilized polysaccharide biopolymer.

### Field Results

The maximum oil recovery was 14 percent of the original oil in place (OOIP). The mean value was close to 4 percent of OOIP in 20 fieldwide projects. In any case, the recovery was greater than with waterflooding alone. In 23 fieldwide projects, the maximum recovery was 143 STB/acre-ft, and the mean recovery value was 34.4 STB/acre-ft.

The injection of high-molecular-weight, water-soluble polymer has become a recognized enhanced oil recovery (EOR) method. In the United States, EOR production from polymer flooding has increased in the last decade. It still represents only approximately 4 percent of total EOR production, with thermal methods comprising approximately 80 percent. There are now well-established guidelines for successfully applying polymer flooding and for obtaining maximum recovery and additional oil (Chang, 1978; Gao, 1987; Needham and Doe, 1987).

## -1 GUIDELINES FOR POLYMER APPLICATION

### Reservoir Characteristics

*Reservoir depth* is a critical factor only when related to reservoir temperature. Temperatures less than 200 °F assure a stable polymer solution.

*Reservoir pressure* is not critical if it permits the injection pressure to be less than the formation parting pressure and is not so high that it requires expensive pumping equipment.

The *porosity* of the reservoir rock must be medium to high (higher than 18 percent) to assure a good storage capacity.

Absolute *permeability* of the reservoir rock is considered to be good between 50 and to 250 md. Moderate permeability values (between 15 and 50 md) cause higher injection pressures. Permeability values considered very good (between 250 and 1000 md) and excellent (higher than 1000 md) assure higher recoveries with conventional waterflood and leave polymer flood expenses difficult to justify.

The *permeability variation* concept connected with reservoir heterogeneity is better than permeability alone for determining areas of polymer flood applicability. Heterogeneous reservoirs are good candidates for polymer flooding for two reasons. First, the polymer solution reduces the rock's permeability. Second, the solution has a tendency to divert toward unswept reservoir areas or areas where waterflooding resulted in an unsatisfactory sweep.

### Fluid Characteristics

The *oil viscosity* that directly controls the water-oil mobility ratio should not be higher than 150 to 200 cp. Viscosity of less than 100 cp is even more preferable. Thermal methods of recovery are competitive when reservoirs' oil viscosities are high. When they are low, less than 5 cp, waterflood and/or other EOR methods may be preferred.

The *water-oil ratio* at the start of the project should be low, even zero. This will mean higher mobile oil saturation. Polymer flooding applied from the beginning as a secondary recovery process instead of waterflooding will afford better chances of success than waterflooding.

### Reservoir Selection

As stated, polymer flooding is successfully applied as a secondary recovery process when the mobile oil saturation in reservoir is still high. The effect of polymer flooding is to improve areal and vertical sweep efficiencies, reducing the water-oil mobility ratio and diverting the injected fluid toward larger reservoir areas. Polymer flooding does not improve the displacement efficiency, since it has no effect on the existing capillary forces and interfacial tensions.

Waterflooded oil reservoirs can also be good candidates for polymer flooding. This is only true, however, when the high WOR values are caused either by a high water-oil mobility ratio (viscous oil reservoirs) and waterconning or by a low vertical sweep efficiency (heterogeneous reservoirs). Under these conditions, the reservoirs still have high mobile oil saturation which can be reduced by the diverting effect of polymer flooding and by the increase of vertical sweep efficiency. Therefore, the use of polymer as a tertiary method (after waterflooding) is attractive only if the high WOR is due to waterconning, high permeability zones or high oil viscosity. Large gas-cap areas and extensive aquifers should be avoided.

### Incremental Oil Recovery

There are several methods to estimate initially the incremental oil recovery from polymer flooding. Johnson's (1956) simplified graphical treatment of the Dykstra–Parsons waterflood calculation method, the fractional flow predictive method of Patton and coworkers (1971), and the empirical correlations published in U.S. DOE Report DOE/ET 12072-2 (1987), among others, can be applied as a screening guide for polymer flood potential.

The technique is to compare the oil recovery expected through using continued waterflooding with the oil recovery expected using the modified flow properties from polymer flood.

Because of its simplicity, Johnson's graphical treatment of the Dykstra–Parsons method is still used throughout the oil industry when a rapid appraisal of a waterflood project is needed.

The technique is based upon the correlation for four variables, namely,

$V$ = permeability variation

$S_{wi}$ = initial water saturation

$M$ = water-oil mobility ratio

$E_R$ = fractional recovery of oil in place, at a specified water-oil ratio, WOR

Figures 6-7, 6-8, 6-9, and 6-10 show plots correlating, $V$, $M$, $S_w$, and $E_R$ for WOR of 1, 5, 25, and 100, respectively.

The permeability variation, $V$ (Craig, 1980) measures the effect of permeability stratification on flood-out performance. Since rock permeabilities usually have a log-normal distribution, the percentages of the total number of permeability values arranged in descending order are plotted on log probability graph paper (Figure 6-11), and a straight line is drawn through the points. The permeability variation is given by

$$V = \frac{\bar{k} - k_\sigma}{\bar{k}} \qquad (6\text{-}6)$$

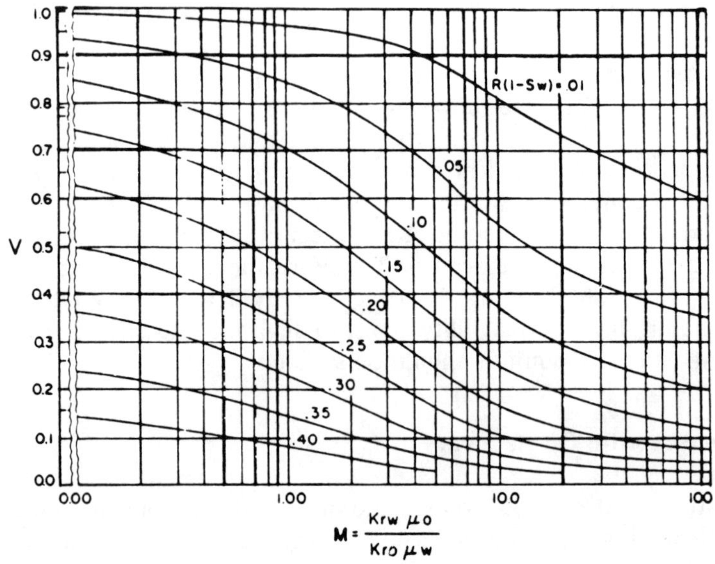


**Fig. 6-7** Correlation of permeability variation, mobility ratio, water saturation, and fractional oil recovery for a producing water-oil ratio of 1 (After Johnson, 1956)

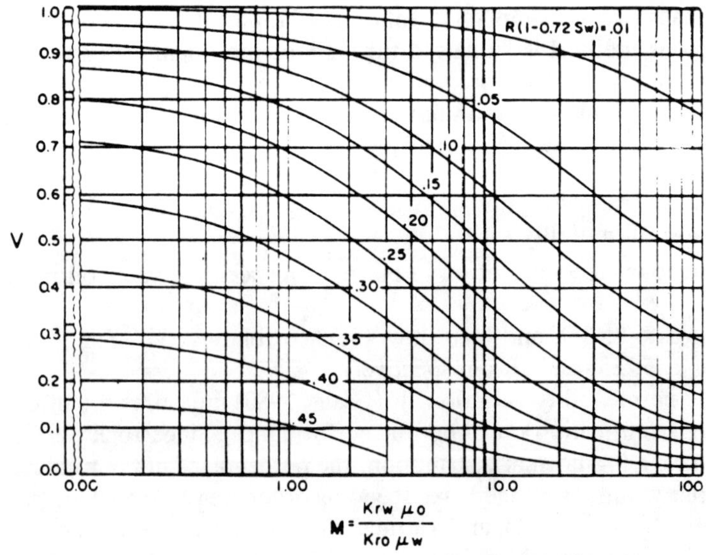


**Fig. 6-8** Correlation of permeability variation, mobility ratio, water saturation, and fractional oil recovery for a producing water-oil ratio of 5 (After Johnson, 1956)

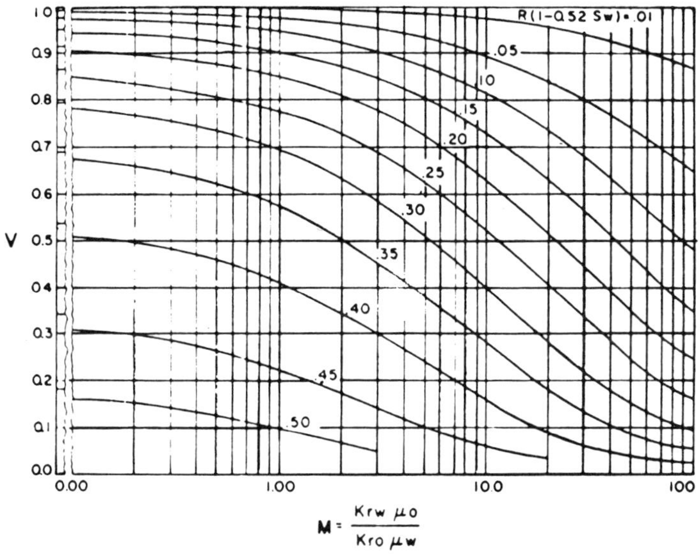


**Fig. 6-9** Correlation of permeability variation, mobility ratio, water saturation, and fractional oil recovery for a producing water-oil ratio of 25 (After Johnson, 1956)

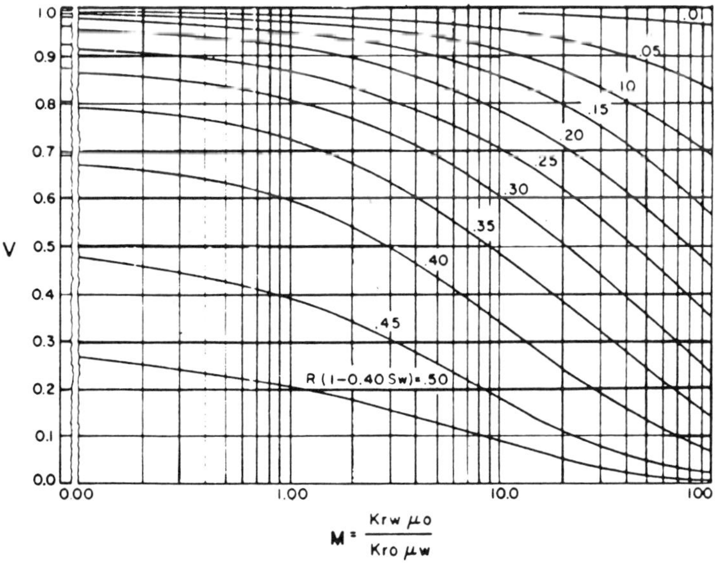


**Fig. 6-10** Correlation of permeability variation, mobility ratio, water saturation, and fractional oil recovery for a producing water-oil ratio of 100 (After Johnson, 1956)

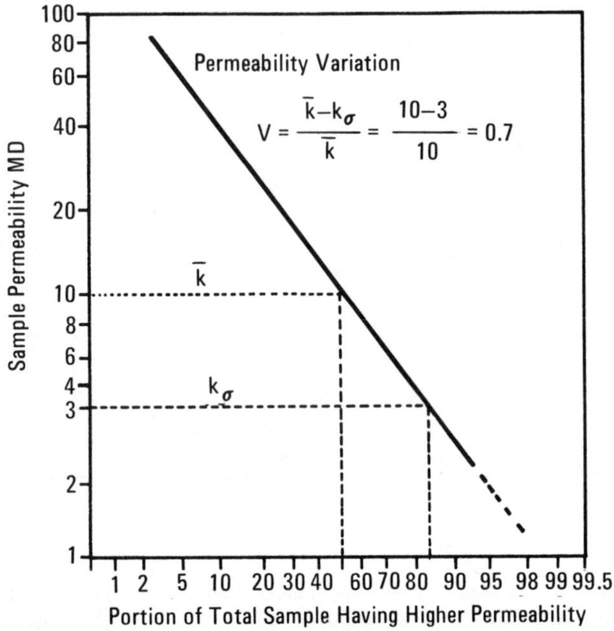


**Fig. 6-11** Log. normal permeability distribution (From Craig, 1980)

where

$\bar{k}$ = mean permeability (the permeability value with 50 percent probability)
$k_\sigma$ = permeability at 84.1 percent of the cumulative sample

The values range from zero to one, a uniform reservoir having the permeability variation of zero.

For a calculated value of the permeability distribution $V$ and a water-oil mobility ratio $M_{w-o}$, the expression $E_R(1 - c \times S_w)$ can be read at the intersection point on the plot for a producing WOR of 1 (Figure 6-6). The same procedure applied for WORs of 5, 25, and 100 makes it possible to know the variation of the fractional recovery of oil in place, $E_R$ for different producing WORs characterizing a conventional waterflood. The modified flow property of polymer flood is the water-oil mobility ratio reduced by the numerical value of the resistance factor $R$. Using this new mobility ratio, the fractional recovery of oil in place, $E_R$, can be obtained for different producing WORs characterizing a continuous polymer flood. The difference in recovered oil for a given WOR value represents the incremental amount of oil due to the polymer project. The method assumes linear flow with pistonlike displacement and with no additional oil being produced after the passage of the front. As was pointed out, the method is a screening device which compares two types of flood under the same flowing conditions and is considered to give good relative answers.

**TABLE 6-2.** Properties of Example Reservoir

| | |
|---|---|
| Irreducible water saturation | 0.20 |
| Relative permeability for water | 0.18 (at $S_{or}$) |
| Relative permeability for oil | 0.60 |
| Water viscosity | 0.473 cp |
| Oil viscosity | 6.4 cp |
| Permeability variation | 0.5 |
| Formation volume factor for oil | 1.05 |
| Resistance factor | 6 |

**Example 6-2.** Compare, at a water cut of 95 percent, the final oil recovery factors expected through conventional water injection and polymer waterflooding. The reservoir properties are shown on Table 6-2.

SOLUTION

**(a)** Mobility ratios:

Water-oil mobility ratio:

$$M_{w-o} = \frac{k_{rw}\,\mu_o}{\mu_w\,k_{ro}} = \frac{0.18(6.4)}{0.473(0.6)} = 4$$

Resistance factor:

$$R = \frac{k_{rw}\,\mu_p}{\mu_w\,k_{rp}} = 6$$

Polymer solution-oil mobility ratio:

$$M_{p-o} = \frac{M_{w-o}}{R} = 0.666$$

**(b)** Oil recovery factors (Table 6-3): The WOR plotted against recovery factor $E_R$ is illustrated in Figure 6-12 for waterflood and for polymer flood. When the water cut, $f_w = 0.95$, the water-oil ratio is

$$\text{WOR} = \frac{f_w}{1 - f_w} = 19$$

and the recovery factor is $E_R = 33.0\%$ for waterflood and $E_R = 41.6\%$ for polymer flood.

**TABLE 6-3.** Oil Recovery Factors

| M | Oil Recovery | WOR = 1 | WOR = 5 | WOR = 25 | WOR = 100 |
|---|---|---|---|---|---|
| 4 | $E_R(1 - C \times S_{wi})$ | 0.11 | 0.205 | 0.32 | 0.38 |
| | $E_R$, % | 13.75 | 23.95 | 35.7 | 41.3 |
| 0.666 | $E_R(1 - C \times S_{wi})$ | 0.20 | 0.30 | 0.38 | 0.425 |
| | $E_R$, % | 25 | 35 | 42.4 | 46.6 |

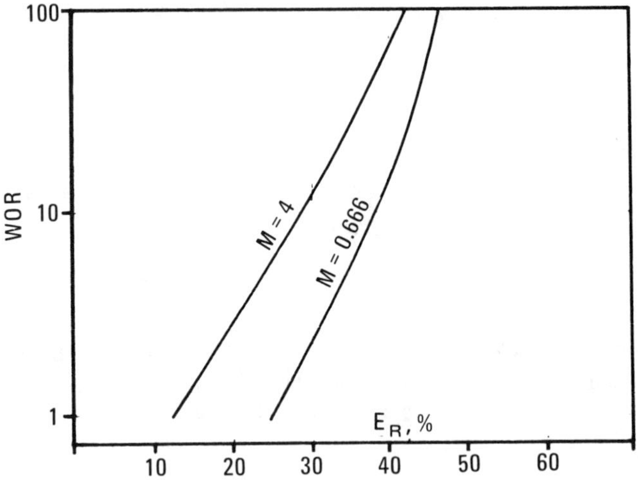


**Fig. 6-12** Water-oil ratio versus oil recovery

Although the method assumes 100 percent volumetric sweep efficiency and continuous polymer solution injection, it still can be said that the additional oil recovery with polymer injection corresponds to an increase of 8 to 9 percent of OOIP over that obtained by waterflood. The oil reservoir can be selected as a polymer flood candidate.

## 6-8 DESIGN CONSIDERATIONS

After the reservoir selection is made, detailed testing is performed on the different polymers available for viscosity, molecular weight, shear degradation, and adsorption. Different polymers' resistance factors are measured using representative core samples. Simulation studies are necessary to determine the optimum polymer concentration, the slug size, and the estimated oil recovery. Field injectivity tests are very useful to determine the polymer's behavior in the porous space of the reservoir and to design the appropriate polymer injection facilities and quality control.

The reservoir engineer should take into account all the existing experience regarding polymer flooding. Thorough consideration should be given to the specific reservoir characteristics and performances before the start of the process.

To illustrate in some detail the work performed after an oil reservoir is identified as a polymer flood candidate, a typical case history is presented.

### Sleepy Hollow Reagan Unit, Nebraska, United States

The Reagan reservoir case history (Christopher et al., 1988) shows that the selection of the reservoir as a polymer flood candidate was made due to the unfavorable mobility ratio under waterflooding (the oil viscosity was 24 cp and injection water viscosity was 0.71 cp).

*Laboratory tests* were performed first to determine the viscosity, screen factor, and shear degradation characteristics of different polymer samples using the low salinity injection water. After choosing the polyacrylamide polymer, which provided the best results, porous media displacement tests using polymer solutions and residual saturation of a refined oil were performed to determine the measurement of the resistance factor and the residual resistance factor. The incremental oil recovery of 10.1 percent of OOIP from polymer flooding was estimated using the fractional flow predictive method (Patton et al., 1971). The 22-cp viscosity of the polymer solution flowing through the porous medium (the apparent viscosity) displaces to the right the fractional flow curve (Figure 6-13).

The value of the producing water cut was reduced as the polymer solution was diverted toward unswept or poorly swept areas of the reservoir. Using computer simulation techniques, the reservoir characteristics were described by history matching the waterflood performance on a three-dimensional black

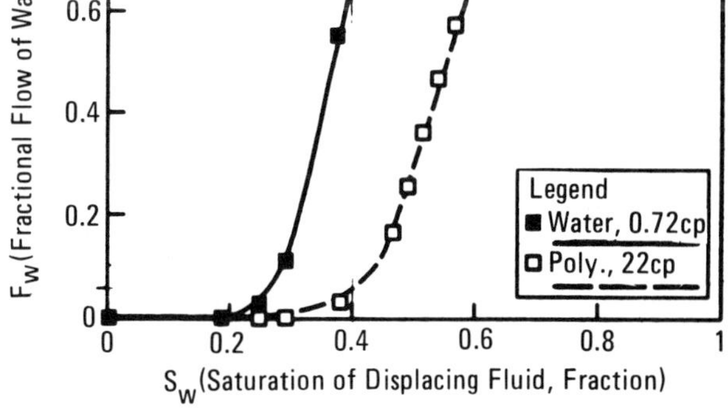


**Fig. 6-13** Fraction flow curves for water and 22-cp polymer solution displacing Reagan crude oil (From Christopher et al., 1988)

oil simulator. The flow of polymer solution in the reservoir was simulated using the results of performance history match, and an optimum polymer concentration of 750 ppm and a 48 percent pore volume slug size were recommended. Using a resistance factor of 28.63 and a reduced mobility ratio of 0.26 at residual oil saturation, the simulator predicted an 8.0 percent OOIP incremental oil recovery.

A *field injectivity test* was initiated with the intent "to develop experience in mixing the polymer in the field, to evaluate injectivity and detect any significant wellbore plugging which could result, and to evaluate the quality of *in-situ* polymer solution . . . " (Christopher et al., 1988). Pressure fall-off tests were performed, swab samples were analyzed, and polymer and tracer concentrations were measured. Since the results of the field test showed injectivity losses but not an excessive polymer degradation, it was decided to use a conservatively lower-molecular-weight polymer with a solution viscosity of 10 cp at 1000 ppm polymer concentration.

*Surface facilities* are shown in Figure 6-14 and were designed for an injection rate of a maximum 25,000 BWIPD at 250 psig and a minimum of 8 days' polymer storage capacity. The highly concentrated polymer solution, transported by tank truck to the site, is injected into the water stream at a rate such that the final concentration desired is achieved in the static mixer and the

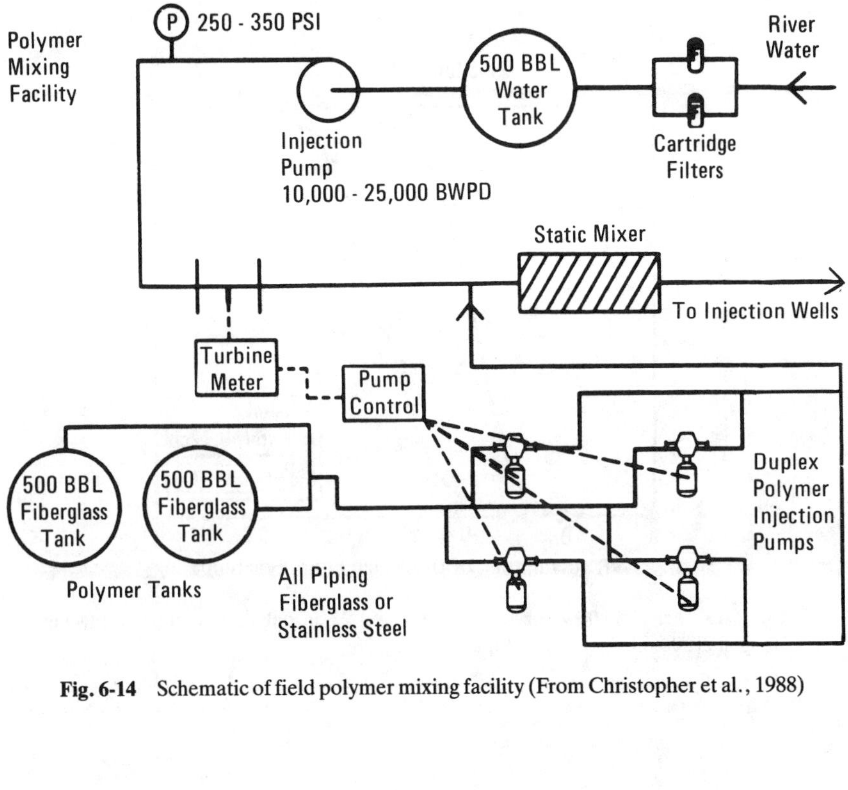


**Fig. 6-14** Schematic of field polymer mixing facility (From Christopher et al., 1988)

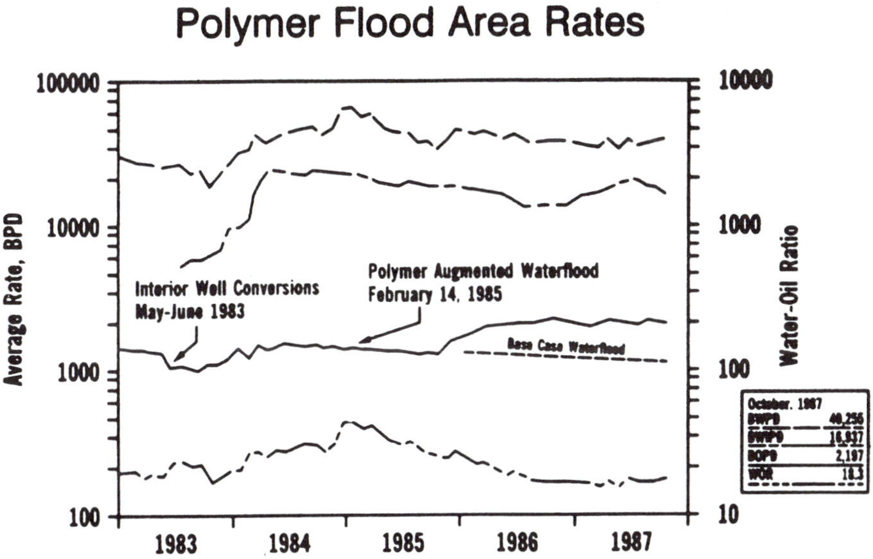


**Fig. 6-15** Injection and production rates from the polymer flood area, showing incremental oil production over the waterflood prediction (From Christopher et al., 1988)

injection lines. Fiberglass injection and source waterlines were installed, 10 producing wells were converted to polymer injection and provided with coated tubing, and a quality control laboratory was set up.

*Quality control* parameters such as the screen factor and the viscosity of the emulsion concentrate and dilute solutions were tested using samples taken at the mixing plant and at each wellhead provided with a sample point. To reduce shear and pressure drop in the piping system, coils of tubing used as flow restrictor and wellhead filters with large filter areas were installed.

*Polymer flood performance* is shown in Figure 6-15.

The rate of oil production has increased from the start of polymer injection, and the WOR has decreased substantially. The results to date are favorable, especially considering the implementation of the polymer flood in the late period of a waterflooded reservoir.

## QUESTIONS AND EXERCISES

**6-1** Find the water-oil mobility ratio if the oil viscosity is 16 times the water viscosity and the formation rock's effective permeability for oil is the same as for water. What is the resistance factor if the mobilities of oil and polymer solution are 2 and 4, respectively?

**6-2** A polymer injection test is conducted on a five-spot pilot with 10 acres (40,000 $m^2$) area, 20 ft (6.1 m) effective thickness, and 25 percent porosity. A total amount of 0.28 PV of 350-ppm polymer solution has to be injected at a daily rate of 377 bbl (60 $m^3$). Since the polyacrylamide is available in both as liquid and powder, calculate how long the test will take and what amount of liquid or powder HPAM should be mixed with water on site. The liquid has 8 wt % active polymer concentration and the powder has 95 wt % concentration.

**6-3** Using the water-oil ratio versus oil recovery, percentage of OOIP (Figure 6-12), calculate the recoveries at breakthrough when $f_w = 50\%$ and after breakthrough when $f_w = 98\%$ for both waterflood and polymer flood.

**6-4** Compare, as a first screening, the oil recovery expected at breakthrough with conventional water injection and with polymer flooding. The reservoir has a permeability variation of $V = 0.5$, the oil-water viscosity ratio $\mu_o/\mu_w = 5$, the resistance factor $R = 4$, and the water-oil relative permeability characteristics are given in Figure 6-1.

**6-5** How much additional oil, percent of OOIP, is produced by continuing the waterflood and the polymer flood up to water cut $f_w = 0.97$?

**6-6** Make comments regarding the recovery obtained at breakthrough and at a water cut of 97 percent with both water and polymer flood. Is the continuous polymer flood economically justified?

## REFERENCES

BUCKLEY, S. E., and M. C. LEVERETT, "Mechanism of Fluid Displacements in Sands," *Transactions of the AIME*, Vol. 146 (1942), pp. 107–16.

BURCIK, J.E., "What, Why and How of Polymers for Waterflooding," *Petroleum Engineering* (August 1968).

CHANG, H. L., "Polymer Flooding Technology—Yesterday, Today and Tomorrow," *Journal of Petroleum Technology* (August 1978).

CHRISTOPHER, C. A., T. J. CLARK, and D. H. GIBSON, "Performance and Operation of a Successful Polymer Flood in the Sleepy Hollow Reagan Unit," Society of Petroleum Engineers/U.S. Department of Energy paper SPE/DOE 17395, SPE/DOE Sixth Symposium on Enhanced Oil Recovery, Tulsa, Oklahoma, April 17–20, 1988.

CRAIG, F. F., *The Reservoir Engineering Aspects of Waterflooding,* third printing (Richardson, TX: Society of Petroleum Engineers, 1980), pp. 45–46.

DAWSON, R., and R. LANTZ, "Inaccessible Pore Volume in Polymer Flooding," *SPE Journal* (October 1972).

DETLING, K. D., "Process of Recovering Oil from Oil Sands," U.S. Patent #2-341-500 (February 1, 1944).

GAO, HONG WEN, *Mobility Control in Oil Recovery by Chemical Flooding, State-of-the-Art Review*, Topical Report NIPER-146 (DE87001210) (Bartlesville, OK: National Institute for Petroleum and Energy Research—U.S. Department of Energy, January 1987).

GOGARTY, W. B., "Mobility Control with Polymer Solutions," *Journal of Petroleum Technology* (June 1967).

GRULA, M. M., and G. W. SEWELL, "Microbial Interactions with Polyacrylamide Polymers," *Proceedings, International Conference Microbial Enhancement Oil Recovery*, E. C. Donaldson and C. J. Bennett, eds., Bartlesville, Oklahoma (1982), p. 219.

JENNINGS, R. R., J. H. ROGERS, and T. J. WEST, "Factors Influencing Mobility Control by Polymer Solutions," *Journal of Petroleum Technology* (March 1971).

JEWETT, R. L., and G. F. SCHURZ, "Polymer Flooding—A Current Appraisal," *Journal of Petroleum Technology* (June 1970).

JOHNSON, C. E., JR., "Prediction of Oil Recovery by Water Flood—A Simplified Graphical Treatment of Dykstra–Parsons Method," *Transactions of the AIME*, Vol. 207 (1956), pp. 345–46.

KNIGHT, BRUCE L., "Reservoir Stability of Polymer Solutions," *Journal of Petroleum Technology* (May 1973).

MAERKER, J. M., "Dependence of Polymer Reaction on Flow Rate," *Journal of Petroleum Technology* (November 1972).

MAERKER, J. M., "Mechanical Degradation of Partially Hydrolyzed Polyacrylamide Solutions," *SPE Journal* (August 1976).

MANNING, R. K., G. A. POPE, L. W. LAKE, et al., *A Technical Survey of Polymer Flooding Projects*, U.S. Department of Energy, Report No. DOE/BC/10327-19, September 1983.

MUNGAN, N., F. W. SMITH, and J. L. THOMSON, "Some Aspects of Polymer Floods," *Journal of Petroleum Technology* (September 1966).

NEEDHAM, R. B., and P. H. DOE, "Polymer Flooding Review," *Journal of Petroleum Technology* (December 1987), p. 1504.

PATTON, J. T., K. H. COATS, and G. T. COLEGROVE, "Prediction of Polymer Flood Performance," *SPE Journal* (March 1971), pp. 72–84.

PYE, DAVID J., "Improved Secondary Recovery by Control of Water Mobility," *Journal of Petroleum Technology* (August 1964), pp. 911–916.

SANDIFORD, B. B., "Laboratories and Field Studies of Water Floods Using Polymer Solution to Increase Oil Recoveries," *Journal of Petroleum Technology* (August 1964), pp. 917–22.

U.S. DEPARTMENT OF ENERGY, *Economics of Enhanced Oil Recovery*, Report DOE/ET/12072-2 (Washington, D.C.: U.S. Department of Energy, 1981), pp. 19–20.

# Chapter 7

# Alkaline Flooding

## 7-1 GENERAL

Alkaline or caustic flooding is another method by which oil displacement efficiency can be improved. The benefits of this process have been known for a long time and were first observed by Squires (1917) and later by others. However, not until 1942 did Subkow offer the explanation that alkaline agents such as sodium hydroxide could react with naturally occurring organic acids in crude oil to produce soaps at the water-oil interface. The effect produced in a reservoir appears to be similar to that of micellar solutions. The difference is that alkaline flooding reduces the interfacial tensions (IFT) with surfactants generated in situ.

Despite the fact that alkaline agents are less expensive, the benefits expected from alkaline flood have not been confirmed by firm field results and still remain a possibility rather than a reality. Indeed, the major difficulty is that the process appears to be highly dependent on minerals on the surface of reservoir rock, which are not chemically inert, and on the crude oil and injection fluid characteristics.

Efforts have been made, especially in the last decade, to understand better the recovery mechanisms generated in alkaline flooding. Since alkaline agents are cost-efficient materials, their use, along with surfactant and/or polymer, could reduce the amount of high-cost surfactant and cosurfactant required in micellar flooding. A reevaluation of alkaline flooding is taking place in order to find ways to reduce the reaction of alkaline agents with reservoir minerals and to take advantage of the combined alkaline/surfactant mixture effect.

## 7-2 DISPLACEMENT MECHANISMS AND METHOD DESCRIPTION

### Displacement Mechanisms

Several mechanisms have been suggested regarding oil displacement by alkaline flooding. The publication of Johnson's review (1976) gives a description of four different mechanisms based on oil emulsification and wettability reversal.

It is known that fluid distribution within the pore spaces of a rock reservoir during (alkaline) waterflood depends upon the wetting and nonwetting phase saturation and upon the direction of the saturation change (Craig, 1980). In a water–wet rock reservoir, the injected water increases the wetting phase saturation, the residual oil being the discontinuous phase. In an oil–wet rock reservoir, the injected water decreases the wetting phase saturation, the residual oil being the continuous phase. It was observed also that residual oil saturation always depends on the dimensionless ratio of viscous to capillary forces defined as the capillary number: $\nu\mu/\sigma\phi$: (velocity × viscosity of the displacing water) ÷ (interfacial tension between water and oil phases × porosity). When the capillary number value can be increased from $10^{-6}$ (conventional waterflood) to $10^{-4}$ or more, the residual oil saturation decreases.

1. The alkaline solutions increase the capillary number value by reacting with the organic acids present in some crude oil to form emulsifying soaps. The petroleum soap or surfactant formed emulsifies oil and water and reduces the interfacial tension by two or three orders of magnitude. This mechanism is referred to as *emulsification and entrainment* because the oil-in-water emulsion formed is entrained by the fluid flow and can then be produced. The residual oil saturation is lowered and an incremental increase in oil recovery can result.
2. When the displacement takes place in an oil-wet reservoir, where residual oil is a continuous phase, the alkaline agent changes the injection water pH and the rock wettability is reversed from oil-wet to water-wet. This mech-

anism is defined as *wettability reversal*. As a consequence the water-oil relative permeability and the water-oil mobility ratio are lowered, with evident benefits on oil recovery (Mungan, 1966).

3. Even in the water-wet reservoirs the discontinuous, nonwetting residual oil phase can be changed to a continuous wetting phase if proper conditions of reservoir temperature, pH, and salinity of the alkaline solution are met (Cooke et al., 1974). The mechanism is referred as *wettability reversal from water-wet to oil-wet*. The presence of water droplets in the continuous oil-wet phase raises the pressure gradient of the flow through porous medium. The capillary forces are overcome and residual oil saturation is reduced.
4. A fourth mechanism, *emulsification and entrapment*, proposed by Jennings et al. (1974), explains that additional oil could be produced because of the entrapment of the oil emulsion droplets by small pores. Because the flow is diverted into poorly swept or unswept areas, it improves the volumetric sweep efficiency, especially in waterflooded viscous oil reservoirs or in heterogeneous reservoir.

Although other displacement mechanisms have been identified, the principal mechanism considered in alkaline flooding is the reduction of the oil-

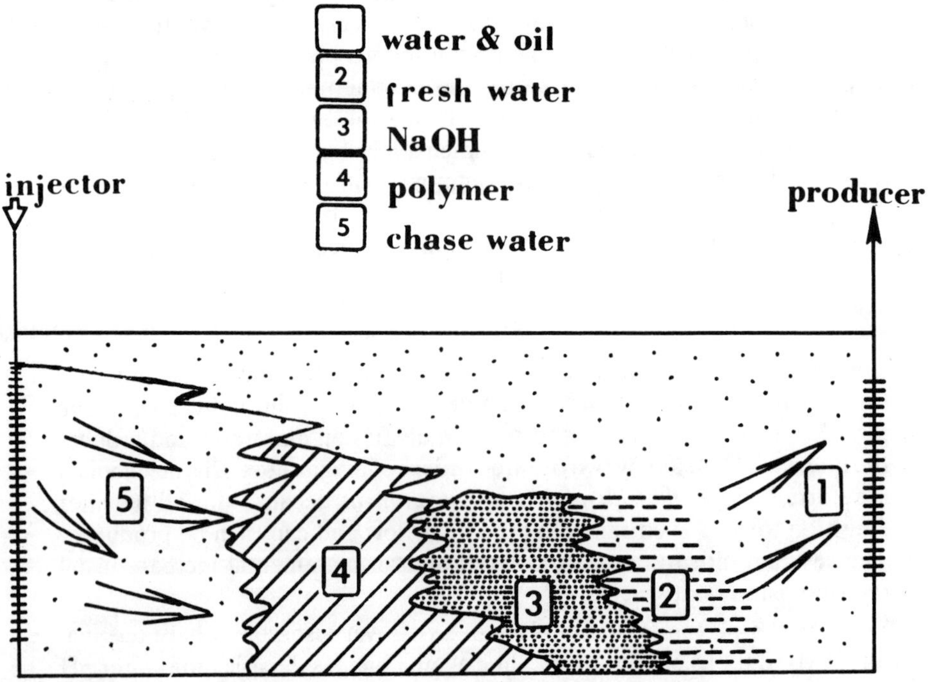


**Fig. 7-1** Schema of alkaline flood injection

water interfacial tension. The chemical reactions involving the alkaline solution and in-place crude oil form a surfactant that reduces the IFT (Surkalo, 1990).

### Method Description

The basic alkaline flooding process starts with a softened water preflush injection followed by the injection of an alkaline solution of about 10 to 30 percent PV and by continuous injection of drive water. Numerous variations have been proposed. The injection of a polymer slug behind the alkaline solution to control mobility and to improve sweep efficiency is desirable (Figure 7-1) if it is cost effective. Because of the complexity of the mineralogy and lithology of petroleum reservoirs the possible reactions between rock–alkaline solution–saline water and oil in the existing conditions of pressure and temperature are considerable. This explains the effort put into laboratory alkaline flooding tests and field trials to design properly the best system for specific reservoir conditions.

The state-of-the-art techniques for alkaline flooding utilize alkaline agents in combination with low concentrations of synthetic surfactant and polymer for mobility control (Ball and Surkalo, 1988).

## 7-3 DESIGN CONSIDERATIONS AND SCREENING CRITERIA

### Design Considerations

The principal goal in designing an alkaline system is to achieve a minimum IFT in the reservoir. The corresponding alkali concentration is then considered optimum. In the laboratory, the alkali concentration yielding the minimum IFT was very low, often less than 0.1 wt % in some of the early flood designs. In the field, this concentration did not survive far from the wellbore because of reaction with the rock and consumption (Surkalo, 1990).

Results from even the earliest laboratory experiments have shown that salinity plays an important role in determining optimum alkali concentration. For instance, minimum IFT could be achieved with distilled water and a wide range of NaOH concentrations, between 0.1 and 0.8 wt % (Figure 7-2).

With brine from Whittier field, California, where a field trial of caustic flooding was conducted from October 1966 to August 1967 (Graue and Johnson, 1974), the lowest IFT was obtained over a very narrow range of NaOH concentrations (0.18–0.27 wt %). Adding alkali to increase concentration and to keep the effect of alkaline solution far from the injection wellbore increased the salinity of the system and the IFT value.

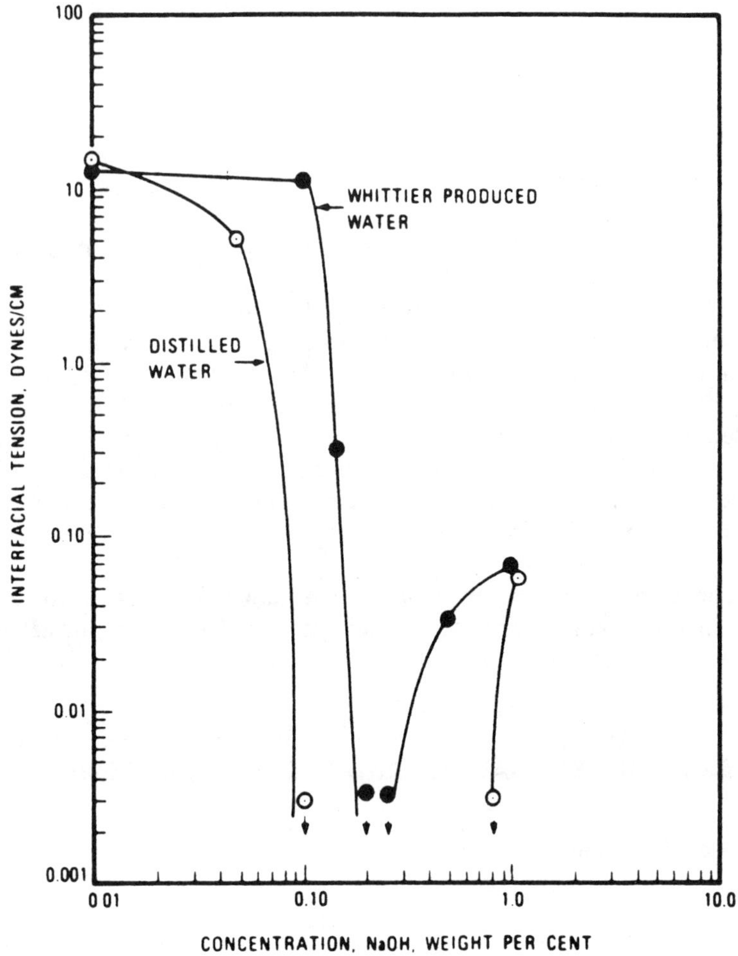


**Fig. 7-2** Interfacial tension of Murphy-Whittier second and third zones crude (From Graue and Johnson, 1974)

Recent laboratory work has been done in an attempt to adjust higher alkali levels without losing the low IFT values. Ball and Surkalo (1988) have found that a low concentration of surfactant, sodium hydroxide (NaOH) or sodium orthosilicate ($Na_4SiO_4$), added to the alkali solution generally yields significant IFT reductions. Table 7-1 shows the IFT obtained from Alberta crude oil and alkali surfactant blends.

Core samples waterflooded to residual oil saturation and then injected with polymer, alkali-polymer, or alkali-surfactant-polymer showed that the last combination was the most efficient (Table 7-2).

As can be observed, the alkali-surfactant-polymer system reduces the residual oil saturation to its lowest value in comparison with the other two

**TABLE 7-1.** IFT Between Alberta Crude Oil and Alkali-Surfactant Blends

| Surfactant at 0.1 wt % | IFT (dynes/cm) +0.8 wt % Naoh | +1.6 wt % NaOH | +0.8 wt % $Na_4SiO_4$ |
|---|---|---|---|
| ENORDET 25-3A | 0.398 | 0.269 | 1.7 |
| ENORDET 45-7 | 0.869 | 0.733 | 0.425 |
| TERGITOL NP-9 | 0.728 | 0.349 | 0.511 |
| PETROSTEP B-105 | 0.020 | 0.016 | 0.010 |
| IPA | 2.0 | 1.2 | 2.8 |

From Ball and Surkalo (1988)

systems. The low amount of surfactant (0.1 wt %) added to an alkaline-polymer system can have the same efficiency in displacing oil that a micellar-polymer system has, but at a lower chemical cost.

**TABLE 7-2.** Tertiary Oil Recovery—Alberta Systems

| System | Final $S_o$ (PV) |
|---|---|
| Polymer | 0.388 |
| Alakli-polymer | 0.251 |
| Alkali-surfactant-polymer (0.1 wt %) | 0.115 |

From Ball and Surkalo (1988)

Although the laboratory tests and studies reported encouraging results, the process will have to be proven by an increasing number of field pilots and by commercial development.

### Screening Criteria

The general screening criteria including temperature (less than 200 °F), permeability (between 50 and 250 md), oil viscosity (less than 150 to 200 cp), and low brine salinity are the same as the limits characterizing chemical flooding using surfactants or polymers. As with other injection solutions, reservoirs with large gas-cap areas and extensive aquifers should also be avoided when considering the use of alkaline flooding.

There are other special aspects to consider when screening reservoirs for alkaline flooding such as the mineralogy of the reservoir rock, the $CO_2$ content of the petroleum reservoir, and the crude oil's acid number. These parameters were identified by Lorenz and Peru (1989) from a detailed study of alkaline field projects conducted since 1960 and were summarized by French and Burchfield (1990) as follows:

**Gypsum.** Reservoirs with gypsum (anhydrite) content greater than 0.1 percent should be rejected as candidates for alkaline flooding. The gypsum reacts and consumes the alkali agent. For instance, 1% gypsum in a reservoir can consume 10 PV of a 1 percent solution of $NaHCO_3$.

**Kaolinite.** Reservoirs that contain appreciable amounts of kaolinite could be candidates for alkaline flooding only with low-pH (about 8.2–10) alkalis. The presence of kaolinite is a serious deterrent to flooding with higher-pH alkalis (Thornton, 1986).

**Montmorillonite.** The montmorillonite content of rock reservoir, because of its large surface area and very high cation exchange capacity, can consume much of the injected alkali by adverse precipitation reactions (Smith, 1978). Reservoirs with a montorillonite content higher than 1 percent and 0.4 wt % divalent ions in the brine are not good alkaline flooding candidates.

**Sandstone.** Sandstone reservoirs are preferred. High-pH alkalis are reactive with limestone.

**Crude oil.** It is important for the crude oil to have a high acid number* to achieve low IFT in alkaline flooding, when the system does not include synthetic surfactant. However, even with crude oil with a low acid number, the alkaline flood can still be successful as a result of mechanisms other than IFT lowering. The detailed study of alkaline field projects mentioned earlier did not show a correlation between oil acid number and project success (French and Burchfield, 1990).

**Carbon dioxide.** The $CO_2$ content of a petroleum reservoir is now considered a strong screening criteria parameter. Reservoirs with high $CO_2$ content, mole fraction of $CO_2$ (in produced gas) higher than 0.01, or a pH less than 6.5 are not good candidates for alkaline flooding (Lorenz and Peru, 1989).

## 7-4 FIELD TRIALS

Although more than 50 alkaline flooding field projects have been tested in the United States in the past 60 years, the method is not yet feasible as a commercial-scale operation (French and Burchfield, 1990). Two field trials of caustic flooding (Whittier oil field and Wilmington Ranger zone, California), where the published field results were combined with laboratory data, are presented next.

* Acid number is the number of milligrams of potassium hydroxide required to neutralize 1 gram of crude oil to pH = 7.0.

### Whittier Oil Field, California, United States

This field had been under a waterflood in progress when an experimental alkaline injection was initiated in 1966. The results were published by Graue and Johnson in 1974.

**Reservoir characteristics.** The project was started as a simultaneous flood in both productive second and third zones of the 63-acre project area at an average subsea depth of 1500 ft and 2100 ft, respectively. Other reservoir and fluid characteristics were

| | |
|---|---|
| Net oil sand thickness | 37 ft and 100 ft |
| Average porosity | 30% |
| Air permeability | 495 md and 320 md |
| Formation dip | 25° to 45° south |
| Oil gravity | 20 °API |
| Reservoir temperature | 120 °F |
| Oil viscosity | 40 cp |
| Well spacing | 1 and 2 acres per well |
| Permeability variation | 0.66 to 0.74 |

**Project implementation.** Approximately 1.6 million bbl of a solution composed of 20 percent PV of 0.2 wt % sodium hydroxide in water were injected through four initial injectors over a period of 10 months. The concentration of 0.2 wt % NaOH is within the range which corresponds with the lowest interfacial tension value, Figure 7-2.

**Results.** Production data are shown in Figure 7-3, and as we observe, the caustic flood produced more oil than could have been expected from the continuation of waterflood. The response to caustic flooding showed drops in WORs also. However, this effect was exaggerated by shutting in a high-volume water producer near injector 135.

**Comments.** What is interesting to point out is that the injectors were located upstructure (Figure 7-4). A high-dip oil reservoir with a very good permeability exhibits a strong gravitational effect even when the crude oil viscosity is 40 cp. When the reservoir is moderately to highly heterogeneous, there are high permeability layers, referred to as thief channels, through which the gravitational flow of fluids rapidly reaches the downstructure producers.

Indeed, tracers were detected a short time (8 to 16 days) after the start of injection into downstructure producers, and injection into Wells 75 and 135 was stopped.

The alkaline flood injection was started upstructure, probably to avoid

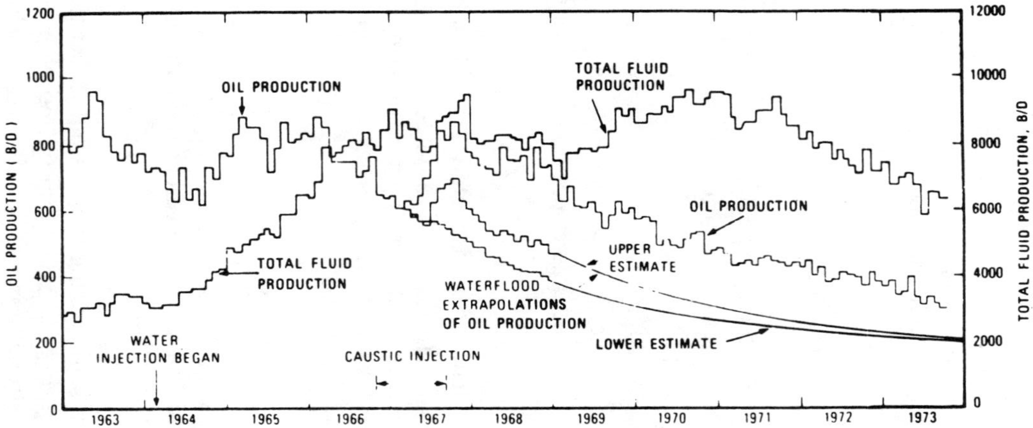


**Fig. 7-3** Production data for wells in caustic flood area (From Graue and Johnson, 1974)

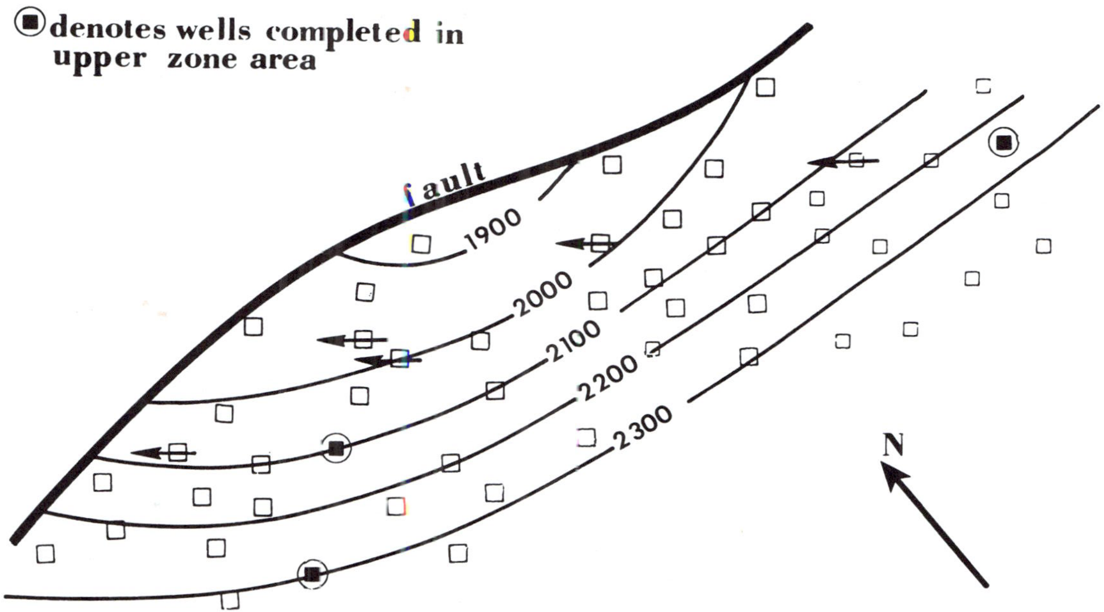


**Fig. 7-4** Wells in field trial area (From Graue and Johnson, 1974)

the high water saturation existing downstructure where the waterflood was supposed to take place. Nevertheless, it is very important to consider carefully the reservoir geometry and recovery mechanisms so that the way in which the new method is implemented will give maximum benefit.

### Wilmington Field Ranger Zone, California, United States

The alkaline demonstration pilot project is comprised of two rows of four injectors each, which enclose the Block 9 pattern area in the Ranger Zone of Wilmington Field. There are two reasons why alkaline flooding was considered for Wilmington—the first is the low recovery efficiency obtained by waterflooding a reservoir which has permeability variations and an unfavorable mobility ratio; the second was the high content of organic acids in the crude oil. The injection was started in the 91 acres pilot project area in January 1979 with 10.2 percent PV softened freshwater with 1 percent added salts, followed by 67 percent PV of alkaline solution (0.39 wt % sodium orthosilicate) and by a postflush of softened water with 0.75 percent added salts. A very detailed study was published by the U.S. Department of Energy (Daube et al., 1987) regarding the reservoir description and project design, implementation, and results evaluation. The main results of this study are described in the paragraphs that follow.

**Reservoir characteristics.** The productive formation lies at a depth of 2225 to 2800 ft and consists of six to eight intervals or subzones of 305 ft total average thickness of interbedded shales and unconsolidated to semiconsolidated sands. The main reservoir and fluid characteristics in the alkaline pilot area are

| | |
|---|---|
| Average net thickness of subzones | between 26 and 52 ft |
| Average porosity between | 0.246 and 0.289 |
| Average permeability between | 131 md and 314 md |
| Average oil gravity, °API | 15–18, 16–23, 18–26, 20–28 |
| Bottom hole temperature | 125 °F |
| Average oil viscosity | 23 cp |

**Laboratory work.** Detailed laboratory studies were performed using core samples recovered in frozen plastic pipe. Special core analysis tests showed an average displacement efficiency with waterflood of 49.4 percent of the oil in place and a cation exchange capacity from 2.6 to 11.5 milliequivalents per 100 grams. The acid number of Ranger Zone crude oil varied from 0.86 to 2.5 milligrams of KOH per gram of oil and measurements of interfacial tension indicated a lower value obtained with sodium orthosilicate than with sodium hydroxide. The consumption tests performed showed alkali consumption ranged from 8.6 to 16.9 milliequivalents of alkali per 100 grams of rock at reservoir temperature and over a test period of 36 days. The reservoir contained a significant amount of clays and also gypsum. The laboratory displacement tests with alkaline solution indicated an average incremental recovery over waterflooding of about 6 percent of OOIP. The concentration of ions in the produced water suggested that during alkaline waterflooding the alkali was consumed and the core material was undergoing dissolution.

**Simulation.** A simulator model with the ability to represent the effects of the alkaline fluids by changes in the relative permeabilities was used in a one-dimensional form to simulate laboratory tests, a two-dimensional form to represent a limited area of the field, and a three-dimensional form to represent the entire project (Dauben et al., 1987). The pilot area simulation (three-dimensional model) predicted an incremental oil recovery of about only 1.9 percent of OOIP compared with 7.25 percent of OOIP predicted from the smaller two-dimensional pattern. This disappointing result appears to be the effect of the long-term consumption of alkaline fluids and of the larger surface area.

**Minifield test.** A two month, one well injectivity test was conducted before the start of the full scale pilot and satisfactory results were obtained regarding fluids injectivity and the operation of the surface facilities in handling the chemicals.

**Project implementation.** The alkaline slug injection started in March 1980. After the 10.2 percent PV preflush injection, 49,927,000 bbl (67 percent PV) of alkaline solution were injected over a period of 34 months at an average rate of 30,000 bbl/day.

**Scale deposition.** Scale deposition in producing wells after the breakthrough of alkaline solutions has been a major problem. Numerous workovers were conducted to overcome the plugging effects of scale in producing wells. The scales composed of calcium carbonate, magnesium silicate and amorphous silica were treated with a combination of acid and scale inhibitor injections.

**Evaluation of results.** The performance of the Block 9 pattern area is shown in Figure 7-5. As we can observe, the alkaline solution injection did not seem to have an impact on production data. The reduced injection rates after mid-1984, caused by the shutdown of two injection wells for repair, along with scale problems in producing wells, were responsible for the decline in gross oil production. The water-oil ratio did not appear to be practically influenced, although a reduced water production was observed in some of the wells. The plot, since 1976, of the oil cut versus cumulative oil recovery indicates that there

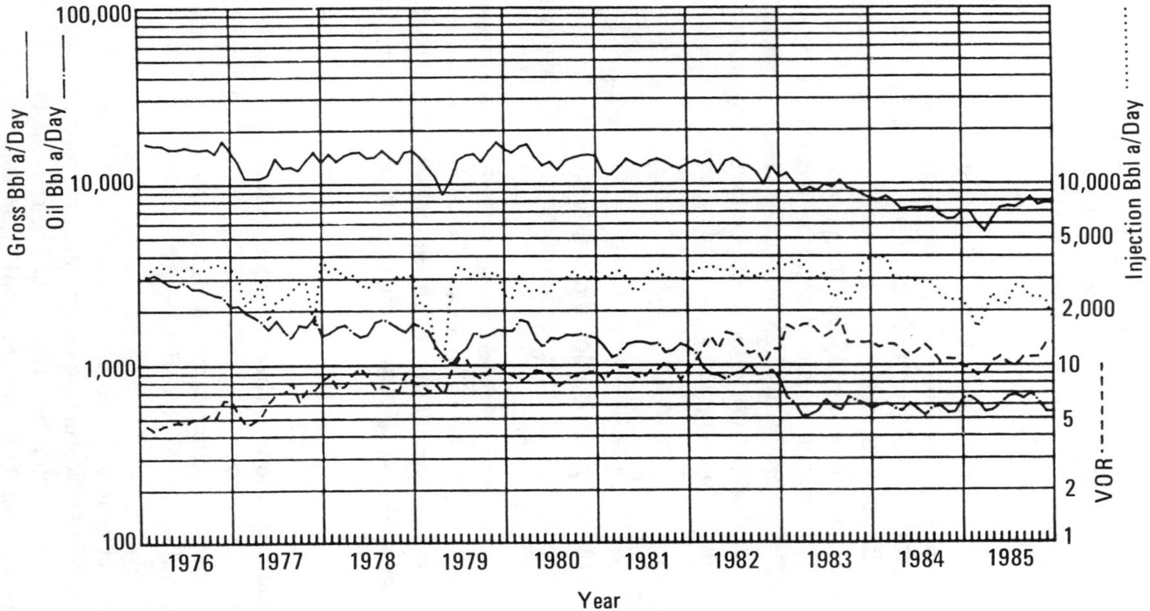


**Fig. 7-5** Production data Ranger economic block 9 (From Dauben et al., 1987)

was no improvement in oil recovery compared to waterflooding. The high consumption of chemicals and the wellbore plugging problems were the main reasons this project proved to be unsatisfactory, not to mention the lower than expected oil recovery (Dauben et al., 1987).

**Comments.** The Wilmington Field Ranger Zone alkaline flood project is an example of how a very-well-prepared, developed, and executed project still did not produce the expected results in the field. It is very important to evaluate the complex reactions of alkaline chemicals with the reservoir rock and fluid content. Even so, losses from chemical consumption observed in laboratory tests, when transferred to the field scale, can be the main reason for failure. Another important aspect is reservoir geometry. The Ranger Zone consists of several subzones separated by impermeable shale barriers. Even if these shale barriers are continuous, they are in hydrodynamic communication through the producers, each opening up at least six subzones. A selective injection schedule starting with the lowest continuous subzone possibly could have been prepared, controlled, and evaluated with less rock surface exposed to chemical reactions.

## QUESTIONS AND EXERCISES

**7-1** Enumerate the mechanisms suggested for oil displacement by alkaline flooding and explain the differences.

**7-2** Describe what can be done to adjust higher alkali levels without losing low IFT values.

**7-3** Enumerate and explain the special aspects to consider as screening criteria for alkaline flooding.

**7-4** What are the merits of the two alkaline field trials—Whittier Oil Field and Wilmington Ranger Zone?

## REFERENCES

Ball, J., and H. Surkalo, "New Variations of Chemicals EOR and Their Economic Potential: The ASP Process (Alkaline-Surfactant-Polymer)," *Proceedings of the Improved Oil Recovery Conference*, Abilene, Texas, September 1988 (U.S. Department of Energy, DOE/BC/14286-1, March 1989), pp. 229–50.

Cooke, C. E., R. E. Williams and P. A. Kolodzie, *Journal of Petroleum Technology*, (December 1974), p. 1365.

Craig, F. F., *The Reservoir Engineering Aspects of Waterflooding*, third printing (Richardson, TX: Society of Petroleum Engineers, 1980), pp. 12–17.

Dauben, D. L., R. A. Easterly, and M. M. Western, *An Evaluation of the Alkaline Waterflooding Demonstration Project Ranger Zone Wilmington Field, California* (U.S. Department of Energy, DOE/BC/10830-5, May 1987).

FRENCH, T. R., and T. E. BURCHFIELD, "Design and Optimization of Alkaline Flooding Formulations" SPE/DOE 20238, presented at the Seventh Symposium on Enhanced Oil Recovery, Tulsa, Oklahoma, April 22–25, 1990.

GRAUE, D. J., and C. E. JOHNSON, JR., "Field Trial of Caustic Flooding Process," *Journal of Petroleum Technology* (December 1974), pp. 1353–58.

JENNINGS, H. Y., JR., C. E. JOHNSON, and C. D. MCAULIFFE, "A Caustic Waterflooding Process for Heavy Oils," *Journal of Petroleum Technology* (December 1974), pp. 1344–52.

JOHNSON, C. E., JR., "Status of Caustic and Emulsion Methods," *Journal of Petroleum Technology* (January 1976), pp. 85–91.

LORENZ, P. B., and D. A. PERU, "Guidelines Help Select Reservoirs for $NaHCO_3$ EOR," *Oil and Gas Journal* (September 11, 1989), pp. 53–57.

MUNGAN, N., "Certain Wettability Effects in Laboratory Water Floods," *Journal of Petroleum Technology* (February 1966), pp. 247–52.

SMITH, F. W., "Ion-Exchange Conditioning of Sandstone for Chemical Flooding," *Journal of Petroleum Technology* (June 1978), pp. 959–68.

SQUIRES, F., "Method of Recovering Oil and Gas," U.S. Patent #1,238,355 (1917).

SUBKOW, P., "Process for Removal of Bitumen from Bituminous Deposits," U.S. Patent #2,288,857 (1942).

SURKALO, HARRY, "Enhanced Alkaline Flooding," *Journal of Petroleum Technology* (January 1990), pp. 6–7.

THORNTON, S. D., *Reaction of Sodium Hydroxide with Silicate Minerals* (U.S. Department of Energy, Report No. NIPER-129, N-115, April, 1986).

# Chapter 8

# Miscible Fluid Displacement

## 8-1 GENERAL

As far back as 1927 Uren and Mahonay showed the existence of a relationship between oil recovery and the presence in the reservoir of interfacial tension among reservoir fluids. A substantial amount of oil (an average of 66 percent of OOIP) remains in the reservoir after water or gas injection. The remaining oil is trapped as a discontinuous phase in the swept zone or forms a continuous phase in the unswept zone of the reservoir. The oil is trapped due to capillary forces and interfacial tensions and remains trapped regardless of how many pore volumes of water or low pressure gas are injected into the reservoir. Only by injecting a miscible agent which reduces the retaining forces to zero is nearly total displacement possible in the pores contacted by the miscible agent.

Miscible oil displacement is the displacement of oil by fluids with which it mixes in all proportions without the presence of an interface, all mixtures remaining single phase.

Beginning in the early 1950s and continuing into the 1960s, hydrocarbon fluids and alcohol were recommended as miscible agents and some were developed and field tested.

*Propane*, *LPG* mixtures, and low-molecular-weight *alcohols* were the solvents extensively investigated in the laboratory and in field tests. Many limiting factors such as high costs, unfavorable mobility ratios, and low volumetric sweep efficiencies restricted their full development. Even if small slugs of solvent were injected, driven through the reservoir with natural gas or with brine in the case of the alcohol slug, dilution and fingering of the solvent into the oil reduced their effectiveness.

*Natural gas*, *flue gas*, and *nitrogen* at high pressure and *enriched hydrocarbon gas* were found to achieve miscibility with the reservoir oil. High injection pressures and the composition requirement for miscibility of enriched gas limit the number of prospective oil reservoirs where this process of miscible fluid displacement can be used. Slugs of fluids containing oil, water, surfactants, and cosolvents (alcohol) in various formulations are known as *micellar-polymer (MP) solutions*. These were found to be technically efficient, resulting in initial miscibility and ultralow interfacial tension (IFT) further from the injection area. A favorable alternate injection fluid for increasing oil recovery is $CO_2$. The *$CO_2$ miscible process* has two main advantages. A relatively lower operating pressure is required for $CO_2$ to reach miscibility with reservoir fluids than with other solvents, and $CO_2$ is widely available from naturally occurring underground sources.

In this chapter, after presenting the miscible displacement principle, miscible flooding with hydrocarbon solvents will be emphasized. Micellar-polymer flooding will be described in Chapter 9 and $CO_2$ miscible flooding in Chapter 10.

## 8-2 PHASE BEHAVIOR

### Phase Change Representations

Review of state-of-the-art miscible displacement has been treated comprehensively by an SPE monograph (Stalkup, 1984). Also, solvent–crude oil properties, phase behavior, and interfacial tension are detailed in the work of Lake (1989).

The miscible displacement mechanism of the EOR processes is understood when the phase behavior of crude oil, water, and enhanced oil recovery (EOR) fluids is known. *A phase is a homogeneous, physically distinct, mechanically separable portion of a material with a given chemical composition and structure* (Barrett et al., 1973). Oil and water represent two distinct liquid phases. Ice cubes in a glass of water is a two-phase system (ice and water). Ice cubes, water, and the vapor above the glass is a three-phase system. Even two regions of a solid with different compositions or crystal structures represents two distinct solid phases. The number of *components* in a system is the number of different kinds of atoms or molecules. An alloy, such as steel, consisting of two elements (iron and carbon) has two components. Water, $H_2O$, is a compound (having two elements); however, there is but one component, namely, the water molecule. Oil is a complex multicomponent system in which the components are the molecules of methane, $CH_4$, ethane, $C_2H_6$, propane, $C_3H_8$, and so on. Thus, if the elements of the system always appear in the form of a compound, we count only the molecule as the component and not the elements themselves (Barrett et al., 1973).

As we observed in Chapter 3, the change from water phase to vapor (steam) phase is dependent on both temperature and pressure. A graphical representation of the corresponding values of temperature and pressure at which phase changes occur is called a pressure-temperature (P-T) or a phase diagram. Other useful methods for displaying phase behavior data are triangular or ternary diagrams and pressure-compositon (P-X) diagrams. The first two methods, P-T and ternary diagrams will be presented here. The third, (P-X) diagram for mixtures of miscible displacement fluids and reservoir oils, is more complex since the exact composition of a reservoir fluid is very difficult, if not impossible, to discern.

### P-T Phase Equilibria Diagram

**One-component system.** The classic example of a one-component system is water. The P-T phase diagram for water is given in Figure 8-1. The lines OC, OF, and OS represent corresponding temperatures and pressures where two phases coexist, respectively, as vapor-liquid, liquid-solid, and solid-vapor and are in equilibrium. At the *critical point* C liquid and vapor exist simultaneously in equilibrium. At a temperature higher than critical temperature $T_c$, the substance cannot be liquified regardless of how much pressure is applied. The gaseous phase above the critical temperature is called a *gas*; below the critical temperature it is called a *vapor*. At the triple point 0—which occurs at a pressure of 4.58 mm of mercury or 0.0882 psi (0.608 kPa) and a temperature of 32.0136 °F (0.0075 °C)—the solid, liquid, and saturated vapor exist simultaneously (Betts, 1981). A phase diagram is of great practical importance because it allows us to predict the state of the system for any particular temperature and pressure.

**Example 8-1.** On the basis of information given in Figure 8-1, show that ice should float in water (Barrett et al., 1975).

SOLUTION

$$\frac{dp}{dT} < 0 \qquad \text{slope of OF = negative}$$

For a given $T$, ice changes to liquid as $p$ increases. This suggests that the volume of the ice decreases on melting because this would relieve pressure. Hence, ice is less dense than water and will float.

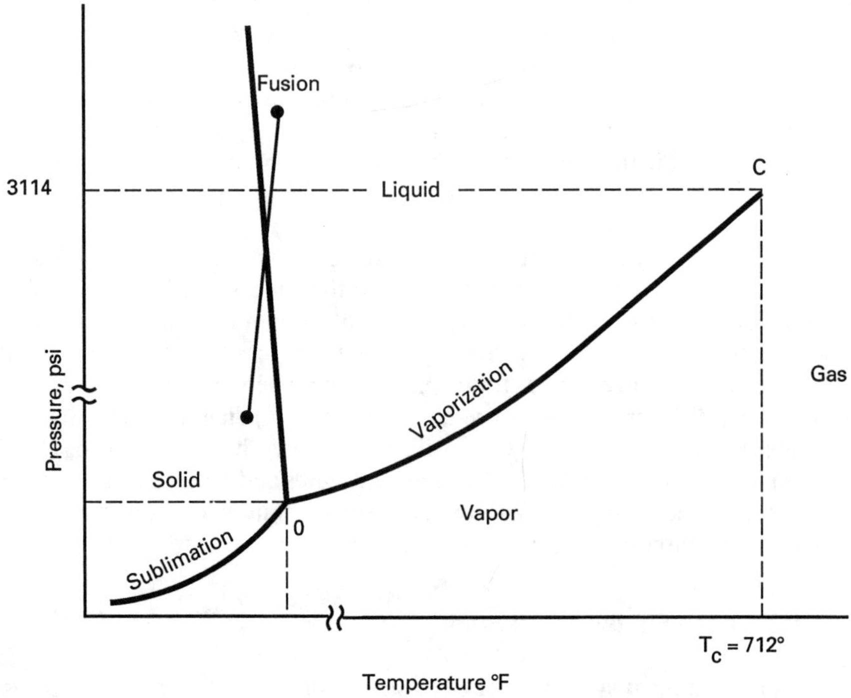


**Fig. 8-1** Phase diagram for water

**Multicomponent system.** Compared to the one-component system where two phases coexist only along a P-T line, in a multicomponent system such as an oil reservoir's, two phases gas and liquid are present in an area enclosed by the P-T phase diagram. Figure 8-2 is the P-T phase diagram of a particular reservoir fluid.

As we observe, the area enclosed by the bubble point and dew point lines which meet at the critical point is the region in which both gas and liquid phases will exist. The maximum two-phase temperature is called the cricondentherm. The curves within the two-phase region show the percentage of the total hydrocarbon volume which is liquid for any temperature and pressure (Craft and Hawkins, 1959). Various types of reservoirs can be defined by the location of the initial reservoir temperature and pressure with respect to the two-phase region.

Let us consider a reservoir where the composition corresponds to the phase diagram given in Figure 8-2, initially at 300 °F (147.4 °C) and 3700 psia (259 atm), point A. Assume that reservoirs containing the same hydrocarbon mixture are located initially at

Point B, or 180° F (81.4° C) and 3300 psia (231 atm)

Point C, or 75° F (23.6° C) and 3000 psia (210 atm)
Point D, or 150° F (64.9° C) and 2000 psia (140 atm)

For each reservoir location the existing phases, initial and during production operations, and the reservoir type are defined with reference to the phase diagram (Table 8-1).

Initially, the existing phases are in

Reservoir A, *gas*; the point A lies outside the two-phase region above the dew point line.
Reservoir B, *gas*; the point B lies the same as above and between the critical point temperature and the cricondentherm.
Reservoir C, *liquid*; the point lies outside the two-phase region above the bubble line.
Reservoir D, *liquid* and *gas*; the point lies within the two-phase region.

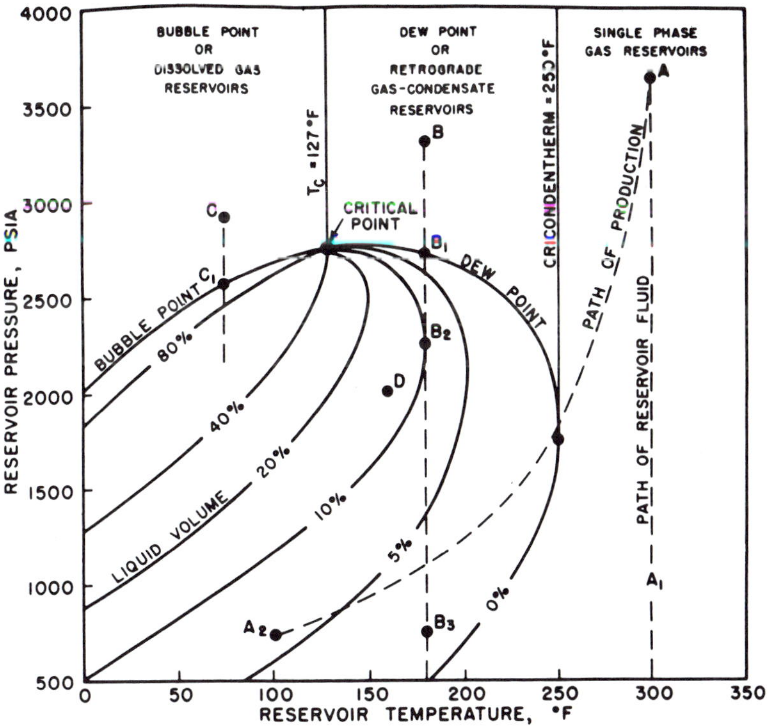


**Fig. 8-2** Pressure-temperature phase diagram of a reservoir fluid (From Craft and Hawkins, 1959)

During production operations, the reservoir temperature remains constant and the pressure declines.

The existing phases are in

*Reservoir A*, *Gas*, as pressure declines along path A-$A_1$.

*Reservoir B*, *Gas*, until the dew point pressure is reached at 2545 psia, point $B_1$. *Gas and Liquid* (*condensate*), below point $B_1$ since a liquid condenses out of the gas up to a maximum volume of 10 percent, reached at 2250 psia, point $B_2$. This is called *retrograde condensation* because declining pressure generally causes vaporization, not condensation. With the continuous decrease of reservoir pressure *vaporization* of the retrograde liquid occurs from $B_2$ to $B_3$.

*Reservoir C*, *Liquid* (*Oil*), until the bubble point pressure is reached at 2550 psia, point $C_1$. *Oil and Gas*, below point $C_1$, since the gas evolves from solution, build up an equilibrium gas saturation and flow with the oil through the reservoir to the producers.

*Reservoir D*, *Oil and Gas*, the oil zone is overlaid by a gas zone or gas cap.

**TABLE 8-1.** Reservoir Types Defined with Reference to Phase Diagrams

| INITIAL LOCATION ($T$ and $p$) | EXISTING PHASES | | | RESERVOIR TYPE | NOTES |
|---|---|---|---|---|---|
| | INITIAL in RESERVOIR | DURING PRODUCTION OPERATIONS | | | |
| | | in RESERVOIR | at the SURFACE | | |
| A | Gas | Gas | Gas | Dry-gas reservoir | No gas condenses |
| | | | Gas + liquid (condensate) | Wet (Rich)-gas reservoir | Gas condenses at the surface |
| B | Gas | Gas + liquid (condensate) | Gas + liquid (condensate) | Retrograde gas-condensate reservoir | Gas condenses at the surface and in the reservoir (retrograde) |
| C | Liquid | Oil + gas | Oil + gas | Oil reservoir | A secondary gas cap is forming in the reservoir |
| D | Liquid + gas | Oil+gas | Oil + gas | Oil reservoir with initial gas cap | Gas may condense at surface; at surface and in the reservoir |

During production operations each reservoir's existing phases appear at the surface as follows:

A, *Gas*. If the temperature of the fluid produced through the wellbore and into surface separators does not enter the two-phase region, the reservoir type in this case is a *dry-gas reservoir*.

*Gas and Condensate*. If the fluid produced enters the two-phase region due to a temperature decline, as along line A-$A_2$, the reservoir type is a *wet-* or *rich-gas reservoir*.

B, *Gas and Condensate*. This reservoir type is a *retrograde gas-condensate reservoir*.

C, *Oil and Gas*. This reservoir type is an *oil reservoir* or *dissolved (solution) gas-drive reservoir*.

D, *Oil and Gas*. This reservoir type is an *oil reservoir with initial gas cap*.

Pressure-temperature diagrams are also useful for studying the miscibility behavior between a solvent and crude oil or between a solvent and dry gas.

Figure 8-3 shows a P-T diagram representing the vapor pressure curves of pure ethane (curve AC) and pure heptane (curve $BC_7$). Curves $A_1C_1B_1$, $A_2C_2B_2$, and $A_3C_3B_3$ trace the bubble and dew point curves for various mixtures of ethane and heptane. For instance, curve $A_2C_2B_2$ represents a mixture containing 50.25 wt % ethane. The dashed line in Figure 8-3 connects the critical (plate) points for all mixtures of ethane and heptane. A constant temperature line representing reservoir conditions when pressure declines during production operations shows that all mixtures are single-phase above the dashed curve, whereas beneath this curve two phases, gas and liquid, may coexist, depending on the mixture composition (Stalkup, 1984).

### Ternary Diagram

The *ternary diagram* represents, on an equilateral triangle, the phase behavior of a three-component mixture (Figure 8-4).

Each corner of the equilateral triangle represents one of the components (100 percent), and each side corresponds to zero percent of the component represented by the opposite corner. This simple three-component system has characteristics as follows (Stalkup, 1984):

- Mixtures can be plotted on the diagram. Thus, mixture A in Figure 8-4 contains 40 percent component 1, 20 percent component 2, and 40 percent component 3.
- A mixture with an overall composition given by point A is a two-phase mixture at a given pressure and temperature: an equilibrium gas phase (assuming component 1 is a gas) with composition $y$ saturated with conden-

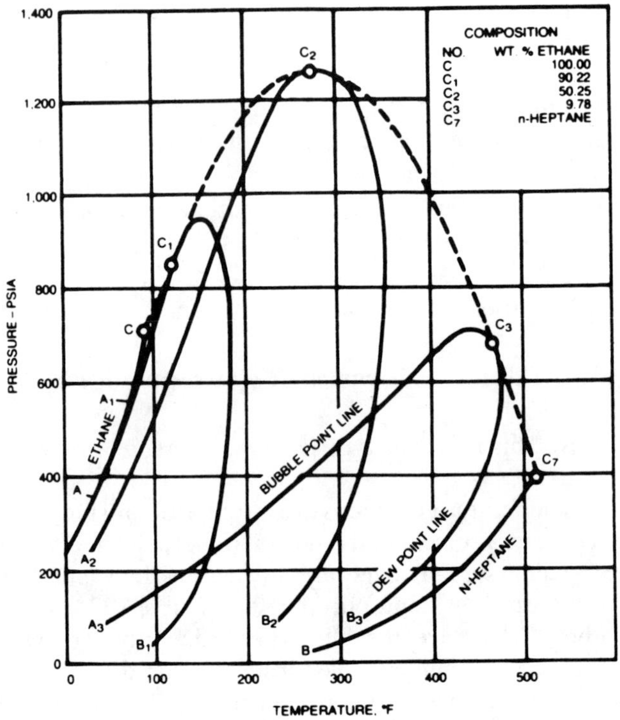


**Fig. 8-3** Pressure-temperature diagram for the ethane/n-heptane system (From Stalkup, 1978)

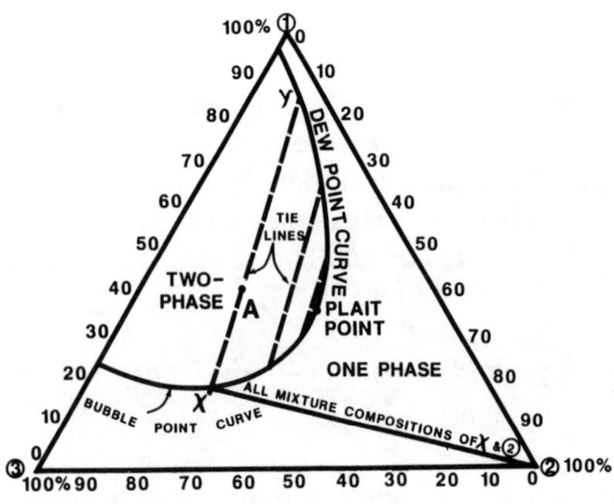


**Fig. 8-4** Ternary diagram of three component mixtures

sible components at its dew point and an equilibrium liquid phase with composition $x$ saturated with vaporizable components at its bubble point.

- The tie line is the dashed line (x-y) connecting the equilibrium gas and liquid compositions. The tie lines must vanish as the Plait point is reached since all phase compositions are equal at this point.
- The dew point curve connects all the dew point compositions.
- The bubble point curve connects all the bubble point compositions.
- The Plait point is the critical point where these curves meet and form the *binodal curve* which separates regions of one-phase behavior with compositions lying outside the curve and of two-phase behavior with compositions lying inside the curve.
- The limiting tie line is the line tangent to the binodal curve through the Plait point.
- The two-phase region's size is reduced when pressure is increased and is increased when temperature increases.
- The compositions of all the mixtures of two fluids are represented by the straight line connecting the composition of the two fluids. For example, along the straight line connecting the bubble point of fluid $x$ with that of component 2 lies the compositions of all mixtures of the two fluids, indicating that 2 is completely miscible with fluid $x$ (one-phase region). Along the straight line connecting the bubble point of fluid $x$ with that of component 1, all mixtures of the two fluids are immiscible since the mixtures fall into the two-phase region.

### Pseudoternary Diagram

The *pseudoternary diagram* represents the phase behavior of a multicomponent system, such as a hydrocarbon reservoir, by grouping the components of the reservoir fluids into three pseudocomponents (Figure 8-5):

- the light component (such as methane) = $C_1$
- the intermediate hydrocarbons (rich gas, light oil, or condensate) = $C_2$–$C_6$
- the heavy hydrocarbons (heavy oil) = $C_{7+}$

The pseudoternary diagram can simultaneously represent

1. phases.
2. component concentrations in mixture.
3. overall composition.
4. relative amount of each phase in the two-phase region.

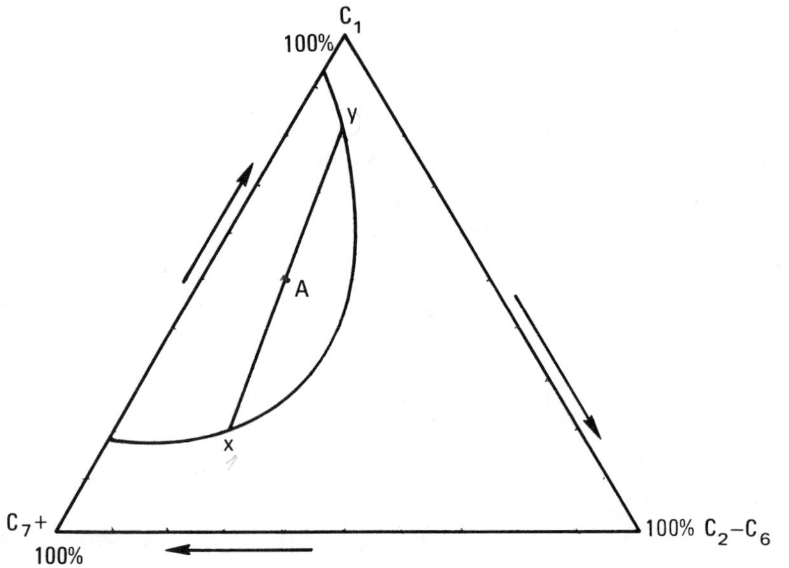


Fig. 8-5 Pseudoternary diagram for a hydrocarbon reservoir

1. *Phases*. Let us assume that the hydrocarbon reservoir is a two-phase mixture which has an overall composition represented on the pseudoternary diagram (Figure 8-5) by point A. The corresponding tie line has

**a.** an equilibrium liquid phase $S_l$ with composition $x$.

**b.** an equilibrium gas phase $S_g$ with composition $y$.

2. *Concentrations*. The component concentrations in the mixture are given in Table 8-2.

3. *Overall Composition*. The overall composition of the reservoir hydrocarbon mixture (point A) can be calculated based on each of the three pseudocomponents $C_1$, $C_{2-6}$, and $C_{7+}$ and on the relative amount of the two phases $S_l$ and $S_g$.

For example, the overall composition (point A) based on the pseudocomponent $C_1$ is equal to the concentration of $C_1$ in phase $S_l$ multiplied by the

**TABLE 8-2.** Component Concentrations in Mixture

| | | Concentration | | |
|---|---|---|---|---|
| Composition | Phase | $C_1$ | $C_{2-6}$ | $C_{7+}$ |
| x | Liquid | 0.20 | 0.20 | 0.60 |
| A | Liquid + gas | 0.50 | 0.15 | 0.35 |
| y | Gas | 0.80 | 0.10 | 0.10 |

relative amount of phase $S_l$ plus the concentration of $C_1$ in phase $S_g$ multiplied by the relative amount of phase $S_g$.

$$C_1 = C_{1,S_l} \times S_l + C_{1,S_g} \times S_g \tag{8-1}$$

or based on

$$C_{2-6} = C_{2-6,S_l} \times S_l + C_{2-6,S_g} \times S_g \tag{8-2}$$

and

$$C_{7+} = C_{7+,S_l} \times S_l + C_{7+,S_g} \times S_g \tag{8-3}$$

**4.** *Relative Amount*. The relative amount of phases $S_l$ and $S_g$ can be found using Eqs. 8-1, 8-2, or 8-3 and Eq. 8-4:

$$S_l + S_g = 1 \tag{8-4}$$

So, the relative amount of phase

$$S_l = \frac{C_1 - C_{1,S_g}}{C_{1,S_l} - C_{1,S_g}} = \frac{C_{2-6} - C_{2-6,S_g}}{C_{2-6,S_l} - C_{2-6,S_g}} = \frac{C_{7+} - C_{7+,S_g}}{C_{7+,S_l} - C_{7+,S_g}} \tag{8-5}$$

and, the relative amount of phase

$$S_g = \frac{C_1 - C_{1,S_l}}{C_{1,S_g} - C_{1,S_l}} = \frac{C_{2-6} - C_{2-6,S_l}}{C_{2-6,S_g} - C_{2-6,S_l}} = \frac{C_{7+} - C_{7+,S_l}}{C_{7+,S_g} - C_{7+,S_l}} \tag{8-6}$$

Finding the relative amounts of phases $S_g$ and $S_l$ of the hydrocarbon mixture represented by point A suggests the use of the inverse lever rule (Lake, 1989). For example, if $x$ represents 100 percent liquid phase $S_l$ and $y$ 100 percent gas phase $S_g$, the relative amount of phase $S_l$ in the mixture represented by point A is the line segment between y and A divided by the length of the tie line between x and y.

$$S_l = \frac{\text{line segment yA}}{\text{tie line xy}} \quad \text{and} \quad S_g = \frac{\text{line segment xA}}{\text{tie line xy}} \tag{8-7}$$

where line segments and tie line are the differences between the concentrations based on each of the three pseudocomponents $C_1$, $C_{2-6}$, or $C_{7+}$.

**Example 8-2.** Using the ternary diagram data given in Figure 8-5 and the data given in Table 8-2, find the relative amount of phases $S_g$ and $S_l$ for the hydrocarbon mixture which has an overall composition represented by point A.

SOLUTION

Equations 8-5 and 8-6 or the inverse lever rule state that

$$S_g = \underbrace{\frac{0.50 - 0.20}{0.80 - 0.20}}_{\text{based on } C_1} = \underbrace{\frac{0.15 - 0.20}{0.10 - 0.20}}_{\text{based on } C_{2-6}} = \underbrace{\frac{0.35 - 0.60}{0.10 - 0.60}}_{\text{based on } C_{7+}} = 0.50$$

$$S_l = \frac{0.50 - 0.80}{0.20 - 0.80} = \frac{0.15 - 0.10}{0.20 - 0.10} = \frac{0.35 - 0.10}{0.60 - 0.10} = 0.50$$

based on $C_1$   based on $C_{2-6}$   based on $C_{7+}$

or

$$S_l = 1 - S_g = 0.50$$

## 8-3 HYDROCARBON-SOLVENT MISCIBLE FLOODING

### Residual Oil Saturation and IFT

The objective of miscible displacement is to increase the oil recovery by reducing the residual oil saturation to the lowest possible value. Since residual oil saturation depends on the capillary number $N_C$,

$$N_C = \frac{u\mu}{\sigma} \tag{8-8}$$

where

$u$ = superficial or actual velocity, ft/day, since only pores and not the full area conduct fluid, ($u = v/\phi$).
$\mu$ = oil viscosity, cp
$\sigma$ = interfacial tension, dynes/cm

and since the residual oil saturation decreases when the capillary number increases (Figure 8-6), the interfacial tension should be reduced to its lowest value by injecting a slug of miscible solvent driven by natural gas until miscibility is achieved. Then, only one phase will result from the mixture of miscible fluids, with no interfaces and consequently no IFT between the fluids.

The injection fluids that can achieve in-reservoir miscibility with reservoir fluids are so-called first-contact miscible and multiple-contact miscible fluids (Stalkup, 1984).

### First-Contact or Direct Miscibility

The first-contact miscible injection fluids used are liquid petroleum gas mixtures. These solvents mix directly with reservoir oils in all proportions and the mixture remains single phase. The phase behavior required for achieving *direct miscibility* in ideal conditions can be illustrated on a pseudoternary diagram where LPG solvent is represented by $C_{2-6}$, the driving gas of the solvent slug by $C_1$, and the heavy hydrocarbons pseudocomponent by $C_{7+}$ (Figure 8-7).

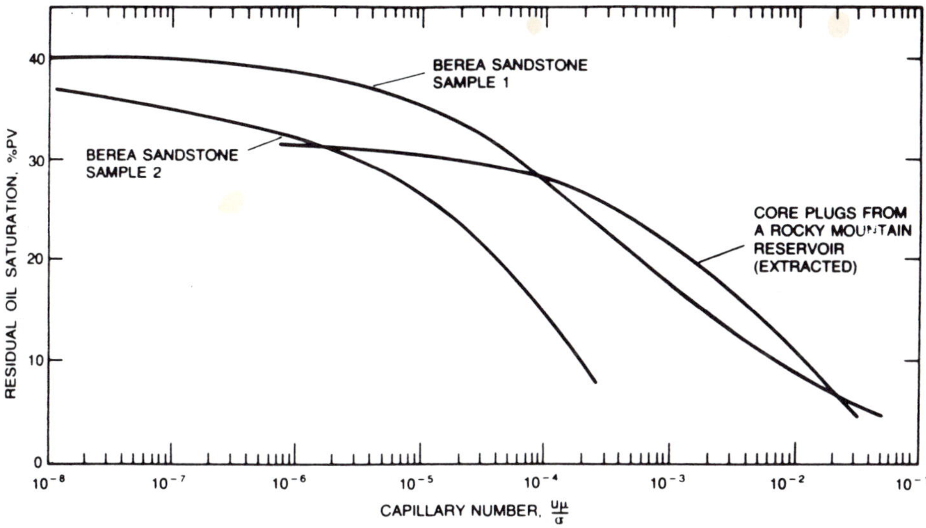


**Fig. 8-6** Dependence of residual oil saturation on capillary number (From Stalkup, 1984)

As we observe, not only is LPG first-contact miscible with reservoir oil, but LPG diluted with $C_1$ up to composition A is also directly miscible with reservoir oil. In fact, the miscibility between reservoir oil, LPG, and $C_1$ is represented by the area of the triangle $OC_{2-6}A$, where OA is the line tangent to the binodal curve. Any higher proportion of $C_1$ existing in composition A will move the line segment OA to intercept the binodal curve and pass through

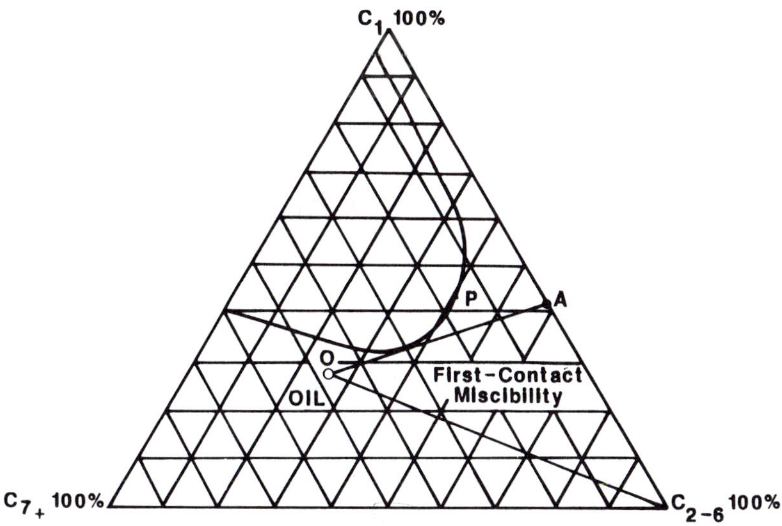


**Fig. 8-7** First-contact miscibility scheme

the two-phase region where the miscibility is lost. In reality, the phase behavior during miscible displacement of the oil by the solvent slug is more complex. As the solvent slug continues to pass through the reservoir it also mixes with the driving gas behind it. A dilution process starts which moves the composition profile to intersect the two-phase region where the direct miscibility is lost.

### Multiple-Contact or Dynamic Miscibility

Multiple-contact miscible injection fluids are natural gas at high pressure, enriched natural gas, flue gas, nitrogen, and carbon dioxide. These fluids are not first-contact miscible and form two-phase regions when mixed directly with reservoir fluids. The miscibility is achieved by the mass transfer of components which results from multiple and repeated contact between the oil and injection fluid during the flow through the reservoir.

There are two processes through which dynamic miscible displacement can be achieved in the reservoir, namely, *condensing gas drive* and *vaporizing gas drive* (Stalkup, 1984). Both processes are related to the location of the reservoir oil's composition on a pseudoternary diagram with respect to the limiting tie line (the tangent line to the binodal curve through the Plait point).

**Condensing gas drive process.** The condensing gas drive process takes place when the reservoir oil composition "O" lies to the left of the limiting tie line PB (intermediates–lean crude oil) on the pseudoternary diagram and when the injected solvent, which is a *mixture of natural gas* ($C_1$) *and intermediates* ($C_{2\text{-}6}$), has a composition lying between A and B in Figure 8-8.

The miscibility results from the in situ transfer (condensation) of intermediate hydrocarbon ethane through butane from the solvent injected into the reservoir oil.

Assuming that natural gas with the minimum concentration of $C_{2\text{-}6}$ (point B) is injected into the reservoir with oil composition "O," the mechanism of dynamic miscibility takes place as follows (Stalkup, 1984):

- The compositions of all mixtures of two fluids, after their first contact, are represented by the straight line OB, and most fall within the two-phase region.
- The resultant two-phase mixture is $M_1$, and its tie line connects the equilibrium liquid phase composition $L_1$ with the equilibrium gas phase composition $G_1$.
- Further injection of solvent B contacts equilibrium liquid phase composition $L_1$ and forms a new overall mixture $M_2$ with an equilibrium liquid phase composition $L_2$ and an equilibrium gas phase composition $G_2$.
- Continued injection of solvent B enriches the reservoir's overall mixtures $M_1, M_2, M_3, \ldots$ with more intermediate components $C_{2\text{-}6}$. A transition zone is created along the bubble point curve with equilibrium liquid phase com-

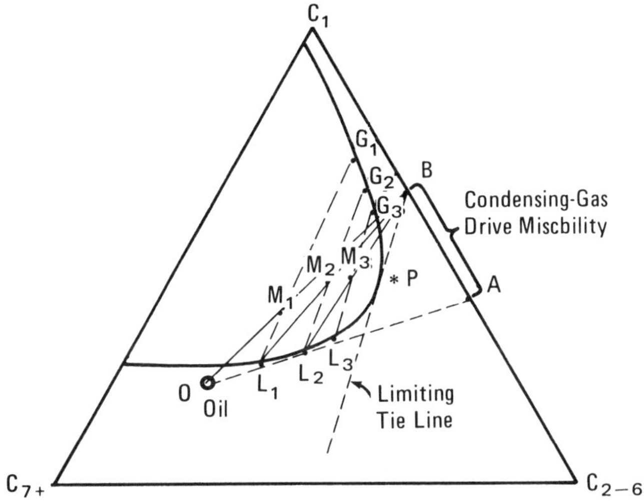


**Fig. 8-8** Condensing gas driving miscibility scheme

positions $L_1$, $L_2$, $L_3$, . . . until the Plait point P composition is reached and the reservoir oil becomes directly miscible with the injected solvent B.

It has to be pointed out that the reservoir oil is also contacted by the equilibrium gas phase compositions and a transition zone is established along the dew point curve from $G_1$, $G_2$, . . . to the Plait point P where direct miscibility is achieved.

For a given solvent composition there is a minimum pressure called the *minimum miscibility pressure* (MMP) above which the dynamic miscibility can be obtained in a condensing gas drive process. Since the two-phase region size of a pseudoternary diagram is reduced when reservoir pressure is increased, lower concentrations of intermediates $C_{2-6}$ in the injected solvent are needed to accomplish miscibility at higher reservoir pressures.

**Vaporizing gas drive process.** The vaporizing gas drive process takes place when the reservoir oil composition "O" lies on or to the right of the limiting tie line PB (crude oil reach in intermediates), and when the injected solvent has a composition lying to the left of the limiting tie line and also to the left of the tangent line OA (Figure 8-9).

The injected solvents used are *natural gas at high pressure* (the high-pressure gas process), *flue gas*, *nitrogen*, and $CO_2$. The miscibility is attained above a minimum miscibility pressure which has different values corresponding to the different gases injected and to different reservoir oil compositions. The mechanism of multiple-contact miscibility results from the in situ mass transfer through vaporization of intermediate hydrocarbon components from the reservoir oil into the injected gas.

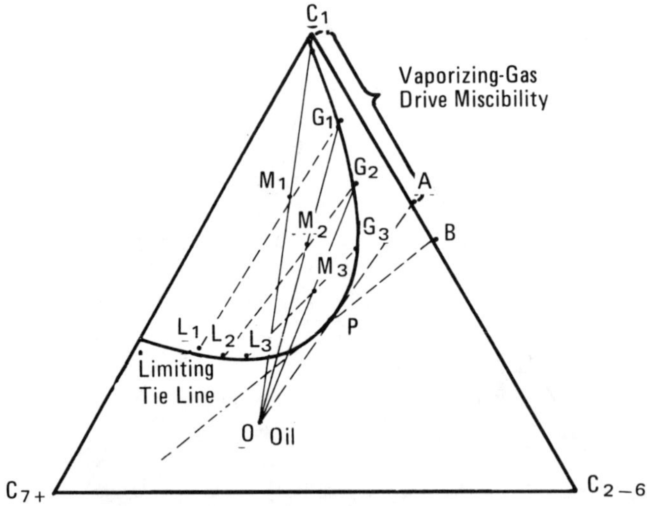


**Fig. 8-9** Vaporizing gas driving miscibility scheme

The $CO_2$ miscible flood is an EOR method commercially developed and, as we pointed out, is presented separately in Chapter 10. The vaporizing gas drive mechanism explained here uses natural gas $C_1$ at high pressure as an injection solvent (Stalkup, 1984). Since natural gas is readily accessible, the process has been successfully developed in the field. Miscibility is achieved in much the same manner using nitrogen gas and flue gas.

- The injected high-pressure gas $C_1$ in contact with the reservoir oil "O" vaporizes, or extracts some $C_{2\text{-}6}$ hydrocarbons from oil. A gas phase $G_1$ results, with intermediate hydrocarbons in equilibrium with the liquid phase $L_1$. The two-phase mixture has the overall composition $M_1$.
- Further injection of gas $C_1$ pushes the equilibrium gas $G_1$ so that it contacts more fresh reservoir oil "O," extracts more intermediates, and reaches a new overall composition $M_2$. At the ends of the tie line are the equilibrium gas phase composition $G_2$ which has more intermediate hydrocarbons and the equilibrium liquid phase composition $L_2$.
- Continued injection of gas $C_1$ causes $G_2$ to contact more fresh reservoir oil "O," enriching the gas phase composition at the displacing front along the dew point curve from $G_2$ to $G_3$ and so on, until the Plait point P composition and direct miscibility is reached.

It has to be pointed out that the vaporizing gas drive method is not a slug process. The high-pressure gas itself serves as both solvent and drive gas (Stalkup, 1984). The continued injection of the high-pressure gas is the major difference between vaporizing gas drive floods and the majority of condensing gas drive and first-contact miscible floods.

The two mechanisms of dynamic miscibility just described (namely, *transfer through condensation* of intermediate components from rich solvent into the intermediates–lean crude oil, and *transfer through vaporization* of intermediate components from crude oil rich in intermediates into the solvent-lean gas at high pressure) are more complex in the fluid flow under actual reservoir conditions. However, the pseudoternary diagram behavior is very useful for recognizing the type of miscibility that can take place with various specific reservoir oil characteristics and solvent compositions.

## 8-4 FIELD DEVELOPMENT OF HYDROCARBON-SOLVENT FLOODING

The available data on field tests and development of miscible hydrocarbon displacement show the use of

- liquid petroleum gas as a first-contact miscible process.
- enriched hydrocarbon gas as a multiple-contact condensing gas drive process.
- high-pressure natural gas as a multiple-contact vaporizing gas drive process.

Beginning in the early 1950s and continuing into the 1960s and beyond, numerous field tests and several projects have been developed in the United States using LPG especially for miscible displacement.

However, the need for excessively large LPG slugs and the high market value of liquid propane made LPG injection economically unattractive. The use of enriched hydrocarbon gas and high-pressure natural gas processes has become more attractive in areas where there is no readily available market for the gas and in deep reservoirs where the high miscibility pressure can be achieved without approaching the parting pressure of the formation.

Several presentations of selected projects using hydrocarbon-solvent flooding are given by van Poollen (1980) and by Stalkup (1984), and recent technical papers containing valuable information regarding their design, implementation, and performance evaluation are available (Da Sie and Guo, 1988; Mazzocchi et al., 1988).

To illustrate reservoir and miscible slug characteristics and their accompanying operating strategies, some case histories are presented next.

### Rainbow Keg River "B" Pool, Alberta, Canada

This is an example of a recently developed direct miscibility process using a tertiary first-contact hydrocarbon miscible flood (Mazzocchi et al., 1988).

**Reservoir characteristics.** The reservoir is a reef of Middle Devonian age discovered in 1965 at 5900 ft (1800 m) and originally containing 272 million STB (43 million $m^3$) of a 38 °API (SG = 0.830) oil. The reef's geometry shows two structural lobes (north and south) separated by an east–west-oriented saddle and characterized by a 360 ft (110 m) formation maximum thickness. A geological and petrophysical model of the reservoir has been developed based on detailed studies and analyses of a large volume of cores, logs, three-dimensional seismic, reservoir, and production data. Within the original thick oil column, the formation above the oil-water contact was characterized as having a predominant limestone zone on the top with mainly vuggy and intergranular porosities and permeabilities; the formation below has two dolomite zones with good to excellent reservoir qualities as a result of the vuggs and fractures system.

**Recovery mechanisms.** Primary production resulted from fluid expansion drive until 1968 since the reservoir saturation pressure of 1572 psig (10,845 kPa) was far below the initial pressure of 2465 psig (17,000 kPa). Water injection into the underlying aquifer started in 1969 and continued until 1984. The oil-water contact rose 203 feet (61 m) in the northern lobe and 221 ft (67 m) in the southern lobe, reaching a level above the saddle (Figure 8-10).

The oil recovery was approximately 37 percent of OOIP and the residual oil saturation range to waterflood was estimated for three petrophysical rock types at less than 0.25 for the vugular, fractured rock type 1, between 0.25 and 0.47 for rock type 2, with average porosities, and higher than 0.47 for tight matrix rock type 3.

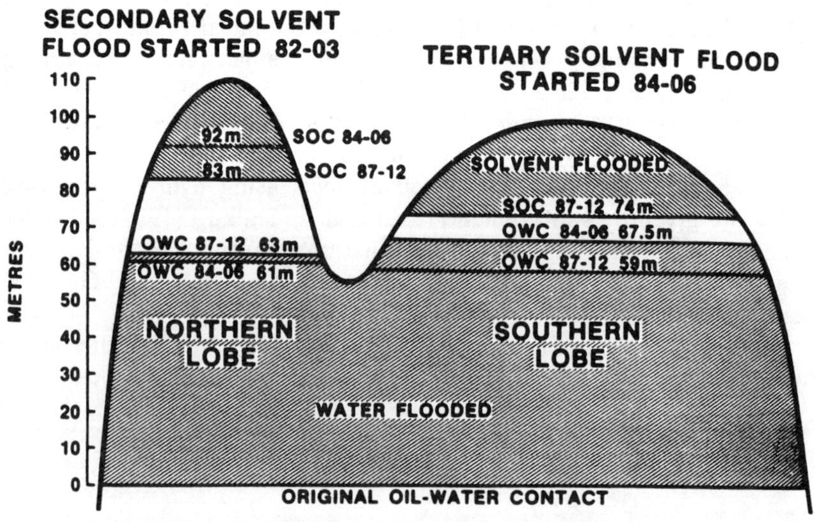


**Fig. 8-10** Rainbow Keg River B pool tertiary scheme. Average fluid levels (From Mazzocchi et al., 1988)

**Secondary and tertiary miscible displacement.** The reservoir geometry and characteristics were considered appropriate for direct miscible or first-contact miscible displacement using hydrocarbon solvent (LPG). The miscibility conditions and the design of the miscible flood were determined based on prediction made from miscibility correlation curves, visual cell phase envelope determinations, slim tube experiments, and core displacement studies. The pseudophase ternary diagram obtained (Figure 8-11) indicated that at 2200 psia (15,169 kPa) and 185° F (85° C) the minimum hydrocarbon solvent $C_{2+}$ requirement of composition "S" was 57 percent to achieve first-contact miscibility with the reservoir oil of composition "O" (Mazzocchi et al., 1988).

**Operating strategy.** A solvent slug of 19.5 percent initial hydrocarbon pore volume injected in the crestal virgin zones of the reef and chased over with a pore volume of dry gas will spread horizontally across the top of the reef and will push about half of the mobilized oil down into the waterswept zone. The oil and solvent banks will form a tertiary common bank below the saddle that moves downward and displaces the residual oil. Water from the waterswept zone is simultaneously produced allowing the formation of the tertiary oil bank and the downward movement. The schematic representation of a tertiary miscible flood (Mazzocchi et al., 1988) is shown in Figure 8-12.

It is interesting to observe the original well completion techniques using sliding sleeves and packer combinations to control tertiary oil bank movement, water, oil and solvent production. The schematic representation of the recompletion strategy is shown in Figure 8-13. Four groups of perforation intervals are sequentially moved downward below the oil bank as the flood progresses (Mazzocchi et al., 1988).

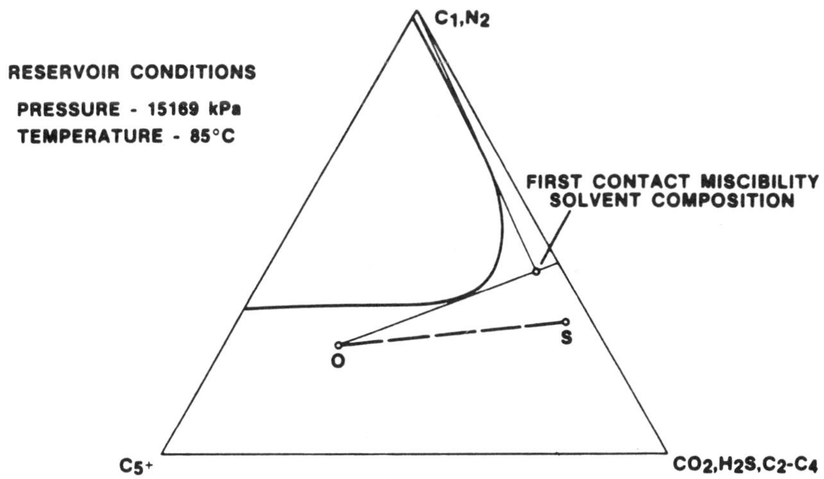


**Fig. 8-11** Rainbow Keg River B pool ternary diagram (From Mazzocchi et al., 1988)

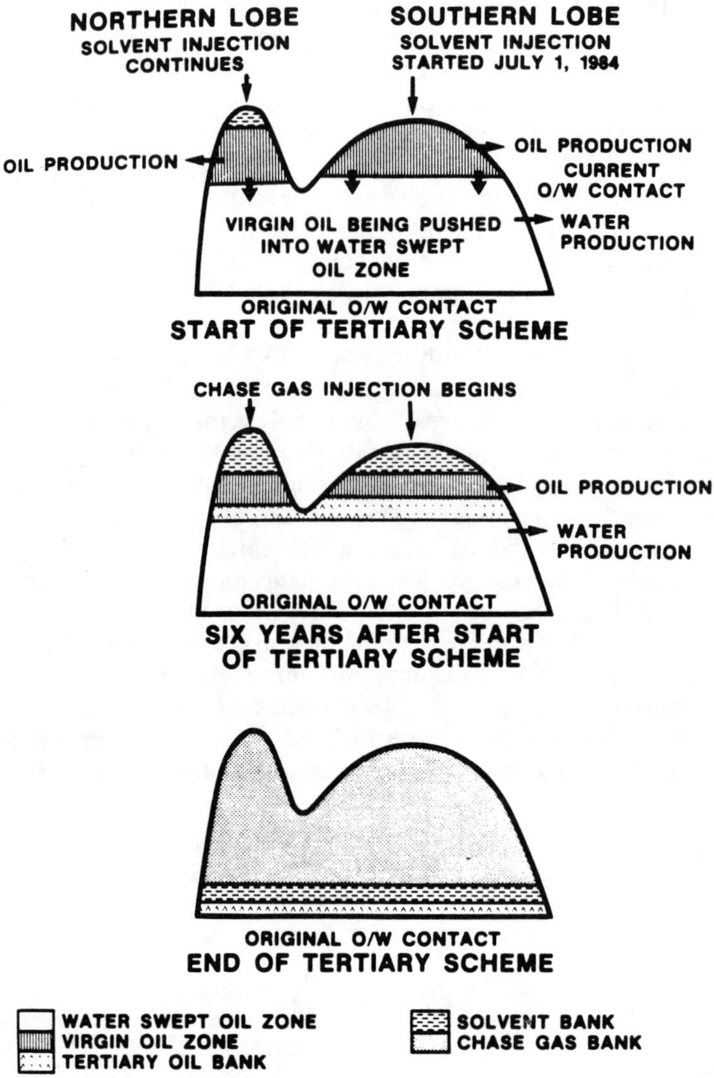


**Fig. 8-12** Rainbow Keg River B pool tertiary scheme. Schematic representation of tertiary miscible flood (From Mazzocchi et al., 1988)

The system provides control of the fluids being produced and of the downward displacement, minimizing drawdown and preventing coning of solvent and chase gas. The producing GOR and WOR values dictate the timing of opening and closing the perforations.

**First results.** Solvent flooding into the crestal zone of the northern lobe began in March 1982. The miscible flood into the crestal area of the southern lobe was started in June 1984. The total solvent injection corresponded to a

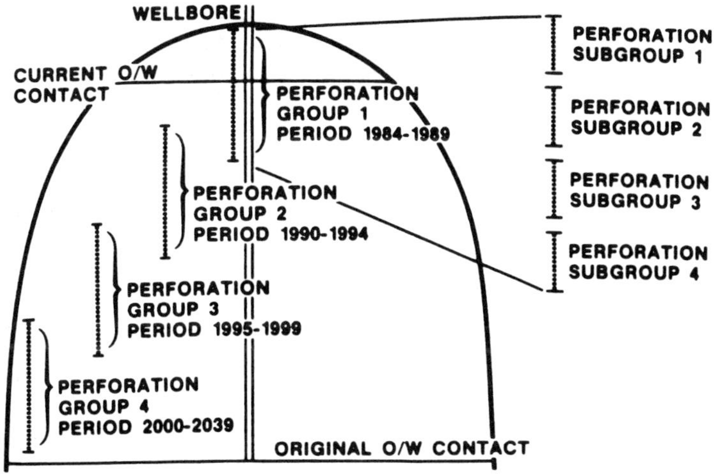


**Fig. 8-13** Rainbow Keg River B pool tertiary scheme. Schematic representation of recompletion strategy (From Mazzocchi et al., 1988)

9.1 percent initial hydrocarbon pore volume (19.5 percent IHCPV was the projected solvent slug injection volume). As can be observed in Figure 8-14, oil production continued to decrease but at a slower rate than before. The water production increased as predicted and the gas production also increased after the breakthrough of solvent at the producers.

The cumulative solvent injected until June 1989 into the "B" pool amounted to 41,727 MMCF ($1.175 \times 10^9$ m$^3$), and the total amount produced was 7.9 MMCF, or 18.9 percent solvent recycled. The oil recovery increased to 43.8 percent of OOIP by the end of June 1987, compared with 37 percent

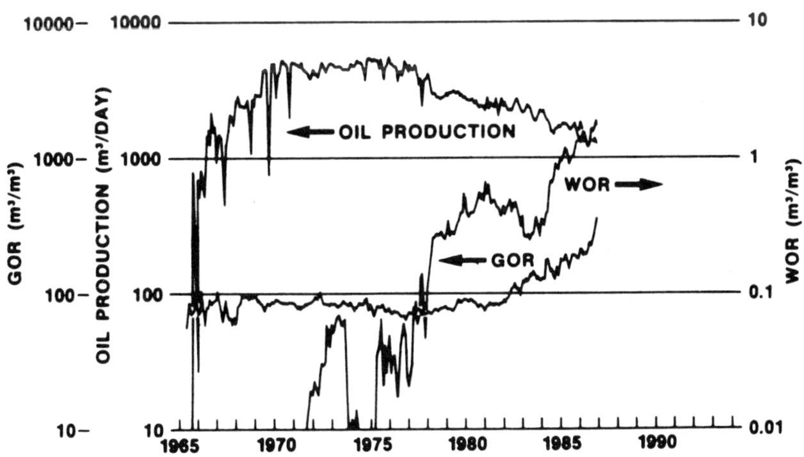


**Fig. 8-14** Rainbow Key River B pool production history (From Mazzocchi et al., 1988)

in 1982 at the start of solvent injection in the northern lobe. Volumetric calculations indicated a 26 ft (8 m) downward movement in the oil-water contact in the southern lobe (Mazzocchi et al., 1988). The target of the project was to displace and produce 55 × $10^6$ STB residual oil or an increased recovery of 20 percent of OOIP.

The asphaltene deposition caused severe plugging problems in the northern lobe producers, and the wells were treated with xylene to regain production rates.

**Comments.** The Rainbow Keg River "B" Pool represents an interesting example of a first-contact miscible displacement process developed for a specific reservoir. The process is at its early stage of implementation, and further publications would update its behavior.

Given the dual porosity system of the reservoir rock, it is apparent that gravity stable oil displacement is very important. The tertiary miscible bank flows into and out of the granular porosity of the matrix and the easier flow fluids through vuggs and fractures is restrained.

### Westpem Nisku D Reef, Alberta, Canada

This is an example of a condensing gas drive process using a multiple-contact hydrocarbon miscible flood (Da Sie and Guo, 1988).

**Reservoir characteristics.** The reservoir is a carbonate pinnacle reef discovered in 1978 at a 10,171-ft (3100 m) depth and originally containing 15.7 million STB (2.5 million $m^3$) of 45° API (SG = 0.80) volatile oil of 0.19 cp viscosity. The main reef, called Zeta Lake Member, has a 250-ft (76-m) maximum thickness, an average porosity of 12 percent, a horizontal permeability of about 1050 md, and an average vertical permeability of 100 md. It is fully dolomitized and has abundant vuggy and pinpoint vuggy porosity as well as brecciation fracturing (Da Sie and Guo, 1988). The initial oil saturation was 89 percent.

**Recovery mechanism and miscibility conditions.** The reservoir, with a temperature of 218° F (102° C) an initial pressure of 5911.5 psig (40,760 kPa) and a saturation pressure of 3966 psig (17,345 kPa), produced by a fluid expansion drive a cumulative amount of 530,000 STB (83,670 $m^3$) oil or 3.37 percent of OOIP by May 1981. The reservoir pressure dropped to 4693 psig (32,355 kPa). Preliminary model study results justified miscibly flooding this type of reservoir.

Using plant residue gas in the form of 15 percent ethane plus and 85 percent natural gas, a combined slug of 1.2 pore volumes was injected in a slim tube experiment to find the miscibility conditions. The hydrocarbon solvent was considered to be miscible at the inflection point on the recovery curve (Figure 8-15).

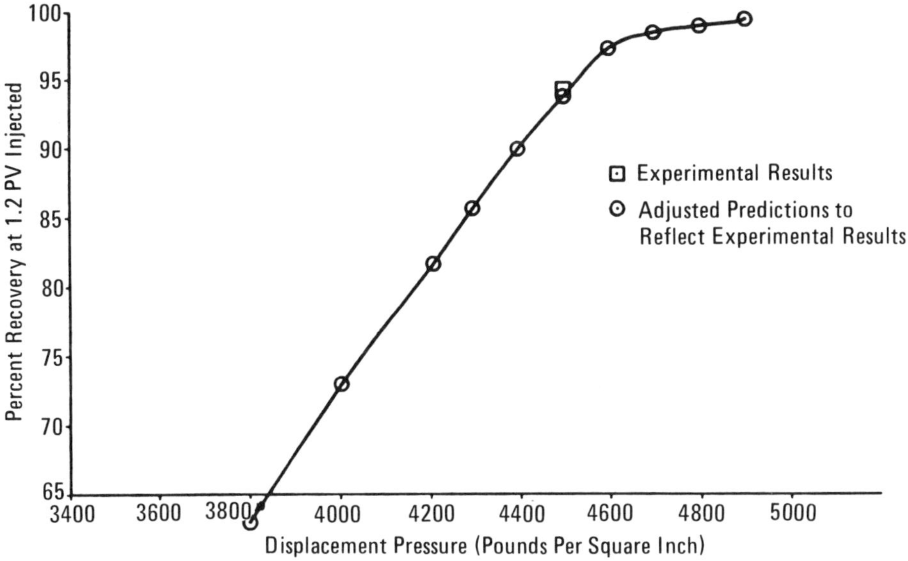


**Fig. 8-15** Slim tube simulation and experimental results (From Da Sie and Guo, 1988)

The pseudoternary diagram given in Figure 8-16 shows that at 4800 psia the enriched gas slug is miscible with the oil on multiple contact, transferring the ethane plus from the injected solvent into the reservoir oil (condensing gas drive process).

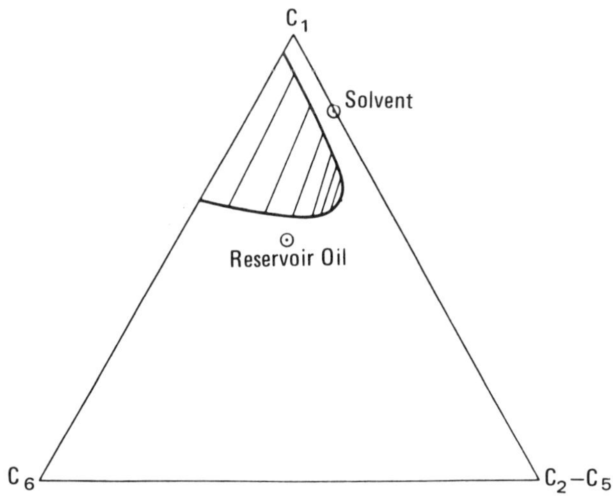


**Fig. 8-16** Termary diagram for reservoir fluid system at 4800 psia and 218 °F (From Da Sie and Guo, 1988)

**Field operations and results.** The vertical hydrocarbon miscible flood was started in May 1981, and the total amount of gas injected by June 1987 was 13 billion scf ($363 \times 10^6$ $m^3$). Cumulative oil production by June 1987 was 5 million STB (803,000 $m^3$) or 34 percent of OOIP. The determination of the solvent-to-oil interface in reservoir conditions was performed comparing pulsed neutron capture logs run before and after the start of the process and analyzing core samples obtained from a new well drilled in the area in early 1987. The results provided clear evidence of the existence of a miscible swept zone with an average residual oil saturation of 5 percent. The gravity-stable solvent-oil flat interface moving downward did not yet break through into the producing well effluent. The producers are perforated at the base of the reef (Figure 8-17) and the production rates are high (2800 bbl/day or 450 $m^3$/day per well). The vertical displacement rate has to be below the gravity-stable critical rate to prevent viscous fingering of solvent into the oil. It is expected that gas breakthrough due to coning will begin when the solvent-oil contact is within 33 ft (10 m) of the existing completion (Da Sie and Guo, 1988).

The Westpem Nisku D Reef condensing gas drive miscible process is applied as a vertical displacement of a highly saturated oil column. It appears to be very efficient and has the potential to reach a high oil recovery estimated at 84 percent of OOIP.

## Block 31 Field, Texas, United States

This is an example of a vaporizing gas drive miscible flood using high-pressure gas injection (Herbeck and Blanton, 1961; Stalkup, 1984).

**Reservoir characteristics.** The reservoir is a Middle Devonian siliceous limestone formation at an 8400 ft (2561 m) depth, having a 350-ft (107-m) thickness and originally containing about 250 million STB ($39.75 \times 10^6$ $m^3$) of

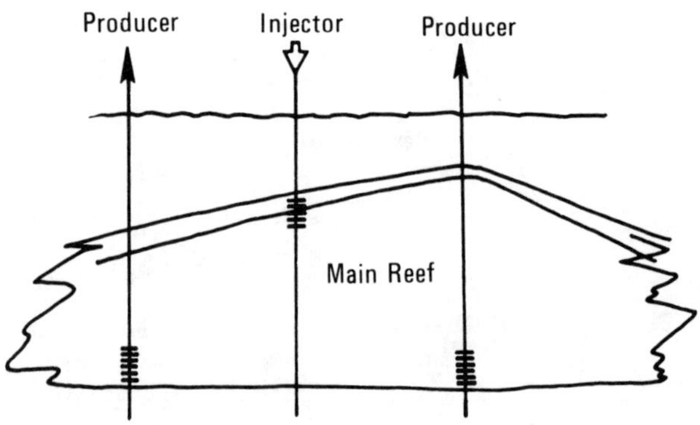


**Fig. 8-17** Schematic cross section of the Westpem Nisku D Reef (Adapted from Da Sie and Guo, 1988)

46° API (SG = 0.797) and 0.2-cp viscosity volatile oil, at a 140° F reservoir temperature. The reservoir rock has 15 percent porosity, 1 md permeability, and an irreducible water saturation of 37 percent.

**The miscible process.** Since the reservoir was highly undersaturated, it was produced primarily by a depletion elastic drive. Beginning in 1952 hydrocarbon gas was injected and flue gas since 1966 through 24 gas injection wells into a 350-acre inverted nine-spot pattern at a pressure of 4200 psia (28,958 kPa) (miscibility pressure of 3500 psia) and at a rate of 135 MM scf/day total gas. Flue gas was produced by steam boilers and compressed to injection pressure. The boilers were fueled with natural gas, avoiding the use of purchased hydrocarbon gas.

**The results.** The results of the project are considered very good since the oil recovery is close to the predicted value of 60 percent of OOIP.

### Hassi-Messaoud Field, Algeria

This is the largest vaporizing gas drive process in the world using high pressure lean gas multiple-contact miscible flood (Pottier et al., 1967).

**Reservoir characteristics.** The reservoir is a domal structure at an 11,000-ft (3353-m) depth with a 355-ft (108-m) average thickness of sandstone characterized by discontinuous intercalations of siltstone and clay beds. The original undersaturated oil in place was 23 billion STB ($3657 \times 10^6$ $m^3$) at an initial pressure of 7200 psi and a 245° F temperature. The rock reservoir has 6 to 10 percent porosity, and the variable permeability ranges between 10 and 180 md.

**The process and results.** The process of gas injection began in 1964 by injecting 175 MM scf/day ($4.9 \times 10^6$ $m^3$/day) lean gas and maintaining the reservoir pressure at 4500 psia, above the miscibility pressure of 3750 psia. Alternate gas-water injection was beneficial in retarding gas breakthrough and/or controlling GOR increase after gas breakthrough. The project was successful and profitable.

## 8-5 SCREENING CRITERIA

Hydrocarbon-solvent miscible flooding can only be applied to certain types of oil reservoirs. The reservoir's characteristics are the screening criteria used to decide if a reservoir is a good candidate for hydrocarbon-solvent miscible flooding. Exceptions are to be expected since the project's profitability depends on the economic factors prevailing at the time of the flood.

The guidelines for identifying potential oil reservoir candidates for the first-contact miscible process, the condensing gas drive process, and the vaporizing gas drive process were well established by Stalkup (1984) and are given in the paragraphs that follow.

### The Oil Viscosity and Gravity

*The viscosity* of the oil under reservoir conditions is perhaps the most important screening criterion since all three types of hydrocarbon-solvent miscible processes work best with oil viscosities at 1 cp or less, especially for horizontal floods. A practical upper viscosity limit may range between approximately 3 and 5 cp. The upper limit for gravity-stable floods depends on the reservoir's vertical permeability.

*Crude oils* with *gravities* above 30° API are most suited to first-contact miscible and to condensing gas drive miscible processes. Volatile undersaturated crude oils, with an API gravity greater than 40° API and rich in intermediate-molecular-weight hydrocarbon components, are required for the vaporizing gas drive miscible process.

### Reservoir Pressure and Depth

*The injection pressure* necessary to maintain the minimum miscibility pressure required by the process must be below the formation parting pressure in reservoir. Since the first-contact miscible process requires a miscibility pressure in the range of 900 psia to 1300 psia to operate above the miscibility pressure and below the parting pressure, the reservoir *depth* has to be between 1500 ft to 2500 ft. The condensing gas drive process requires *pressures* in the range of 1500 to 3000 psia and reservoirs with *a minimum depth* between 2000 ft and 3000 ft.

Vaporizing gas drive miscibility is achieved at high pressures of 3500 psia (minimum) and up to 6000 psia (maximum); thus the method is consequently restricted to *deep* reservoirs.

### Reservoir Geometry

The *direction of flow* is important in all three types of hydrocarbon-solvent miscible processes. Better operating conditions and results occur in reservoirs with substantial relief and high and uniform permeabilities, favoring gravity-stabilized flooding. In horizontal reservoirs, vertical permeability has to be restricted to avoid or reduce gravitational segregation.

### Oil Saturation at Start of the Project

Hydrocarbon-solvent miscible methods reduce the residual oil saturation after waterflooding, and therefore are tertiary methods. In this case a minimum of 25 percent PV residual oil saturation at the start of the project is desirable. A higher percentage of oil in place at the start of the project will afford even better results. Some of the hydrocarbon-solvent miscible processes, particularly vaporizing gas drive, have been initiated in large and nonwaterflooded reservoirs as secondary EOR methods with good success to date.

### High-Risk Factors

The project risk is increased by the presence of extensive fracturing, a gas cap, a strong water drive, or high permeability contrasts.

Although hydrocarbon-solvent miscible flooding has proven to be a technically effective EOR process, hydrocarbon injection fluids are in short supply and not practically available in the quantities required for large-scale oil field flooding in the United States (DOE, 1984). However, hydrocarbon miscible flooding with the vaporizing gas drive process using lean gas is attractive because lean hydrocarbon gas is available at relatively low cost and the produced gas can be compressed to miscibility pressure and reinjected.

## QUESTIONS AND EXERCISES

**8-1** Explain what miscible oil displacement means.

**8-2** Enumerate some of the agents that will achieve miscibility with reservoir oil.

**8-3** Describe, with reference to the P-T phase diagram given in Figure 8-2, the existing phases—initial and during production operations—of the oil reservoir located at $p_i$ = 2850 psia and $T$ = 150° F.

**8-4** Considering the pseudoternary diagram of Figure 8-5, find the relative amount of phases $S_l$ and $S_g$ for the hydrocarbon mixture with the overall composition $C_1$ = 40 percent, $C_{2-6}$ = 20 percent, $C_{7+}$ = 40 percent.

**8-5** Show where the reservoir oil composition should be located on the pseudoternary diagram of Figure 8-18 to have first-contact miscibility with dry-gas composition A (Figure 8-18).

**8-6** Determine the lowest $C_{2-6}$ wt % in gas composition to have direct miscibility with oil composition $O_1$ (Figure 8-18).

**8-7** Find the types of miscible processes possible with oil compositions $O_1$, $O_2$, and $O_3$ at reservoir pressures $p_1$ and $p_2$ (Figure 8-18).

**8-8** Explain the vaporizing gas drive mechanism, and state the important difference between this method and the majority of condensing gas drive and first-contact miscible floods.

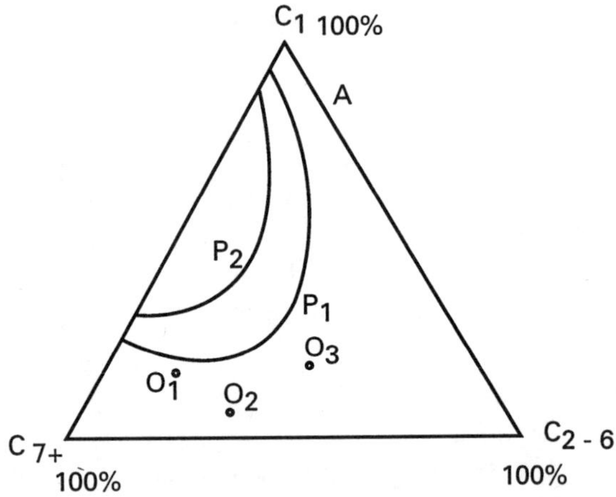


**Fig. 8-18** Pseudoternary phase diagram

## REFERENCES


BARRETT, C. R., W. D. NIX, and A. S. TETELMAN, *Solutions Manual*, *The Principles of Engineering Materials* (Englewood Cliffs, NJ: Prentice-Hall, 1975), p. 28.

BARRETT, C. R., W. D. NIX, and A. S. TETELMAN, *The Principles of Engineering Materials* (Englewood Cliffs, NJ: Prentice-Hall, 1973), p. 117.

BETTS, JOHN E., *Physics for Technology* (Reston, VA: Reston, 1981), p. 192.

CRAFT, B. C., and M. F. HAWKINS, *Applied Petroleum Reservoir Engineering* (Englewood Cliffs, NJ: Prentice-Hall, 1959), p. 64.

DA SIE, W., and D. S. GUO, "Assessment of a Vertical Hydrocarbon Miscible Flood in the Westpem Nisku D Reef," SPE/DOE 17354, *Proceedings*, *Enhanced Oil Recovery Symposium*, Tulsa, Oklahoma, April 17–20, 1988.

HERBECK, E. E., and J. R. BLANTON, "Ten Years of Miscible Displacement in Block 31 Field," *Journal of Petroleum Technology* (June 1961).

LAKE, LARRY W., *Enhanced Oil Recovery* (Englewood Cliffs, NJ: Prentice-Hall, 1989), Chapters 7 and 9.

MAZZOCCHI, E., et al., "Tertiary Application of a Hydrocarbon Miscible Flood: Rainbow Keg River 'B' Pool," SPE/DOE 17355, *Proceedings*, *Enhanced Oil Recovery Symposium*, Tulsa, Oklahoma, April 17–20, 1988.

POTTIER, J., et al., "The High Pressure Injection of Miscible Gas at Hassi-Messaoud," *Proceedings*, *Seventh World Petroleum Congress,* Mexico City, Vol. 3 (1967), pp. 533–544.

STALKUP, F. I., "Carbon Dioxide Miscible Flooding: Past, Present, and Outlook for the Future," *Journal of Petroleum Technology* (August 1978), pp. 1102–1112.

STALKUP, FRED I., *Miscible Displacement*, Society of Petroleum Engineers Monograph Series (Richardson, TX: SPE, 1984), Chapters 2 (pp. 6–16), and 5–7.

U.S. DEPARTMENT OF ENERGY, *Enhanced Oil Recovery* (Washington, D.C.: National Petroleum Council, June 1984), p. 2.

VAN POOLLEN, H. K.., and associates, *Fundamentals of Enhanced Oil Recovery* (Tulsa, OK: PennWell, 1980), pp. 121–129.

# Chapter 9

# Micellar-Polymer Flooding

## 9-1 GENERAL

It is well known that water and oil will not mix until a third component, soap or surfactant, which has affinity for both water and oil, is added. The use of an aqueous soap solution to reduce the interfacial tension of the oil-water system in order to displace the residual oil was recommended by Atkinson in 1927. During the following years laboratory studies and reported research have shown that it is necessary to reduce and maintain the interfacial tension at 0.01 to 0.001 dyne/cm to have an effect on the residual oil. These very low interfacial tension values can be attained with petroleum sulphonates derived from crude oil (Jones, 1964). These sulphonates have high interfacial activity, are less expensive, and are potentially available in large supply. Research on the use of surfactants in micellar solutions to recover more oil was carried out by many investigators including Gogarty (1967), Gogarty and Tosch (1968), Holm (1971), and Healy et al. (1975) among others.

## 9-2 PRINCIPLE AND METHOD DESCRIPTION

### Principle and Characteristics

The micellar solution composition which assures a gradual transition from the displacement fluid water to the oil displaced, without the presence of an interface, is as follows:

| | % |
|---|---|
| Surfactant | 10–15 |
| Oil | 25–70 |
| Water | 20–60 |
| Cosurfactant | 1–4* |

Water-soluble electrolytes such as inorganic salts may be used in preparing micellar solutions to gain better solution viscosity control. In order to have control of mobilities, the micellar solution slug is driven by a polymer slug. This process is called micellar-polymer flooding or MP flooding.

The micellar solution operates miscibly with reservoir fluids including oil and water without phase separation, thus assuring that nearly 100 percent of the residual oil can be displaced. In the field, however, this high percentage is reduced due to reservoir rock nonuniformity. For instance, oil recovery may actually be 64 percent when areal sweep efficiency is 80 percent and invasion efficiency also is assumed 80 percent. The micellar solutions are different from emulsions due to the microscopic size of the discontinuous phase. The internal phase of a micellar solution is in the form of extremely small droplets of $10^{-6}$ to $10^{-4}$ mm compared with $10^{-4}$ mm and higher for water-oil emulsions. The micellar solutions are also referred to in the literature as surfactant slugs, microemulsions, soluble oils, and so on. They are translucent, homogeneous, and thermodynamically stable.

### Pseudoternary Diagram

The three major components of micellar solutions which are oil, surfactants, and water can be represented on a ternary diagram (Figure 9-1).

In the two-phase region one phase is oil external and the other is water external, both being in equilibrium and lying at opposite ends of a tie line A-B. In the one-phase region, all components are miscible and no interfaces are present. The pseudoternary diagram for practical micellar-polymer displace-

* This is a fourth component, usually alcohol, which can be added to enhance the possibility for the micellar solution to include oil or water.

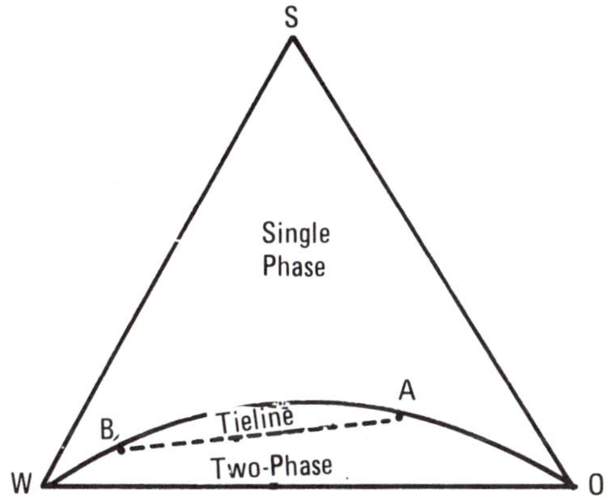


**Fig. 9-1** Ternary diagram

ment in a specific reservoir should have a reduced two-phase region. The surfactant slug moving through the reservoir changes its composition by including oil and water in a miscible displacement.

In reality, the surfactant-oil phase behavior and the changes which take place in micellar-polymer processes are far more complicated. Laboratory studies have shown that the reduction of the interfacial tension required for a miscible displacement can be achieved only in a very narrow range of salt concentration. Indeed, the salinity of the brine influences the phase behavior of micellar solutions which in turn directly correlates with the interfacial tension. Surfactant-brine-oil phase behavior measurements can replace the more difficult measurements of the interfacial tension and also provide a basis for micellar-polymer flooding design (Lake, 1989).

### Method Description

The MP flooding process is applied in general after waterflooding, as a tertiary process following the sequence shown in Figure 9-2.

**Preflush.** When the reservoir water salinity is too high, direct contact with the micellar solution is avoided by first injecting a low-salinity brine slug which adjusts salinity and hardness.

**Micellar solution.** This is prepared either as a low- or high-surfactant concentration solution. In the first case, large pore volumes of 15 to 60 percent or more of solution are injected into the reservoir, and the residual oil is displaced gradually from the porous space contacted. In the second case, a small pore volume (about 3 to 10 percent) of high concentration surfactant solution rapidly displaces the oil contacted. The process later becomes low concentration due to dilution with formation fluids.

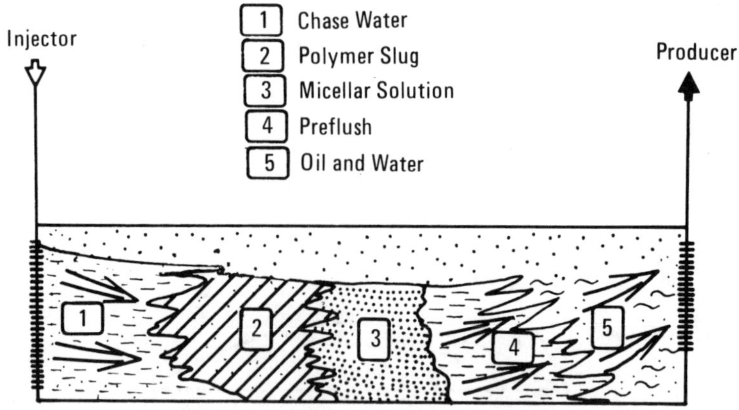


**Fig. 9-2** Schematic micellar-polymer flood

**Polymer slug.** This is necessary to control the mobilities, backing up the micellar slug and preventing it from being penetrated by the chase water.

**Chase water.** This represents the driving energy necessary to propel the MP solution through the reservoir to the producers.

The MP flood process applied in the field may differ in many details since the chemical system for a specific reservoir needs to be established by laboratory experiments.

## 9-3 LABORATORY EXPERIMENTS

MP flooding is an attractive process because the displacement efficiency can be almost 100 percent in zones swept by surfactant. Laboratory experiments have tried to formulate micellar solutions that are highly efficient and stable in reservoir conditions.

### Experimental Conditions

Berea sandstone core samples of 10 to 48 in. in length (25 to 122 cm) and 2 in. (5 cm) in diameter with 18 to 21 percent porosity and 140 to 450 md permeability have been used. When possible, the flooding tests utilized samples of reservoir rock. All the experiments were carried out at temperatures less than 120° F (49° C) and in tertiary displacement conditions. Core samples saturated with water were flooded with crude oil to irreducible water saturation and then flooded with water to residual oil saturation (values between 28 and

56 percent). Following this, the MP system was injected. The frontal velocities of 5 to 10 ft/day (1.5 to 3.0 m/day) were changed after the 1970s to a frontal advance of 1 ft per day or less, corresponding with reservoir flow velocities (Healy et al., 1975). The chemical system formulation varied from aqueous surfactant (68 to 92 percent water) to soluble oil (68 to 76 percent hydrocarbon). The mobility buffer consisted in most cases of partially hydrolized polyacrylamide.

**MP behavior in porous medium.** The laboratory experiments performed in various flood conditions and with different MP system formulations contributed useful information about MP behavior.

**Mobilities.** In water-wet porous media the micellar solution disperses the water film from the grains surface and partially carries it away (Davis et al., 1969). A stable oil bank with a higher mobility than water is formed ahead of the micellar solution slug. The polymer solution, which protects the micellar slug from behind, displaces the remaining irreducible water and forms a water bank. The micellar solution mobility must be equal to or less than the mobility of the stable bank ahead and its slug size sufficient to prevent dispersion. The polymer slug should be less mobile than the micellar slug and oil bank.

**Viscosity.** The micellar solution viscosity which controls mobility is influenced by the water content, the electrolyte concentration, the hydrocarbon and surfactant qualities, and the cosurfactant presence (Gogarty, 1968).

For instance, the presence of electrolyte in the micellar solution maintains the same viscosity of the solution at higher water percentages.

**Adsorption.** The adsorption of surfactants on the grains' surface has been observed. Although the petroleum sulphonates with high equivalent weights cause the greatest reduction in interfacial tension, they are insoluble in water and are readily adsorbed (Latil, 1980). Since the lower equivalent weight sulfonates show very little adsorption and are water soluble, a compromise is achieved by mixing these sulphonates with those of high equivalent weights. In addition, the chemical system is provided with various mineral compounds which are adsorbed in preference to the surfactant. Other mineral additives (ammonia or sodium carbonate) protect the surfactant slug against mineral salts in the formation water. The permeability reduction of the porous medium after the flow of a micellar slug was observed by Trushenski (1974) and explained as a beneficial effect caused by the interaction with the clay content of reservoir rock. The adsorption of polymers which follow the micellar solution is diminished and a less concentrated polymer slug can then be utilized.

## 9-4 FIELD TEST PROJECTS

The first micellar-polymer flood test was initiated in Dedrik Field, Illinois, in 1962. Despite industry interest the complexity of the method and its cost application restricted the number of field tests and the productive area involved. Two projects are described here: the largest conducted on 407 acres and started in 1977 by Marathon in the M-1 Field, Illinois, and the most successful pilot conducted in a 0.68-acre five-spot pattern by Exxon in the Loudon Field, Illinois, beginning in 1981.

### The M-1 Project, Illinois, United States

The M-1 Project was a commercial-scale test of the Maraflood-enhanced oil recovery process. The Maraflood technique was developed by the Marathon Oil Company. It consisted of the use of water-oil microemulsions of which the discontinuous phase was in the form of droplets of microscopic size achieved by the use of a large percentage of a suitable surfactant and a certain amount of alcohol. The evaluation (Cole, 1988) and the final report (Stover, 1988) of the M-1 Project included project design and operations, production history, performance evaluation, and economic evaluation.

**Reservoir characteristics.** The oil reservoir is a Robinson sandstone characterized by a meandering stream deposition with a net average thickness of 27.8 ft at an average depth of 950 ft. An average porosity of 18.9 percent and an arithmetic mean permeability of 103 md, with a Dykstra–Parsons permeability variation coefficient of 0.59 and 0.085 clay fraction content, characterize the rock reservoir. The oil viscosity is 7 cp and the reservoir water salinity is 16,575 ppm TDS.

The reservoir was developed to primary production under a dissolved gas drive mechanism from 1906 to 1941. The secondary recovery consisted of air repressuring (1941 to 1958) and was followed by waterflooding from 1958 to the start of the M-1 Project in 1976. The combined primary and secondary recovery averaged approximately 30 percent, and the most representative oil saturation at the start of the project was estimated to be 40 percent.

**Project design.** The scope of the project was to demonstrate economic recovery of tertiary oil in a mature waterflood project. The crude oil sulfonation/micellar-polymer technology used evolved from 15 years of field test experience with MP floods in the Robinson Sand (Cole, 1988).

The micellar slug design was based on the results of radial corefloods through $\frac{1}{8}$-in.-diameter "wellbores" formed in 2-in.-thick disks cut from 6-in.-diameter cores in the project area. After cleaning and vacuum drying, the cores were resaturated with reservoir crude, waterflooded with synthetic brine to residual oil, and injected with the micellar slug, mobility control buffer, and chase water. With the designed slug, recovery in the lab exceeded 70 percent of the residual oil after waterflooding in cores with permeability greater than 100 md.

In the field, 60 percent of 407 acres of the M-1 Project area was developed using 2.5-acre five-spot pattern spacing. The remaining area was developed with 5.0-acre pattern spacing.

The micellar-polymer fluid injection started in February 1977 (Cole, 1988) in the following sequence:

- The 10 percent pore volume micellar slug consisted of

  | | |
  |---|---|
  | Active sulfonate | 10.0 wt % |
  | Inorganic salt | 2.5 wt % |
  | Supply water | 80.0 wt % |
  | Unreacted oil | 7.5 wt % |

  The micellar slug was compatible with reservoir water so no preflush was needed. To prevent hydroxide precipitation of any iron in solution, 500 ppm of citric acid were added to the initial 2000 bbl of injected slug. Cosurfactants added at the field site were hexyl alcohol 0.8 wt % and isopropyl alcohol 0 to 1.5 wt %.
- 105 percent pore volume of tapered mobility buffer (injection began in November 1978) consisted of partially hydrolized polyacrylamide in concentrations from 1156 ppm 11 percent, to 800 ppm 19 percent, 625 ppm 32 percent, 411 ppm 12 percent, 200 ppm 11 percent, 100 ppm 10 percent, and 50 ppm 10 percent. The polymer injection was completed in the 2.5-acre pattern area in August 1984 and in the 5.0-acre pattern area in 1989.
- 35 percent pore volume of produced water has to be injected until the injected fluids will total 1.5 pore volume.

**Field Operations.** Due to the advanced depletion of the reservoir, 101 of 114 injectors and 122 of 132 producers were drilled for the M-1 Project and were completed "open hole" through the productive formation. Extensive hydraulic fracturing with small fracture lengths was the most effective method of injection and production well stimulation. Treatment of the injectors with hydrogen peroxide was effective against the accumulation of concentrated or cross-linked polymer at sand face (Stover, 1988).

**Production history.** The graphs of the oil cut versus time for the 2.5-acre spacing area and for the 5-acre spacing area are shown in Figures 9-3 and 9-4, respectively.

As can be observed, the oil cut started to increase in the 2.5-acre spacing area in March 1979 after the injection of 18.6 percent PV and in the 5-acre

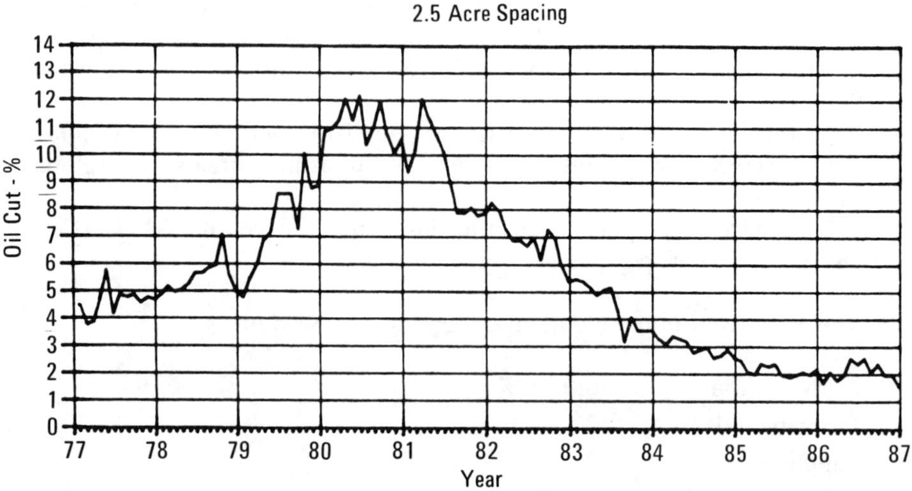


**Fig. 9-3** Oil cut versus time (From Stover, 1988)

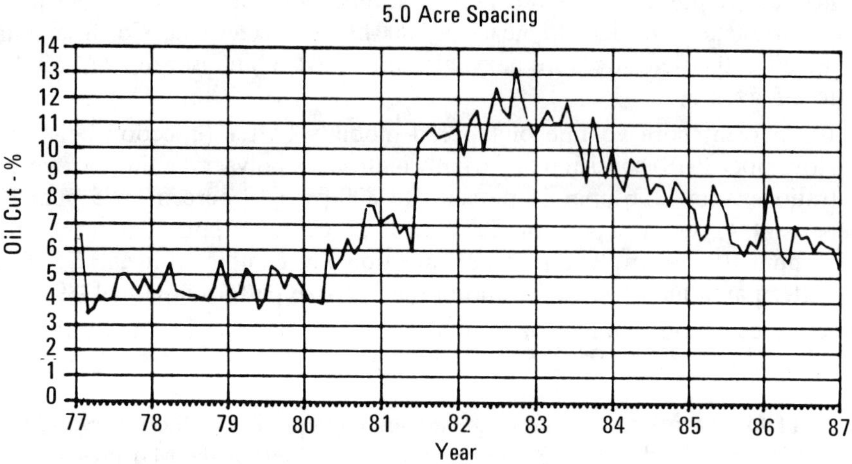


**Fig. 9-4** Oil cut versus time (From Stover, 1988)

spacing area in May 1980 after 24.1 percent PV injected. The peak period of 11 to 12 percent oil cut was reflected in the total oil production in both areas reaching over 700 bbl/day in 1980 with an average of 550 bbl/day between late 1979 and 1982 (Figure 9-5). Since then, the production rate has steadily declined to a total of 145 bbl/day in December 1986 representing 42 bbl/day at a 2 percent oil cut on the 2.5-acre pattern area and 105 bbl/day at a 6.2 percent oil cut on 5-acre pattern area.

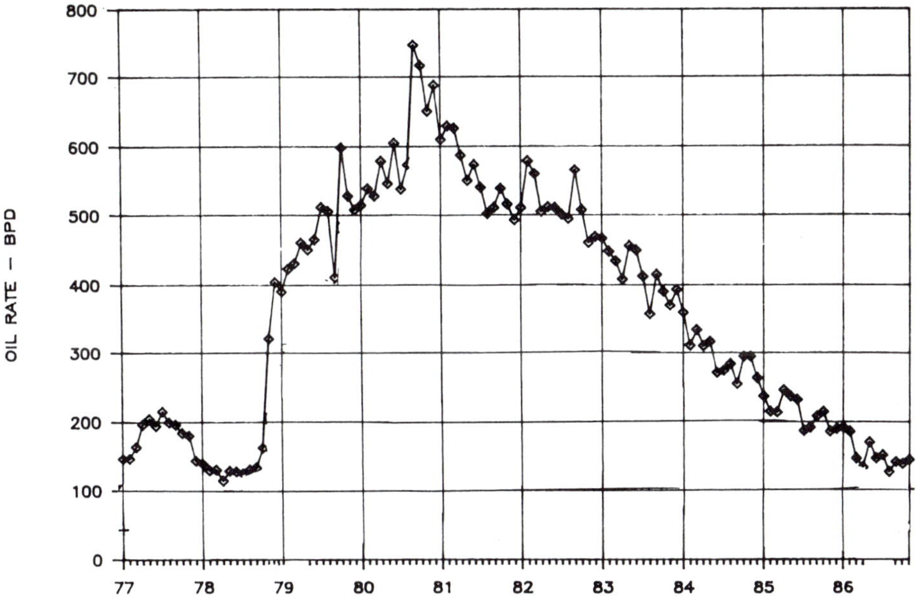


**Fig. 9-5** Oil rate versus time (Adapted from Stover, 1988)

**Performance evaluation.** The 2.5-acre pattern area became uneconomical in December 1986 after recovering 805,500 bbl of oil or 22.2 percent of the 40 percent postwaterflood oil saturation instead of the 38 percent predicted. The 5-acre area was estimated to produce a total of 583,000 bbl of oil or 19.2 percent of the 40 percent postwaterflooded oil saturation instead of the 35 percent predicted.

The lower than anticipated oil recovery was primarily due to poor vertical and areal sweep efficiencies caused by reservoir heterogeneities (Stover, 1988). Three observation wells were drilled in the central 100 acres of the project for the purpose of performing dual induction logs to measure changes in resistivity caused by the moving injected fluids. The logs showed the presence of stacked sand bodies which adversely affected vertical sweep efficiency.

Three core wells also drilled in the same zone showed the presence of at least three distinct zones and provided indication of the displacement efficiency. The best oil displacement took place in the upper zone with residual oil saturations of only 5.3 to 9.8 percent PV. The middle zone was moderately swept and the lower zone had the poorest displacement efficiency with residual oil saturations between 11.0 and 27.5 percent PV. A flow trend was observed parallel to the direction of deposition of the sandstone reservoir when radioactive tritium was injected in 10 injectors prior to the micellar slug.

**Economic evaluation.** Because of the lower than predicted recovery (55 percent of original prediction) the project's revenue and profit were lower also.

The project's operation expenses were much higher than anticipated due to the cost for water pumping (10 miles from source) and for oil treating. Of the total project investment, 63 percent was spent for micellar slug and polymer and 25 percent was spent to drill and complete 233 new injection and production wells.

The M-1 Project demonstrated the technical viability to design, implement on a large scale, conduct, and evaluate the most complex of the EOR methods. The project was well designed and executed and provided valuable information and experience. Future micellar-polymer projects should attempt to reduce the cost of the chemical system, utilize more of the existing wells if possible (instead of drilling new ones), and improve the reservoir description.

### The Loudon Pilot, United States

The Loudon Pilot was a surfactant flood pilot test conducted by Exxon in a 0.68-acre five-spot pattern, using four injection wells, one central producer (Well 13), five observation wells, and one postproject cored well. The Loudon test was conducted under the cost-shared and tertiary incentive programs under a contract with the Department of Energy. It was presented by Bragg et al. (1982) and reviewed by Dauben (1988).

**Reservoir characteristics.** The rock reservoir having 13 ft average thickness is a Mississippian Chester sandstone at 1550-ft depth and with average permeability of 150 md, a porosity of 19 percent, and a Dykstra–Parsons permeability variation of 0.42. An average oil saturation of 25.5 percent pore volume was obtained after waterflooding and characterized the pattern area at the start of the pilot. The oil reservoir had 5-cp viscosity at reservoir temperature of 78° F, and the formation water was highly saline having 104,000 ppm $T_{DS}$.

**Project design.** The chemical system was formulated to avoid the need for a preflush to remove salinity and hardness. To be effective in high-salinity reservoirs, the micellar fluid should not contain a petroleum sulfonate surfactant. The 40 percent pore volume microemulsion of 28-cp viscosity was formulated by mixing a 2.3 wt % surfactant (other than petroleum sulfonate) and biopolymer at a 1000 ppm concentration with the produced brine. The polymer drive water of 38-cp viscosity contained biopolymer at a concentration of 1400 ppm. An oxygen scavenger was added to the system to prevent the oxidation of dissolved ferrous iron together with biocide to inhibit bacterial growth, acid to adjust pH, and well tracers (Dauben, 1988).

**Field operations.** Before the start of the pilot an injectivity test was conducted and useful information was obtained regarding mixing and quality control of chemicals, injectivity, wellhead pressure that would not induce fracturing, and so on. The surface facilities for storage injection and production were either epoxy-coated steel or plastic to prevent corrosion. Nitrogen gas blankets were also maintained on all tanks to avoid the contact with air (oxygen). The fluids produced by Well 13 were equivalent to one-fourth of the total injection into the four wells and were separated using electrical heat treatment, settling tanks, and/or deemulsifiers.

It is important to point out that to detect any net migration of fluids into or away from the pilot area, downhole pressure measurements were made in three shut-in wells around the pilot and continuously in Well 13.

**Pilot Results.** The oil response was observed after 0.25 of the pore volume was produced and when the oil cut started to increase. The oil cut reached a maximum of 0.26 after 0.70 of the pore volume was produced (Figure 9-6).

After 2.25 pore volume of production, about 60 percent of the oil in place was recovered ($E_R = 0.6$). This represents a very high level when compared to other surfactant projects, especially considering the formation's high-salinity water conditions.

The induction log's response in the observation wells showed that the lower part of the formation thickness was generally better swept than the upper zone. Also, samples collected from observation wells indicated loss of polymer viscosity and the existence of bacteria degradation of the biopolymer. The measurement of the concentration of surfactant and polymer in produced field showed that 60 percent of the injected surfactant was produced.

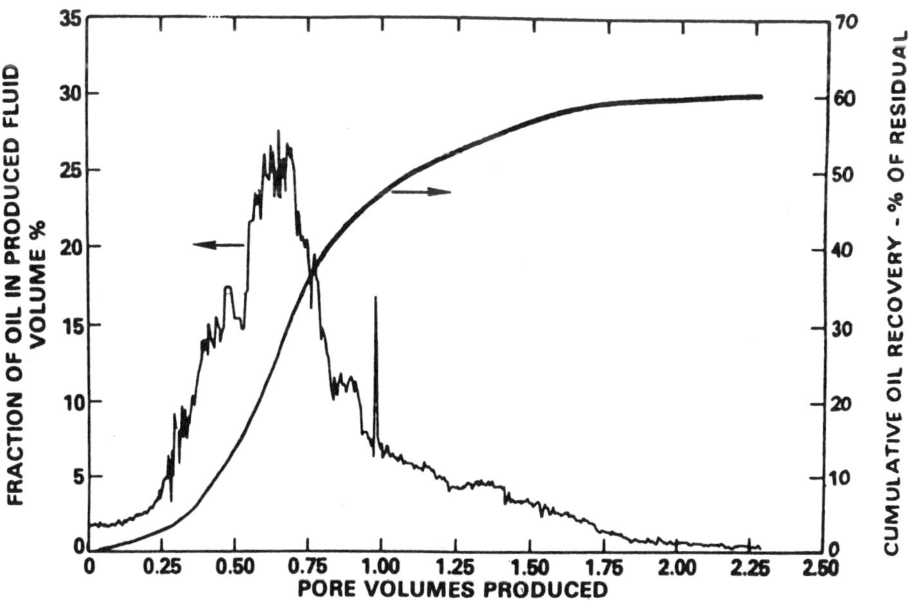


**Fig. 9-6** Pilot oil production from well 13 (From Dauben, 1988)

**Comments.** The Loudon Pilot Test demonstrated the success of a special chemical system formulated for high-salinity formation brine conditions yet avoiding the need for a preflush. The next step to larger, semicommercial- or commercial-scale production is possible since less surfactant is needed to penetrate to the producing well. The use of biopolymers will necessitate improved protection from microbial degradation. The Loudon Pilot is an example of how a very clear interpretation was possible when useful information was provided from sources such as core and well log analysis, interwell tracers, bottom hole pressure measurements, postproject wells cored, injection profile surveys, and others.

## 9-5 SCREENING CRITERIA AND CRITICAL QUANTITIES

### Screening Criteria

Screening parameters for micellar-polymer flooding seem to reach a consensus. A given reservoir will be a micellar-polymer candidate if it is a sandstone, has a temperature of about 200° F or less, a permeability greater than 20 md, a residual oil saturation at the start of the project higher than 20 to 25 percent, and a formation water salinity of less than 200,000 ppm TDS, among other things. This screening approach is useful when considering the MP tertiary-enhanced recovery technique, for eliminating reservoirs that are technically unfeasible or have low economic potential.

For a large geological area or a particular country, it is very important to rank the reservoirs which pass all the criteria according to their EOR potential and estimated contribution to production. A preliminary economic evaluation program, PECϕN, was developed to identify first-rank reservoir candidates (Lake and Pope, 1979; U.S. Department of Energy, 1980; Paul et al., 1982). PECϕN has its limitations however, as it does not consider the time of oil breakthrough, the oil production versus time curve, and the time value of money. Another ranking system called the chemical flood prediction model (CFPM) was developed (Paul et al., 1982; Goldburg et al., 1983). The first step in the development of these models was to identify the critical reservoir and micellar quantities (CRMQ).

### Critical Reservoir and Micellar Quantities

Based on an extensive volume of data from currently proposed, completed and ongoing MP/EOR field tests, the 29 quantities identified as affecting the micellar-polymer process were consolidated into 13 groups constituting the CRMQ (U.S. Department of Energy, 1980). These groups are

1. *Residual oil saturation and distribution*. This is the oil saturation left behind by a pistonlike waterfront in the waterswept area. Its value and its distribution determine the amount of MP/EOR target oil and the high permeability zones, respectively. Higher residual oil saturation means more MP/EOR target oil but also a low permeability zone poorly swept by water.
2. *Reservoir confinement*. A given reservoir is a hydrodynamic unit characterized by confinement at reservoir boundaries and by continuity of the pay zone in the interior of the boundaries.
3. *Natural fractures*. As a system characterizing the reservoir rock is an unfavorable factor for an MP/EOR candidate reservoir. Induced fracturing in the injectors to increase injectivity can be beneficial.
4. *Temperature and depth*. Temperature is a limiting factor when trying to avoid mobility agent degradation, although IFT improves at elevated temperatures. Temperature is directly proportional to depth. This, in turn, controls the formation parting pressure.
5. *Permeability and heterogeneity*. It is important to know the areal and vertical distribution of permeability if possible. Evaluation of core sample analyses (Dykstra–Parsons coefficient) and the well's production-rate history may reveal vertical and areal flow trends.
6. *PVT analysis of crude oil*. The crude oil properties at reservoir conditions, especially the viscosity, are related to the design of the MP chemical system.
7. *Makeup water composition*. This represents total dissolved solids and total cations if full analysis is not available. The mobility buffer efficiency depends on the makeup water's composition for which the designation "fresh" is not sufficient.
8. *Resident water composition*. This is an important parameter to define since it determines the necessity for either an effective preflood or a chemical system tailored to the existing brine.
9. *Relative permeabilities*. These depend upon the microdistribution of fluids within the rock pores. This distribution is controlled by the rock wettability, which in turn, affects mobility requirements.
10. *Pattern type and size*. These seem to play an important role. There are indications that oil recovery is higher in the smaller pattern areas (Lowry et al., 1986).
11. *Clay mineralogy and composition*. These influence the surfactant adsorption, and an empirical link for this dependence might be established.
12. *Rock mineralogy and composition*. The level of carbonate minerals particularly affect the surfactant's electrolyte environment.
13. *Volumetric waterflood data*. This refers to the material balance equation which states that the original fluid in place equals the fluid produced plus the remaining fluid.

The remaining oil, particularly the residual oil saturation in the water-swept zone, is the most important group of the CRMQ and heads the list.

The results of 20 field tests with reported recoveries were analyzed to find correlations between MP oil recovery and the various CRMQ.

## 9-6 PRELIMINARY ECONOMIC EVALUATION MODEL

The predictive model developed by the U.S. Department of Energy (1980) determines for each reservoir if micellar-polymer flooding is technically feasible, and if so, what its economic potential will be. The flow diagram of the proposed screening and evaluation procedure is given in Figure 9-7.

The following example illustrates the method for calculating the micellar-polymer feasibility and potential of an oil reservoir.

**Example 9-1.** The oil reservoir bordered on the north, east, and west by faults and on the south by an aquifer is a monocline, with an 8° dip to the south and an average net thickness $h$ = 24 ft (Figure 9-8). The productive formation is a sandstone at $D$ = 5500-ft depth with $\phi$ = 28% porosity, $S_{wi}$ = 30% irreducible water saturation, $k$ = 400 md average permeability, and Dykstra–Parsons permeability variation $V_{DP}$ = 0.5. The 34 °API crude oil has $\mu_o$ = 3.4 cp viscosity at $T$ = 92° F reservoir temperature and initial and present formation volume factors $B_{oi}$ = 1.15 RB/STB and $B_{of}$ = 1.1 RB/STB, respectively. After a short period of using a gas-dissolved drive mechanism, the reservoir was developed for waterflooding. This took place in a staggered line drive $A$ = 80 acres/pattern (Figures 9-8 and 9-9). The cumulative oil produced under solution gas drive and waterflooding was 17.2 × 10⁶ STB or 48 percent of the original oil in place, the water-oil ratio reached WOR = 21, the residual oil saturation in the area swept by water $S_{orw}$ was estimated at 26 percent pore volume, and the oil saturation in unswept regions of the reservoir after waterflood, $S_o'$, was estimated at 65 percent pore volume. Other data regarding the reservoir include water salinity of 65,000 ppm TDS, water viscosity $\mu_w$ = 0.55 cp, clay content of the rock $W_{clay}$ = 0.05, rock density $\rho_r$ = 156 lb$_m$/ft$^3$, and surfactant density $\rho_s$ = 62.3 lb$_m$/ft$^3$.

Determine whether or not a micellar-polymer system can be formulated for this particular reservoir and how economically viable the MP process would be in the reservoir.

SOLUTION

**(a)** *Technical Feasibility*: This is determined from Figure 9-7 by using reservoir temperature, reservoir brine salinity and by the capillary number. Both the *temperature* and *salinity* of the reservoir are below the feasibility screen limits:

$$T = 92° \text{ F is less than } 229° \text{ F}$$

$$\text{salinity} = 65{,}000 \text{ ppm } T_{DS} \text{ is less than } 200{,}000\ T_{DS}$$

The *capillary number*, which as we know is the ratio of viscous to capillary forces, is the best available guide to whether the residual oil saturation after

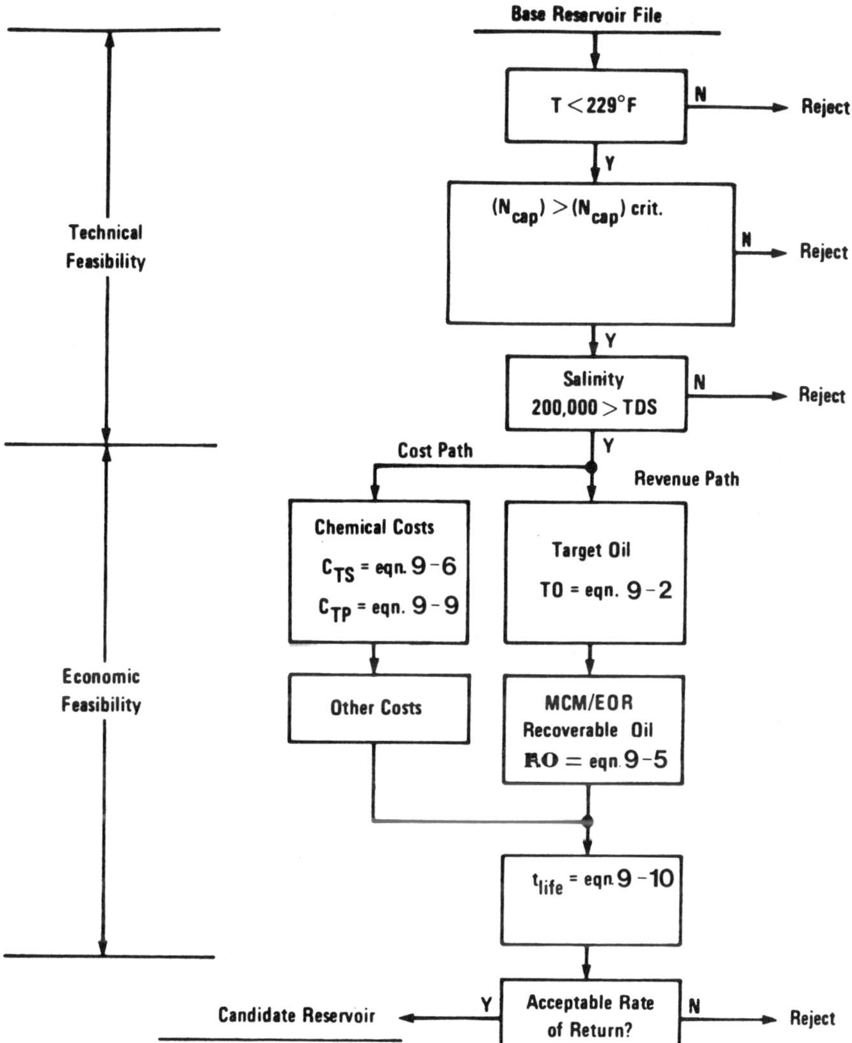


**Fig. 9-7** Schematic of proposed screening and evaluation procedure (Adapted from U.S. Department of Energy, 1980)

waterflooding, $S_{or}$, can be lowered by a micellar system. A reservoir will be an MP/EOR candidate if its characteristic capillary number calculated by Eq. 9-1 for line drive pattern

$$(N_{\text{cap}})_{\text{LD}} = \frac{(4.3 \times 10^{-10})C_p' \, kD/\sigma}{\sqrt{A}(4.56 + 1/\pi \ln A)} \tag{9-1}$$

is greater than $(N_{\text{cap}})$ critical, which is $10^{-5}$ for a water-wetting reservoir and $10^{-4}$ for an oil-wetting reservoir. Assuming the injection pressure gradient of

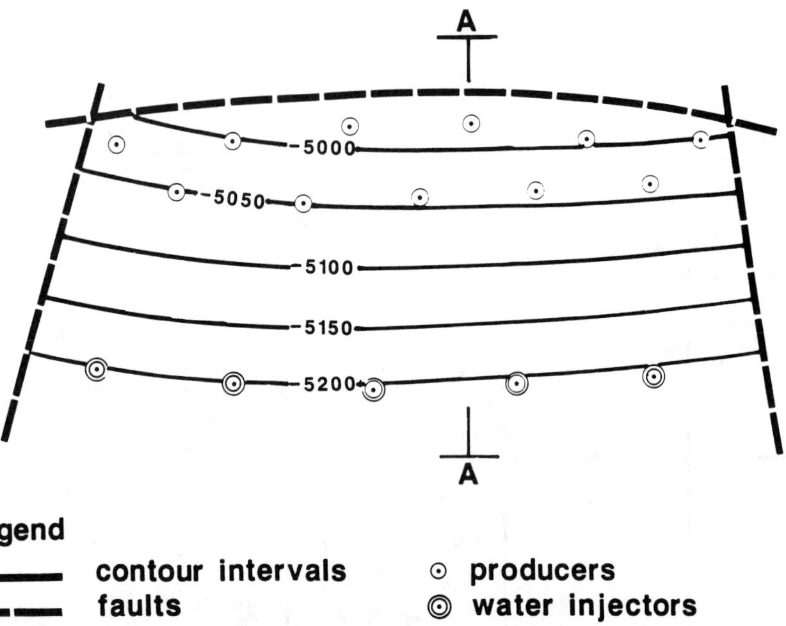


**Fig. 9-8** Subsurface structural map scheme with waterflood pattern development

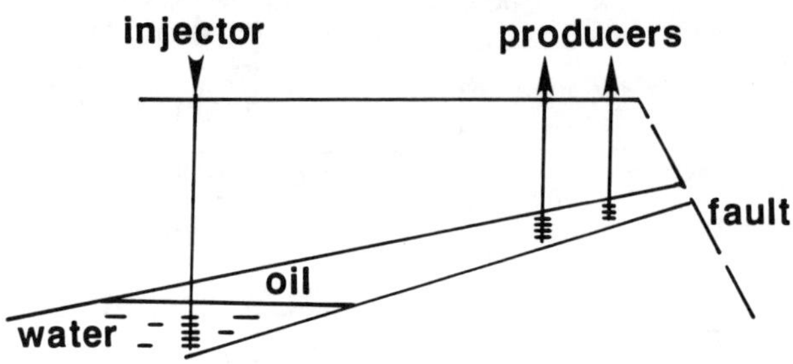


**Fig. 9-9** Structural cross-section scheme A-A′

$C_p' = 0.5$ psi/ft and the interfacial tension $3.33 \times 10^{-3}$ dynes/cm, the capillary number is

$$(N_{\text{cap}})_{\text{LD}} = \frac{1.29 \times 10^{-7}(0.5)400 \times 5500}{\sqrt{80}(4.56 + 1/\pi \ln 80)} = 2.67 \times 10^{-3}$$

Not having available reservoir wettability data, the $(N_{\text{cap}})_{cr}$ for water-wet media will be used:

$$\text{field } (N_{\text{cap}})_{\text{LD}} = 2.67 \times 10^{-3} > 10^{-5} = (N_{\text{cap}})_{cr}$$

Since the reservoir meets temperature, salinity, and capillary number criteria, a chemical system can be designed that will mobilize residual oil.

**(b)** *Economic Feasibility*: This is illustrated by calculating target oil, recoverable oil, surfactant and polymer requirements, project life, field development, cost data, and the revenue to expense ratio.

**1.** *Target Oil*. Target oil, TO, is defined as the oil remaining in the water-floodswept portion of the reservoir at the start of the MP process. Using the same notations as in Appendix II of the 1980 U.S. Department of Energy report, the target oil is

$$TO = \mathrm{PV} \times \frac{S_{\mathrm{orw}}}{B_{of}} E_{\mathrm{v}} \tag{9-2}$$

where the volumetric efficiency $E_{\mathrm{v}}$ is given by a material balance equation applied at the end of waterflooding, for one pore volume PV

original oil in place = remaining oil in place + oil produced

$$\mathrm{PV}\frac{S_{oi}}{B_{oi}} = \mathrm{PV}(1 - E_{\mathrm{v}})\frac{S_o'}{B_{of}} + \mathrm{PV} \times E_{\mathrm{v}}\frac{S_{\mathrm{orw}}}{B_{of}} + N_{\mathrm{p}} \tag{9-3}$$

Solving (9-3) for $E_{\mathrm{v}}$ after replacing

$$\mathrm{PV} = \frac{\mathrm{OOIP} \times B_{oi}}{S_{oi}}$$

gives

$$E_{\mathrm{v}} = \frac{S_{oi}\,B_{of}/B_{oi}(1 - N_{\mathrm{p}}/\mathrm{OOIP}) - S_o'}{S_{\mathrm{orw}} - S_o'} \tag{9-4}$$

or

$$E_{\mathrm{v}} = \frac{0.7 \times 1.1/1.15(1 - 0.48) - 0.65}{0.26 - 0.65} = 0.7738$$

and the target oil

$$\mathrm{TO} = \frac{\mathrm{OOIP} \times B_{oi}}{S_{oi}} \times \frac{S_{\mathrm{orw}}}{B_{of}} E_{\mathrm{v}} = \frac{17.2 \times 10^6 \times 1.15}{0.48 \times 0.7} \times \frac{0.26}{1.1} \times 0.7738$$

$$= 10.767 \times 10^6 \text{ STB} \qquad \text{or} \qquad 1.712 \times 10^6 \text{ m}^3$$

which can be expressed as 57.5 percent of the oil in place at the start of the process.

**2.** *Recoverable Oil*. Recoverable oil, RO, is estimated from the ratio

$$\frac{\text{produced oil}}{\text{displaced oil}} = \frac{\mathrm{RO}}{E_{\mathrm{D}} \times \mathrm{TO}} = 0.8 \tag{9-5}$$

where $E_{\mathrm{D}}$ is the micellar-polymer displacement efficiency (U.S. Department of Energy, 1980) as a function of slug size ($V_{ps}$), surfactant retention ($D_s$), micellar-polymer swept zone residual oil saturation ($S_{\mathrm{orc}}$), and heterogeneity.

From Figure 9-10 (Gupta and Trushenski, 1979) for water-wet rock and capillary number $N_{\mathrm{c}} = 2.67 \times 10^{-3}$

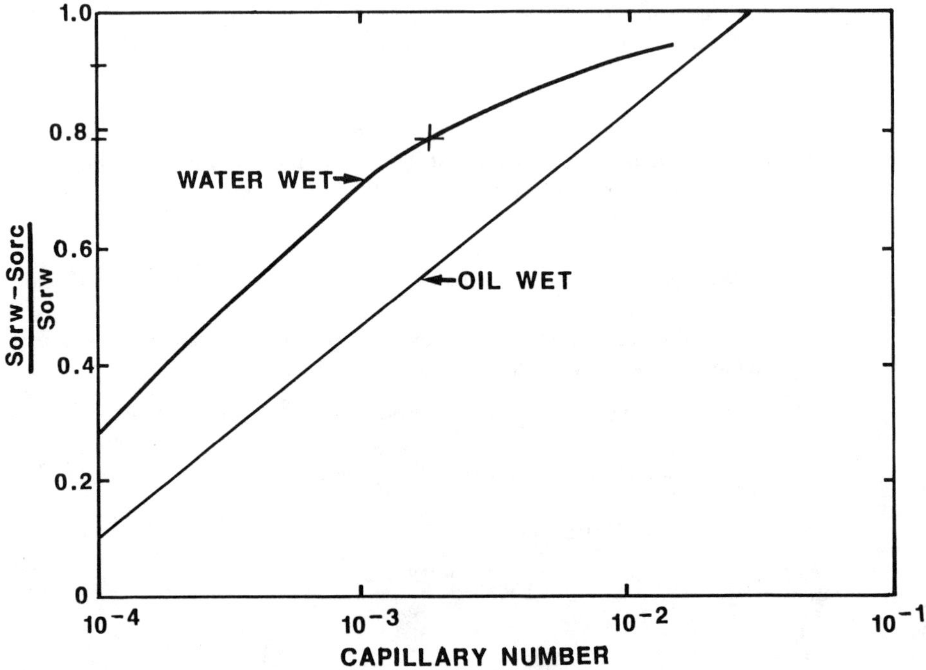


**Fig. 9-10** Capillary desaturation (From U.S. Department of Energy, 1980)

$$\frac{S_{orw} - S_{orc}}{S_{orw}} = 0.79$$

and represents the degree to which the surfactant slug can mobilize residual oil.

The micellar-polymer displacement efficiency depends on reservoir heterogeneity ($V_{DP} = 0.5$), surfactant retention $D_S$, and surfactant slug size $V_{PS}$ required to satisfy all adsorption.

Assuming $V_{PS}/D_S = 1.3$ for one unit of floodable pore volume FPV = PV × $E_v$ (from Figure 9-11)

$$\frac{E_D}{(S_{orw} - S_{orc})/S_{orw}} = 0.82$$

so

$$E_D = 0.79 \times 0.82 = 0.65$$

and

$$\text{RO} = 0.8 \times 0.65 \times 10.767 \times 10^6$$
$$= 5.6 \times 10^6 \text{ STB} \quad \text{or} \quad 0.89 \times 10^6 \text{ m}^3$$

3. *Surfactant Requirement*. The surfactant requirement, $C_{TS}$, is estimated from Eq. 9-6

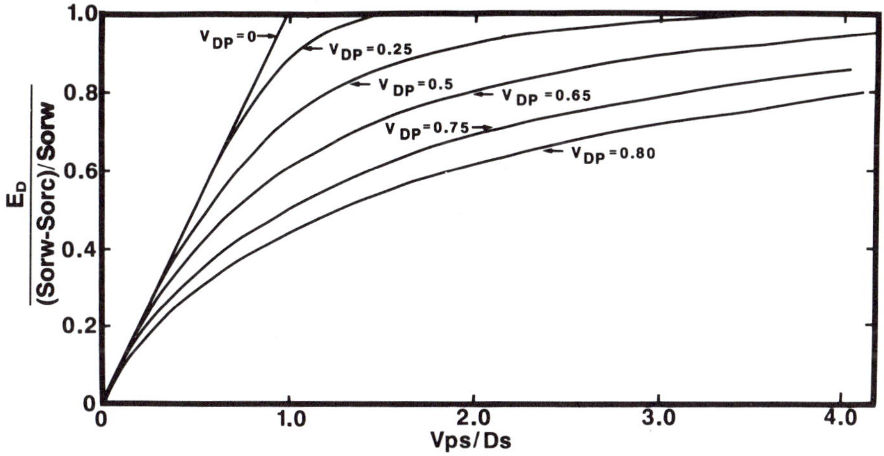


**Fig. 9-11** Effect of slug size/retention ratio on vertical sweep (From U.S. Department of Energy, 1980)

$$C_{TS} = C_S V_{PS} = C_S D_S \times 1.3 \text{ FPV} \tag{9-6}$$

where $C_S$ is the concentration of active surfactant in the injected slug (volume fraction) and is given by

$$C_S D_S = \left(\frac{1-\phi}{\phi}\right)\left(\frac{\rho_r a_S}{\rho_S}\right)\frac{1}{1000} \tag{9-7}$$

The surfactant retention $a_S$ (in milligrams of surfactant per gram of rock) is determined in principal by adsorption on clay surfaces and may be estimated with $W_{clay} = 0.05$.

$$a_S = 3.3(0.05) = 0.165 \text{ mg surf/g rock}$$

and

$$C_S D_S = \frac{1-0.28}{0.28} \times \frac{156}{62.3} \times \frac{0.165}{1000} = 0.0010624$$

The total active surfactant required

$$C_{TS} = 10.624 \times 10^{-4} \times 1.3 \times \frac{10.767 \times 10^6}{0.26} \times 1.1$$

$$= 62{,}913 \text{ STB} \qquad \text{or} \qquad 10{,}000 \text{ m}^3 \text{ active surfactant}$$

4. *Polymer Requirement*. When relative permeability data are available, a plot of total relative mobility for oil and water versus water saturation is represented. The initial mobility of the polymer buffer is made equal to the $\lambda_{min}$, the minimum total mobility of oil and water (Figure 9-12). Then, the viscosity of the mobility buffer is graded down to that of the chase water. A simplified

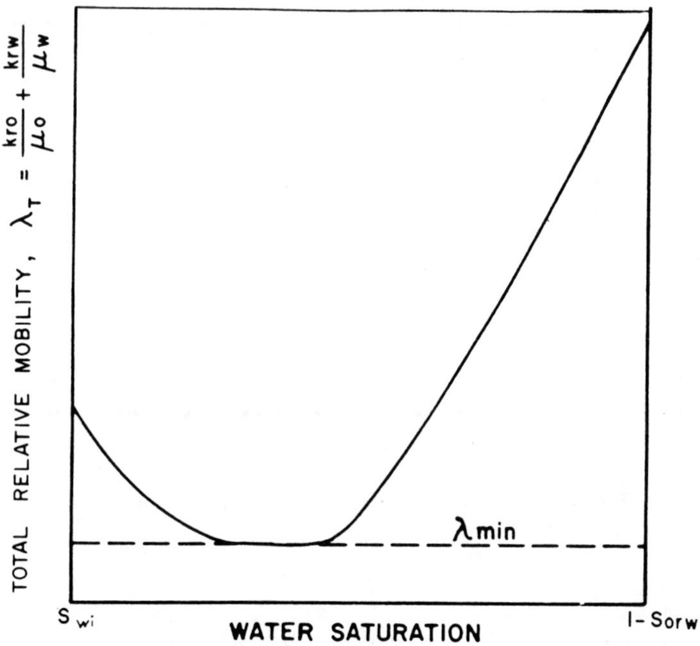


**Fig. 9-12** Total relative mibility versus water saturation curve (From U.S. Department of Energy, 1980)

procedure has been developed for use when such data are not available. Using field test information (U.S. Department of Energy, 1980), this procedure plots the polymer concentration in the initial portion of the drive against the ratio of oil to water viscosity (Figure 9-13). The average concentration of the polymer buffer, $\overline{C}_p$, is given by

$$\overline{C}_p = \tfrac{1}{2} C_p \times V_{\mathrm{MB}} \tag{9-8}$$

assuming that the total volume of the polymer buffer, $V_{\mathrm{MB}}$ is 1 FPV. Therefore, the total polymer requirement

$$C_{\mathrm{TP}} = 5.614 \times 10^{-6} \times \overline{C}_p \times \mathrm{FPV} \times \rho_{\mathrm{MB}}\ \mathrm{lb_m} \tag{9-9}$$

From Figure 9-13 with $\mu_o/\mu_w = 6.18$

$$C_p = 1076 \text{ ppm} \qquad \text{and} \qquad \overline{C}_p = \tfrac{1}{2} \times 1076 \times 1 = 535 \text{ ppm}$$

Taking the density of the mobility buffer $\rho_{\mathrm{MB}} = 62.3\ \mathrm{lb_m/ft^3}$, the total polymer requirement

$$C_{\mathrm{TP}} = 5.614 \times 10^{-6} \times 535 \times 62.3 \times 45.55 \times 10^6$$

$$= 8.523 \times 10^6\ \mathrm{lb_m} \text{ polymer}$$

5. *Project Life*. For a line drive pattern and 0.5-psi/ft injection pressure gradient, the project life is given by

$$t_{\text{life}} = 9430A\phi \frac{4.56 + 1/\pi \times \ln(A)}{0.5 \times kD/\mu_o} Q_D \tag{9-10}$$

where $Q_D$ is the total injection volume in pore volumes ($Q_D = V_{PS} + V_{MB} + V_{CHASE\,W}$) and can be taken as 1.5

$$t_{\text{life}} = 9430 \times 80 \times 0.28 \frac{4.56 + 1/\pi \times \ln(80)}{0.5 \times 400 \times 5500/3.4} \times 1.5 = 5.83 \text{ years}$$

6. *Field Development*. Field development is based on the number of injection-production wells in the field required for the micellar-polymer project.

The area needed to be developed is given by

$$DA = \frac{\text{FPV}}{7758\phi h} \text{ acres} \tag{9-11}$$

$$= \frac{45.55 \times 10^6}{7758 \times 0.28 \times 24} = 873.7 \text{ acres}$$

and the number of repeated patterns

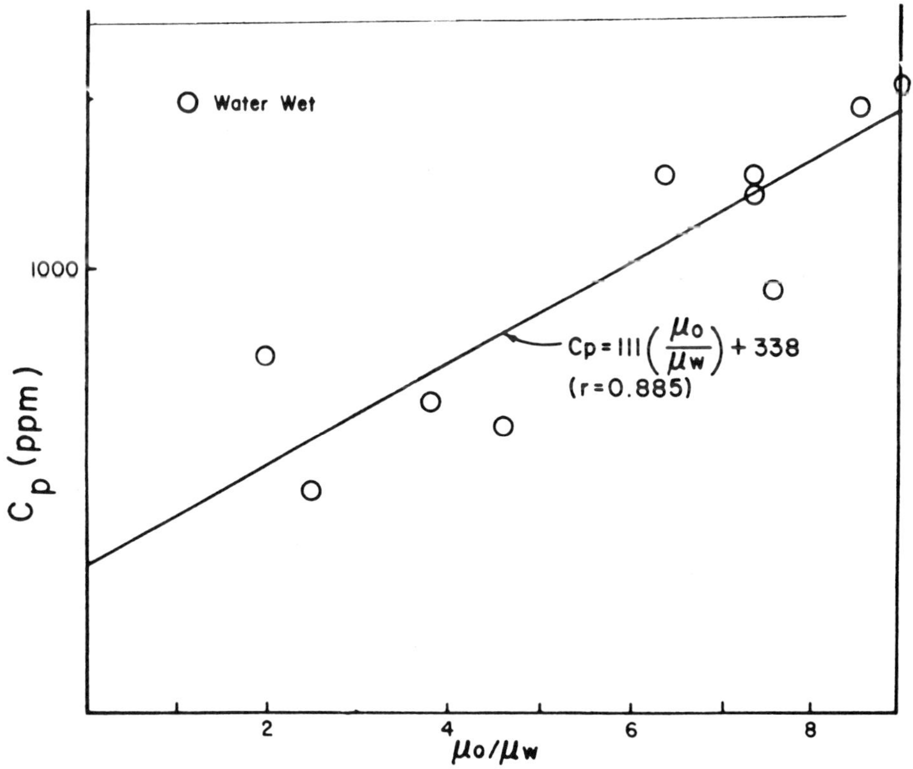


**Fig. 9-13** Initial polymer concentration versus oil-water viscosity ratio (Adapted from U.S. Department of Energy, 1980)

$$NRP = \frac{DA}{A} = \frac{873.7}{80} = 11 \tag{9-12}$$

For direct line drive patterns there is one injector and one producer per pattern, so the project requires 11 injectors and 11 producers.

7. *Cost Data*. These are data on oil price, development costs, operating costs, and chemical costs.

- *The relationship for oil price* (U.S. Department of Energy, 1980) is

$$\$_o = [S_o' - 0.02(40 - °API)]X_s \ \$/STB \tag{9-13}$$

where

$S_o'$ = a base oil price
$X_s$ = 1.0, if no sulphur is present in crude
$X_s$ = 0.9, if crude contains significant sulphur

Assuming $S_o'$ = \$22 with no sulphur content, the oil price is

$$\$_o = 22 - 0.02(40 - 34) = \$21.72/STB$$

- *Development and operating costs* are the expenses incurred in drilling the new wells, purchasing the equipment required, working over and converting the existing wells, and operating the pattern's wells each year. Considering the reservoir geometry and the existing pattern, 11 new injection wells should be drilled. The location of the 11 injectors corresponds to the contour line of the maximum initial oil saturated thickness (Figures 9-14 and 9-15). The existing five water injection wells have to be converted into five supplementary producers for the second half of the project life when they will be put into production from the top of the formation interval. Four existing producers (30 percent) need workover operations. The development cost data are assumed to be

| | | |
|---|---|---|
| drilling: | \$120/ft × 11 wells × 5500 ft/well | = \$ 7.26 × 10^6 |
| equipping new wells: | \$80,000/well × 11 wells | = \$ 0.88 × 10^6 |
| workovers and conversions: | 100,000/well × 9 wells | = \$ 0.90 × 10^6 |
| operating expenses: | 50,000/pattern × years × 11 × 6 | = \$ 3.3 × 10^6 |
| Total Development and Operating Costs | | = \$12.24 × 10^6 |

- *Chemical costs* are assumed to be

\$160/bbl surfactant
\$1.7/lbm polymer

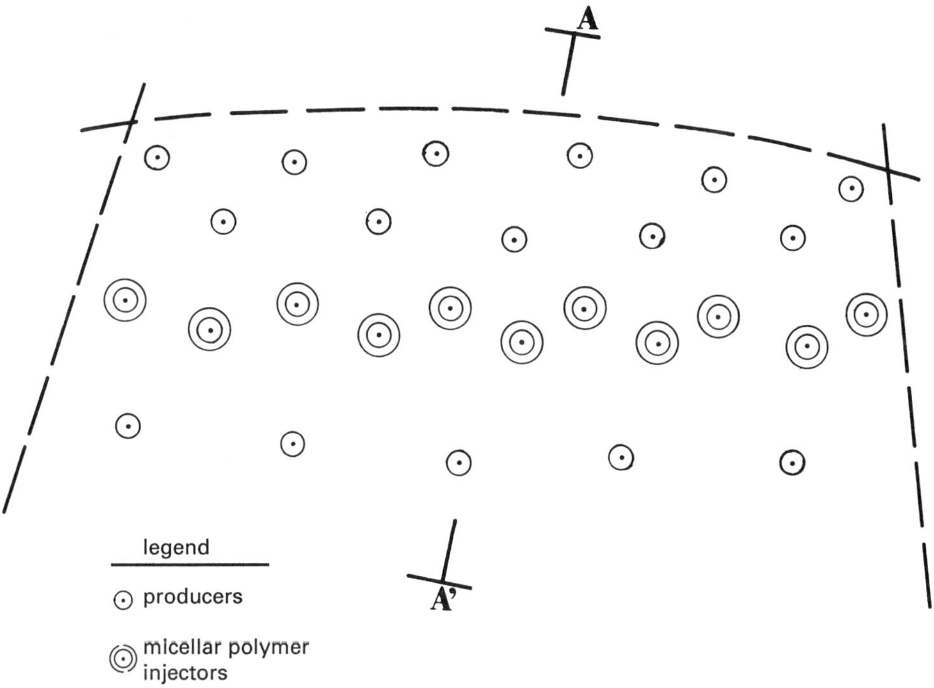


**Fig. 9-14** Subsurface structural map scheme with MP pattern development

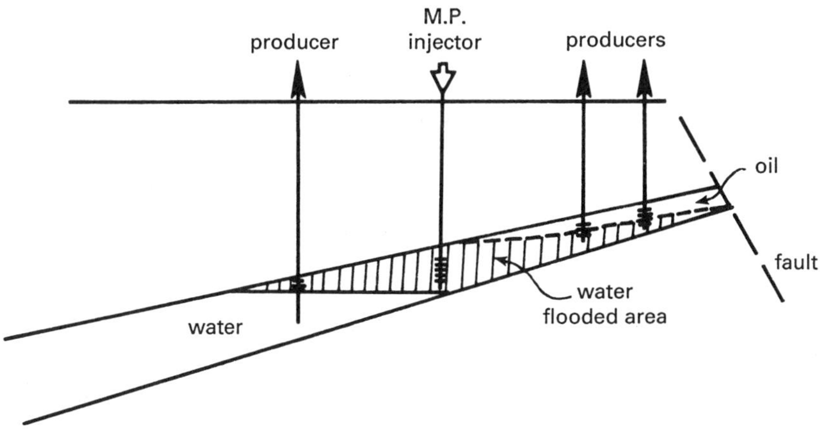


**Fig. 9-15** Structural cross-section scheme A-A′

| | | |
|---|---|---|
| surfactant cost: | $62{,}913 \times 160$ | $= \$10.066 \times 10^6$ |
| polymer cost: | $8{,}523 \times 10^6 \times 1.7$ | $= \$14.489 \times 10^6$ |
| total chemical costs | | $= \$24.555 \times 10^6$ |

**8.** *Economic Calculation*

$$\text{total expenses are } (12.24 + 24.555)10^6 = \$36.795 \times 10^6$$

$$\text{total revenue is } 5.6 \times 10^6 \times 21.72 = \$121.632 \times 10^6$$

and the revenue-to-expense ratio is

$$\frac{R}{E} = \frac{121.632 \times 10^6}{36.795 \times 10^6} = 3.30$$

**9.** *Oil Recovery* as a percentage of the original oil in place is

$$E_R = \frac{\text{RO}}{\text{OOIP}} = \frac{5.6 \times 10^6}{35.833 \times 10^6} = 0.15628 \text{ or } 15.6\%$$

and as a percentage of the oil in place at start of the project

$$E_R = \frac{\text{RO}}{\text{OOIP} - N_p} = \frac{5.6 \times 10^6}{(35.833 - 17.200)10^6} = 0.30 \text{ or } 30\%$$

As it was pointed out, the preliminary economic evaluation model (PECϕN) is a simplified method to screen the national reservoir potential for micellar-polymer flooding and to provide a selection of the better reservoirs. The model does not provide criteria for estimating the time of oil breakthrough, the oil production versus time curve and the time value of money. A new predictive model for micellar-polymer processes that does include the foregoing considerations and circumvents the limitations of the other models was developed.

## 9-7 THE CHEMICAL FLOOD PREDICTIVE MODEL

This model was developed for the U.S. Department of Energy and was designed to find the better micellar-polymer candidates of waterflooded sandstone reservoirs in midcontinent and California (Paul et al., 1982; Goldburg et al., 1983). Apart from the same technical screen and a more detailed economic and cash flow analysis extended over the life of the project, the model also provides the oil production curve, oil breakthrough, and the "peak" oil cut. Example 9-2 illustrates how to determine the oil production curve following the calculation for the Loudon pilot test given in Appendix C of the DOE Final Report (Goldburg et al., 1983).

**Example 9-2.** Given the fractional flow curve (Figure 9-16) constructed using the relative permeability data from the DOE report (Appendix C), determine the oil production curve and compare it with the field results using these data:

| | |
|---|---|
| capillary number | $N_c = 2.7 \times 10^{-3}$ |
| MP displacement efficiency | $E_D = 0.81$* |
| surfactant retention | $D_S = 0.156$ |
| residual oil saturation after chemicals | $S_{orc} = 0.05$ |

SOLUTION

**The Specific Velocity.** The micellar-polymer flood generates an oil bank of constant saturation $S_{ob}$ and fractional flow $f_{os} = f_o(S_{ob})$. The oil bank is driven by a surfactant front having a specific velocity

$$V_s = \frac{1}{1 + D_s - S_{orc}} \tag{9-14}$$

* Assuming water-wet reservoir, and using $N_c$ and Figure 9-11.

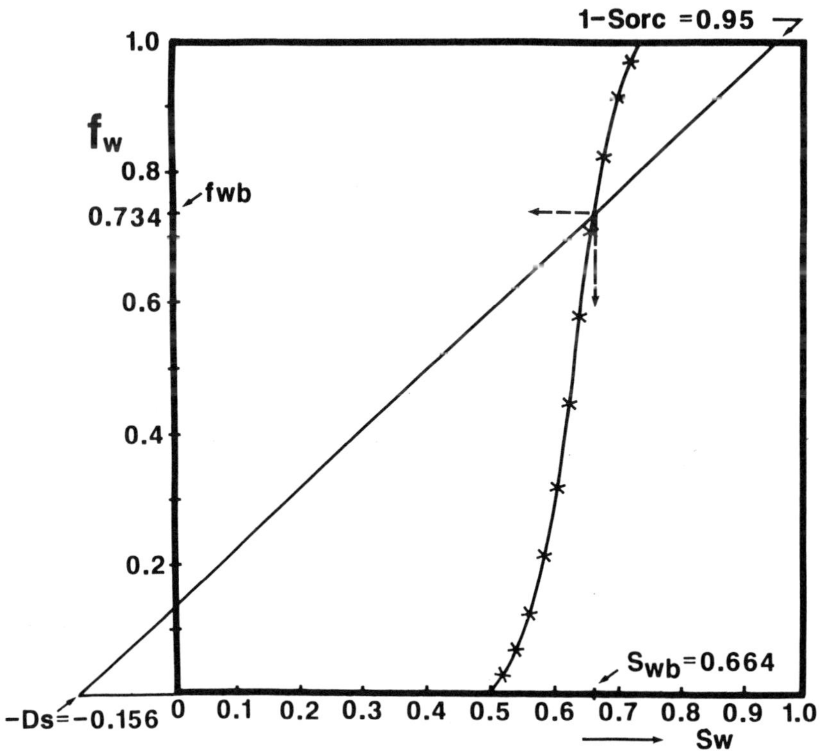


**Fig. 9-16** Fractional flow curve—Loudon Pilot (Data from Goldburg et al., U.S. Department of Energy, 1983)

which can also be expressed in terms of the oil saturation and fractional flow change at the rear of the oil bank (Goldburg et al., 1983)

$$V_s = \frac{f_{ob}}{S_{ob} - S_{\text{orc}}} = \frac{1 - f_{wb}}{1 - S_{wb} - S_{\text{orc}}} \tag{9-15}$$

Equations 9-14 and 9-15 give a relationship for the oil bank saturation which must be solved simultaneously with the water-oil fractional flow curve $f_w = f_w(S_w)$. The graphical solution (Pope, 1980) is given by the intersection of a straight line passing through the points $f_w = 0$, $S_w = -D_s$ and $f_w = 1$, $S_w = 1 - S_{\text{orc}}$ with the fractional flow curve (Figure 9-16). The intersection point coordinates are $S_{wb} = 0.664$ and $f_{wb} = 0.734$. The specific velocity of the surfactant front

$$V_s = \frac{1}{1 + 0.156 - 0.05} = 0.90$$

The specific velocity of the oil bank front

$$V_{ob} = \frac{1 - f_{wb} - f_{oi}}{1 - S_{wb} - S_{oi}} \tag{9-16}$$

where for tertiary applications $S_{oi} = S_{or}$ and $f_{oi} = 0.015$ (Loudon Test)

$$V_{ob} = \frac{1 - 0.734 - 0.015}{1 - 0.664 - 0.255} = 3.09$$

To consider reservoir heterogeneity, an empirical function $M_e$ of the Dykstra–Parsons coefficient ($V_{\text{DP}}$) was developed

$$\log(M_e) = \frac{V_{\text{DP}}}{(1 - V_{\text{DP}})^{0.2}} \tag{9-17}$$

$$M_e = 2.94 \qquad \text{for } V_{\text{DP}} = 0.42$$

The oil production curve is triangular since it can be described with four variables: the oil breakthrough time, $t_{D_{ob}}$, the time of peak oil rate, $t_{D\text{S}}$, the peak oil rate, $f_{\text{opk}}$, and the time of zero rate, $t_{D\text{SW}}$.

The four variables are given by the equations

$$t_{D_{ob}} = (V_{ob} M_e)^{-1} \tag{9-18}$$

$$t_{D\text{S}} = (V_s M_e)^{-1} \tag{9-19}$$

$$f_{\text{opk}} = f_{ob} M_e \left[ \frac{1 - (V_s/V_{ob})^{0.5}}{M_e - 1} \right] \tag{9-20}$$

$$t_{D\text{SW}} = t_{D_{ob}} + 2E_{\text{R}} \frac{S_{\text{orw}}}{f_{\text{opk}}} \tag{9-21}$$

So the *oil breakthrough time* is

$$t_{D_{\text{ob}}} = (3.09 \times 2.94)^{-1} = 0.11 \text{ PV}$$

the *time of peak oil rate* is

$$t_{D\text{S}} = (0.9 \times 2.94)^{-1} = 0.38 \text{ PV}$$

and the *peak oil rate* (*cut*) is

$$f_{opk} = (1 - 0.734)2.94\left[\frac{1 - (0.9/3.09)^{0.5}}{2.94 - 1}\right] = 0.1855 \sim 0.19$$

The calculated triangular oil production curve plotted along with the field curve (Figure 9-6) is given in Figure 9-17.

As we can observe, the comparative results are good but miss on the peak rate location and value. The CFPM might be used to perform sensitivity analysis prior to more costly, fully compositional simulations.

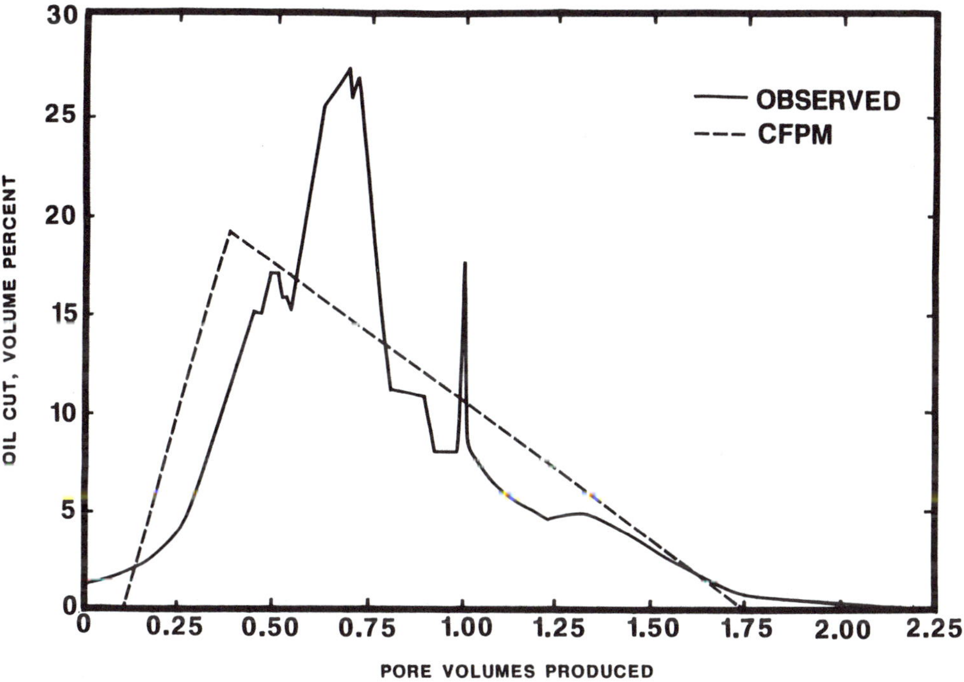


**Fig. 9-17** Oil production—Loudon Pilot (From Goldburg et al., 1983)

## QUESTIONS AND PROBLEMS

**9-1** What are the components of a micellar solution and how does it operate with reservoir fluids?

**9-2** Describe the micellar solution behavior—regarding its mobility—in the porous medium.

**9-3** Explain why the oil recovery obtained in the M-1 Field micellar-polymer project was lower than anticipated.

**9-4** Find the target oil and the recoverable oil of a waterflooded reservoir which has characteristics as follows:

| | |
|---|---|
| Productive area | 720 acres ($2.9 \times 10^6$ $m^2$) |
| Net thickness | 100 ft |
| Porosity | 26% |
| Irreducible water saturation | 29% |
| Formation volume factor for oil | |
| Initial | 1.26 |
| Present | 1.15 |
| Oil recovery (present) | 46% |
| Capillary number | $3 \times 10^{-3}$ |
| Oil saturation in | |
| Waterflooded areas | 25% |
| Unswept regions | 66% |

**9-5** Determine whether or not an MP system can be formulated for the reservoir whose characteristics are given in Example 9-1, assuming the residual oil saturation after waterflood is 31 percent, the oil saturation in unswept regions is 60 percent, and the oil viscosity in reservoir conditions is 2 cp.

**9-6** Rework Example 9-2 using the same characteristic values.

## REFERENCES

ATKINSON, H., U.S. Patent #1651311 (1927).

BRAGG, J. R., W. W. GALE, JR., et al., "EXXON Production Company: Loudon Surfactant Flood Pilot Test," Society of Petroleum Engineers, SPE/DOE 10862 Paper, presented at the 1982 SPE/DOE Joint Symposium on Enhanced Oil Recovery, Tulsa, Oklahoma, April 4–7, 1982.

COLE, L. E., "An Evaluation of the Robinson M-1 Commercial Scale Demonstration of Enhanced Oil Recovery by Micellar-Polymer Flood," Prepared by K&A Technology for the U.S. Department of Energy under Contract No. AC-19-85BC10830-10 (Bartlesville, OK: U.S. Department of Energy, December 1988).

DAUBEN, DWIGHT L., "A Review of the Loudon Surfactant Flood Pilot Test," work performed under contract No. AC-19-85BC10830-8 for U.S. Department of Energy (Bartlesville, OK: U.S. Department of Energy, November 1988).

DAVIS, J. A., and S. C. JONES, "Displacement Mechanisms of Micellar Solutions," *Journal of Petroleum Technology* (December 1969).

GOGARTY, W. B., "Miscible-Type Waterflooding: The Maraflood Oil Recovery Process," SPE of AIME: SPE 1847-A, 42nd Annual Fall Meeting, Houston, Texas, October 1967.

GOGARTY, W. B., and W. C. TOSCH, "Miscible-Type Waterflooding: Oil Recovery with Micellar Solution," *Journal of Petroleum Technology* (December 1968).

GOLDBURG, A., H. PRICE, G. W. PAUL, and T. C. WESSON, "Selection of Reservoirs Amenable to Micellar Flooding," Final Department of Energy Report, October 1978–December 1982, work performed by Department of Energy (U.S. DOE/BC00048-29), August 1983.

GUPTA, S. P., and S. TRUSHENSKI, "Micellar Flooding—Compositional Effects on Oil Displacement," *Society of Petroleum Engineers Journal*, Vol. 19 (1979), p. 116.

HEALY, R. N., R. L. REED, and C. W. CARPENTER, JR., "A Laboratory Study of Microemulsion Flooding," *SPE Journal* (February 1975).

HOLM, L. W., "Use of Soluble Oils for Oil Recovery," *Journal of Petroleum Technology* (December 1971).

HOLM, L. W., "Status of Micellar-Polymer Field Tests—Another View," *Petroleum Engineering International* (April 1980), pp. 100–16.

JONES, L. W., U.S. Patent #3126952 (March 31, 1964).

LAKE, LARRY W., *Enhanced Oil Recovery* (Englewood Cliffs, NJ: Prentice-Hall, 1989), Chapter 9.

LAKE, L. W., and G. A. POPE, "Status of Micellar-Polymer Field Test," *Petroleum Engineering International* (November 1979), pp. 38–60.

LATIL, M., *Enhanced Oil Recovery* (Paris: Edition Technip; distributed in United States by Gulf, Houston, TX, 1980), pp. 212–13.

LOWRY, P. H., H. H. FERRELL, and D. L. DAUBEN, "A Review and Statistical Analysis of Micellar-Polymer Field Test Data," Topical Report DOE/BC/10830-4 (Washington, D.C.: U.S. Department of Energy, November 1986).

PAUL, G. W., L. W. LAKE, and G. A. POPE, "A Simplified Predictive Model for Micellar-Polymer Flooding," SPE 10733, presented at the 1982 California Regional Meeting of the Society of Petroleum Engineers, San Francisco, California, March 24–26, 1982.

POPE, G. A., "The Application of Fractional Flow Theory to Enhanced Oil Recovery," *SPE Journal* (June 1980), pp. 191–205.

STOVER, D. F., "Commercial Scale Demonstration Enhanced Oil Recovery by Micellar-Polymer Flood," Final Report, prepared by Marathon Oil Company for the U.S. Department of Energy under Contract No. AC-19-78ET13077-130 (Bartlesville, OK: U.S. Department of Energy, November 1988).

TRUSHENSKI, S. P., D. L. DAUBEN, and D. R. PARRISH, "Micellar Flooding-Fluid Propagation, Interaction and Mobility," *SPE Journal* (December 1974).

U.S. DEPARTMENT OF ENERGY, "Selection of Reservoir Amenable to Micellar Flooding," First Annual Report DOE/BC/00048-20 (Bartlesville, OK: U.S. Department of Energy, December 1980).

# Chapter 10

# Carbon Dioxide Flooding

The use of carbon dioxide ($CO_2$) to increase the recovery of oil has received considerable attention since 1950. Laboratory research has been conducted and field applications have been initiated and performed indicating a great interest in $CO_2$ flooding.

To understand why $CO_2$ emerged as an important injection agent in oil reservoirs, we should note its principal properties and the factors that make it a useful tool in enhancing oil recovery.

## 10-1 PROPERTIES OF $CO_2$

Carbon dioxide is a colorless, odorless, inert, and noncombustible gas. It has a molecular weight of 44.01, which is one and a half times higher than that of air. The phase behavior of pure $CO_2$ is shown in Figure 10-1 on a P-T diagram.

$CO_2$ is solid at low temperatures and pressures. The solid carbon dioxide (dry ice) evaporates directly to gas at $-78.5°$ C ($-110.7°$ F) and is used primarily as a refrigerant. By increasing the pressure the liquid phase appears for the first time and coexists with the solid and vapor phases at the triple point:

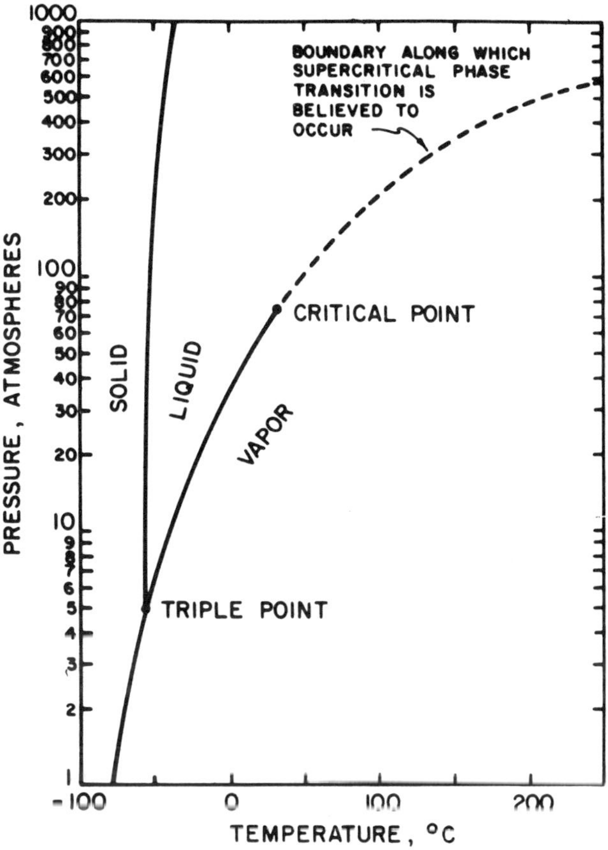


**Fig. 10-1** Carbon dioxide phase diagram (From van Poollen, 1980)

triple point temperature $T_{tr} = -56.6°$ C ($-70.9°$ F), triple point pressure $P_{tr} = 5.28$ atm (77.6 psia). $CO_2$ is usually transported as a liquid in refrigerated trucks or tank cars when it can be utilized in small amounts. The liquid and the vapor phases of $CO_2$ coexist at the critical point: critical point temperature $T_c = 30.7°$ C (87.8° F), critical point pressure $P_c = 73$ atm (1073 psia). Below the critical temperature $CO_2$ can be either a liquid or a gas over a wide range of pressures. Above the critical temperature of 87.8° F, $CO_2$ (pure) will exist as a gas regardless of the pressure applied. However, at increasingly higher supercritical pressures the vapor becomes and behaves more like a liquid. Most $CO_2$ pipelines operate in the supercritical range (van Poollen, 1980).

$CO_2$ density varies with pressure and temperature as does its viscosity and compressibility factor (Sage et al., 1955; Kennedy et al., 1961).

The density of

$CO_2$ (solid) $\rho_s = 1.512$ g/cm$^3$ (12.59 lb/gal) at triple point

$CO_2$ (liquid) $\rho_l = 0.914$ g/cm$^3$ at 0° C (32° F) and 34.3 atm (504.2 psia)

$CO_2$ (gas) $\rho_g = 1.9768$ g/L (0.0164 lb/gal) at 0° C and 1 atm (14.7 psia)

The specific gravity (relative density) of $CO_2$ (gas) $d = 1.529$ (air = 1) at 0° C and 1 atm. Under many reservoir conditions the densities of $CO_2$ and oil are similar.

The viscosity of $CO_2$ is 0.0335 cp at the critical conditions of pressure and temperature, the critical compressibility factor is 0.275, and the specific heat (liquid) capacity at 300 psi is 0.5 Btu/lb-°F. The $CO_2$ is more soluble in oil than in water (2 to 10 times more). Water also is soluble in $CO_2$ and must be removed by drying to prevent condensation and corrosion in the pipelines.

In solution with water $CO_2$ increases water viscosity and forms carbonic acid, which has a beneficial effect on shaley rocks (reduction in pH stabilizes) and on calcareous rocks (dissolving effect).

$CO_2$ is, from a chemical point of view, a final oxidation product of organic compounds with carbon, has acid character being the anhydrite of carbonic acid, and reacts with bases to form carbonates and bicarbonates. The principal $CO_2$ properties such as its acid function, its chemical inertia (protection gas, inert gas, pressure gas agent), its refrigerant function, and its high specific heat capacity explain the multiple utilization of $CO_2$ in different industries and processes.

$CO_2$ has many uses. It is frozen to produce dry ice, it is used to sterilize organic liquids, and it is used in cryogenics, foodstuffs manufacturing, refrigeration, and beverage carbonation. In the area of industry, $CO_2$ is used in heat transfer in nuclear reactors, in welding, in the manufacture of fertilizer and plastics, in neutralization of wastes, and in the manufacture of fire extinguishers. In medical and pharmaceutical applications it is found in salycilic acid for aspirin, mineral waters, and aerosol propellants, and it is used in cryogenic surgery. $CO_2$ is also used in pneumatic conveyor systems for coal and grain slurry lines, in the manufacture of white lead, for oil and gas stimulation, and for tank cleaning.

## 10-2 FACTORS THAT MAKE $CO_2$ AN EOR AGENT

Carbon dioxide is highly soluble in oil and to a lesser extent in water. This results in the following factors which contribute to enhanced oil recovery:

- Reduction in crude oil viscosity and increase in water viscosity
- Swelling of crude oil and reduction in oil density
- Acid effect on carbonate and shaley rocks
- Miscibility effects

*Oil viscosity is reduced* significantly when $CO_2$ is dissolved in crude.

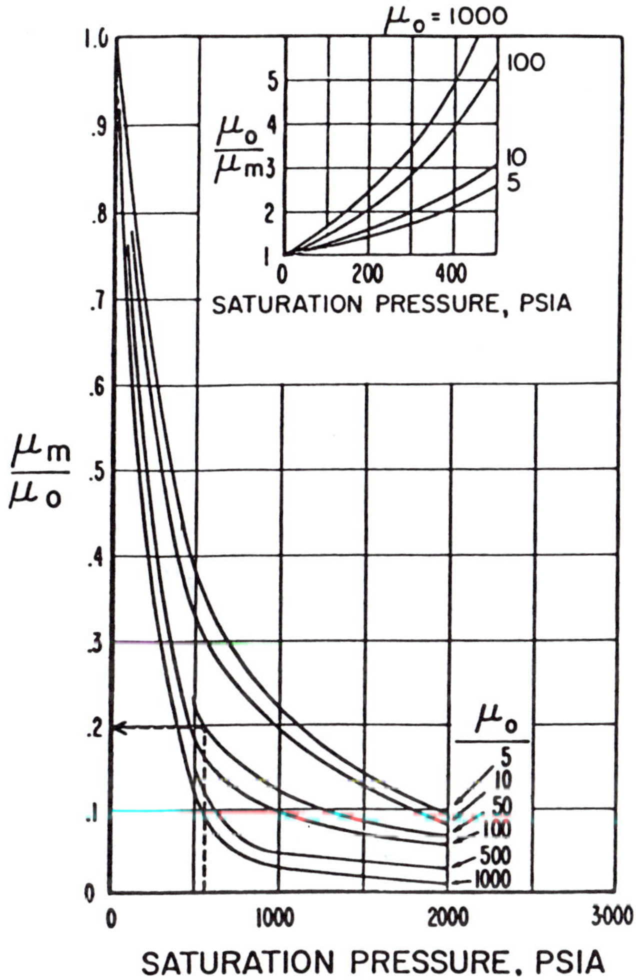


**Fig. 10-2** Viscosity reduction versus saturation pressure (From Simon and Graue, 1965)

Figure 10-2 shows data from the work of Simon and Graue (1965). The symbol $\mu_o$ refers to the original viscosity of the oil while $\mu_m$ refers to the viscosity of that oil after saturation in $CO_2$ at equilibrium. The viscosity reduction, for example, is 10-fold at 2000 psia saturation pressure for an oil with original viscosity of 5 cp (5 mPa·s). This reduction in crude oil viscosity and an accompanying small increase in water viscosity reduces the water-oil mobility ratio.

**Swelling of crude oil.** As a result of $CO_2$ dissolved in the crude, the oil's volume will increase from 10 to 20 percent or more. Figure 10-3 gives the crude oil swelling factor (volume of $CO_2$-saturated crude at saturation pressure and

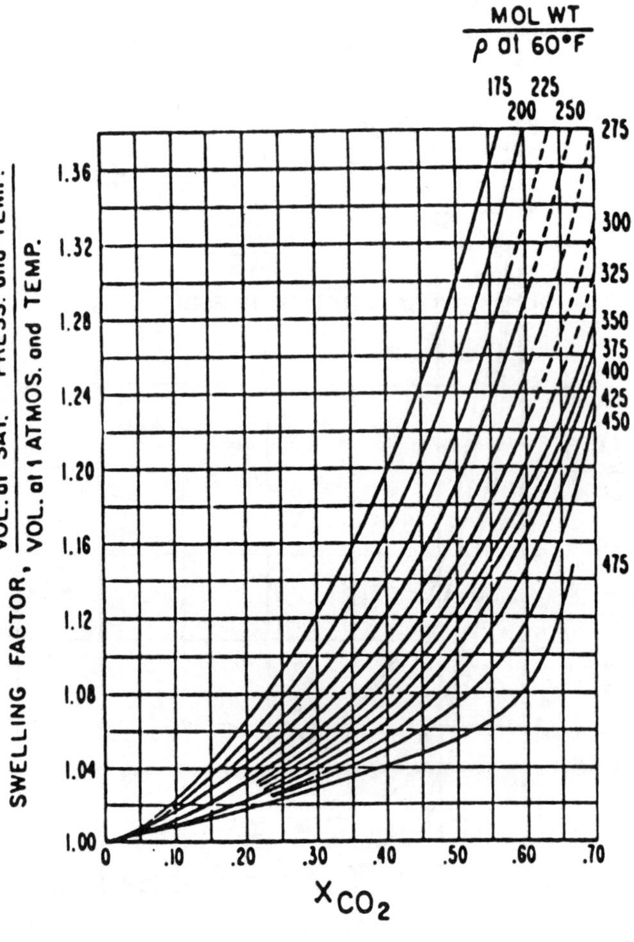


**Fig. 10-3** Swelling factor for oils (From Simon and Graue, 1965)

temperature/volume of $CO_2$-free crude at the same temperature) as a function of the mole fraction of $CO_2$ dissolved $XCO_2$ and the molecular weight of the oil (Simon and Graue, 1965).

An example of how the volume of a West Texas reservoir fluid increased when saturated with $CO_2$ at various temperatures is shown in Figure 10-4 (Holm and Josendal, 1974).

Oil swelling increases the recovery factor since, for a given residual oil saturation, the mass of the oil remaining in the reservoir and expressed in standard conditions is lower than if the abandoned oil was $CO_2$ free.

**Acid effect on carbonate and shaley rocks.** Carbon dioxide in solution with water forms carbonic acid, which in turn, dissolves the calcium and magnesium carbonates. This action increases the permeability of the carbonate rock, improving the well injectivity and in general the fluid flow through the

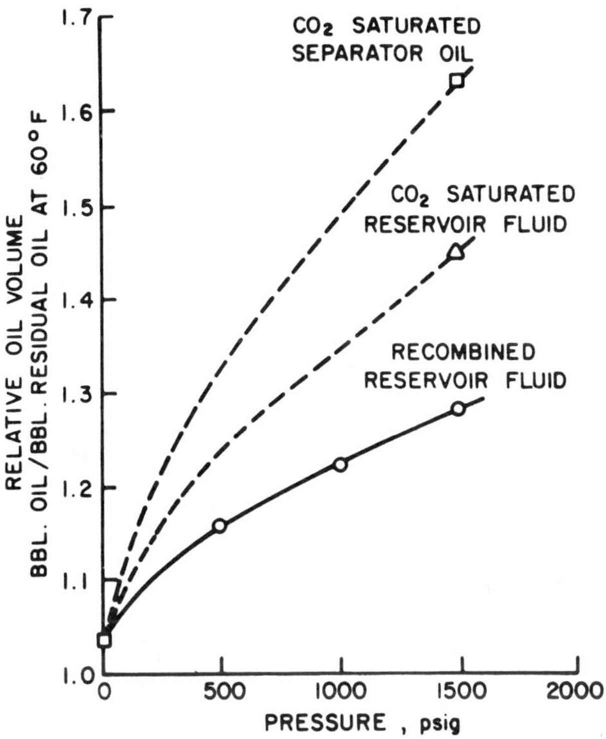


**Fig. 10-4** Relative oil volume versus pressure at 144 °F, west Texas reservoir fluid (From Holm and Josendal, 1974)

reservoir. $CO_2$ has a stabilizing effect on shaley rocks, reducing the pH and preventing the shales from swelling and causing blockage of the porous medium.

**Miscibility effects.** Carbon dioxide is not first-contact miscible with reservoirs oils. $CO_2$ may develop miscibility through multiple contacts under specific conditions of pressure and temperature and with specific oil compositions.

## 10-3 $CO_2$ MISCIBLE FLOODING

### Multiple-Contact Miscibility

Dynamic (multiple-contact) miscibility of $CO_2$ with light- and medium-gravity crude oils is generated as a vaporizing gas drive mechanism. $CO_2$ at appropriate pressures vaporizes or extracts heavier hydrocarbons ($C_5$ through $C_{30}$) from the oil and concentrates them at the displacement front where miscibility is

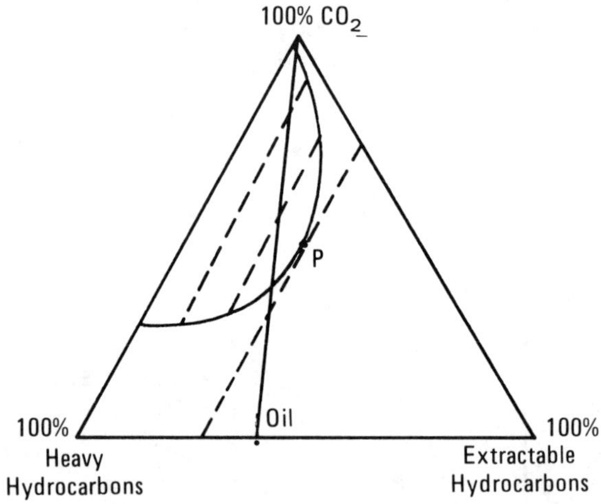


**Fig. 10-5** Pseudoternary diagram vaporizing gas drive process with $CO_2$ (Adapted from Stalkup, 1984)

achieved. Dynamic miscibility with $CO_2$ is possible through a vaporizing gas drive mechanism for reservoir fluid compositions lying to the right of the limiting tie line on a pseudoternary diagram (Figure 10-5).

The difference between the vaporizing gas drive mechanism with $CO_2$ and with natural gas (methane) is that dynamic miscibility with $CO_2$ does not require the presence of intermediate-molecular-weight hydrocarbons in the reservoir fluid. The extraction of a broad range of hydrocarbons from the reservoir oil often causes dynamic miscibility to occur at attainable pressures which are lower than the miscibility pressure for dry hydrocarbon gas (Stalkup, 1984).

### Miscibility Pressure

The minimum miscibility pressure (MMP) above which dynamic miscible displacement with $CO_2$ is possible can be determined from displacement techniques and miscibility experiments. A review of methods for determining miscibility conditions is given by Stalkup (1984). The recommended methods for conducting and interpreting displacement techniques for determining miscibility conditions are gravity-stable experiments, slim tube experiments, and visual cell observations. Also, there have been several attempts to develop correlations from experimental studies of factors affecting $CO_2$ miscibility pressure.

*Gravity-stable experiments* use a vertically sandpacked and oil-saturated test column in which $CO_2$ injected at the top displaces the oil vertically downward at a rate slow enough to maintain a gravity-stable flow, overcoming the $CO_2$'s tendency to protrude into the oil. The experiment is run at different

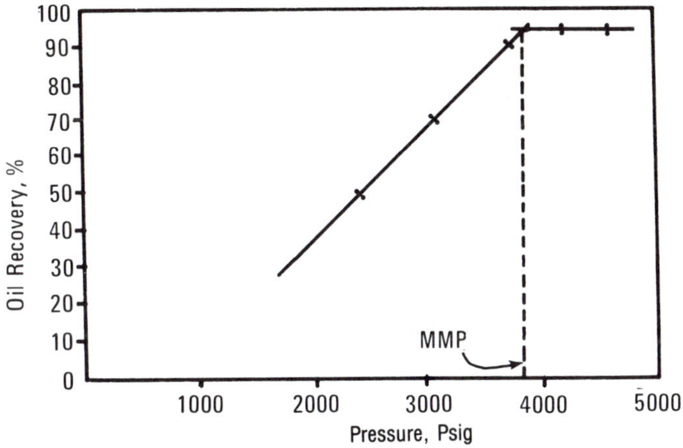


**Fig. 10-6** The minimum miscibility pressure reached in a vertically test column

increasing pressures, and the ultimate oil recoveries (percentage of OOIP) obtained are plotted versus the respective pressures on the same graph (Figure 10-6). The minimum miscibility pressure is the pressure corresponding to the breakover point on the ultimate oil recovery curve.

*Slim tube experiments* are performed in a 40-ft-long, $\frac{1}{4}$-in.-diameter coiled stainless steel tube sandpacked and saturated with oil at a given pressure and temperature. The slim tube apparatus is provided with a capillary tube sight glass (Yelling and Metcalfe, 1980). The combination of small tube diameter, long tube length, and low flow velocity minimizes and suppresses $CO_2$ fingering. The results obtained for a series of $CO_2$ flood experiments in a slim tube apparatus are given in Figure 10-7. The plot of oil recovery versus pressure, after 1.2 PV of injection, shows a sharp break in the recovery curve, indicating the change from immiscible to miscible displacement as pressure is increased.

*Visual cell observations* can describe the gradual color change of the single-phase effluents as the transition zone changes from displaced oil to injected $CO_2$. The interpretation of visual cell observations for $CO_2$ floods is not simple, requires an experienced observer, and may not describe all situations (Stalkup, 1984).

*Correlations* for estimating miscibility pressure have been made since reservoir temperature, oil composition, and characteristics are factors affecting this pressure. As stated by Stalkup (1984), there is a consensus that temperature is an extremely important parameter and that light ends in the oil (methane and nitrogen) and intermediate-molecular-weight hydrocarbons (ethane, propane and butane) have a small effect on $CO_2$ miscibility pressure.

Figure 10-8 shows the Holm and Josenthal correlation (1974) as extended by Mungan (1981). The oil composition, which is characterized by the molecular weight of the pentanes and heavier fractions, and reservoir temperature

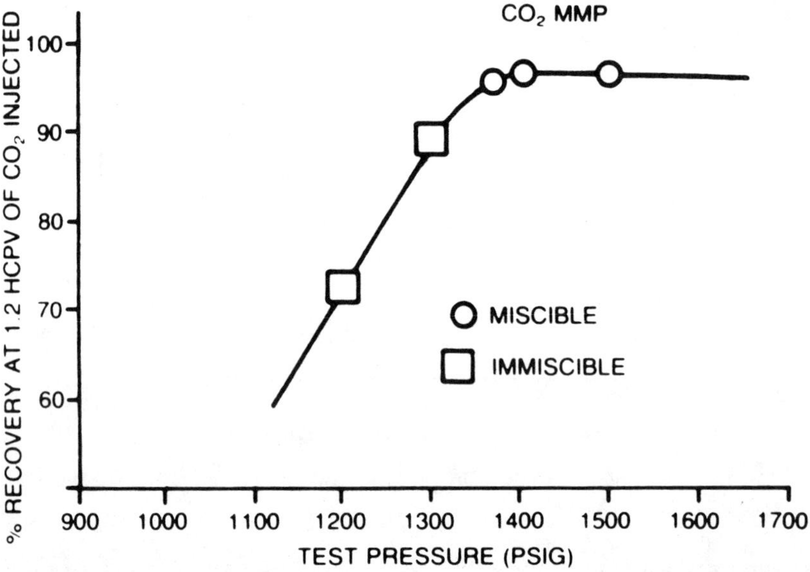


**Fig. 10-7** Test results for fixed oil composition and fixed temperature (From Yelling and Metcalfe, 1980)

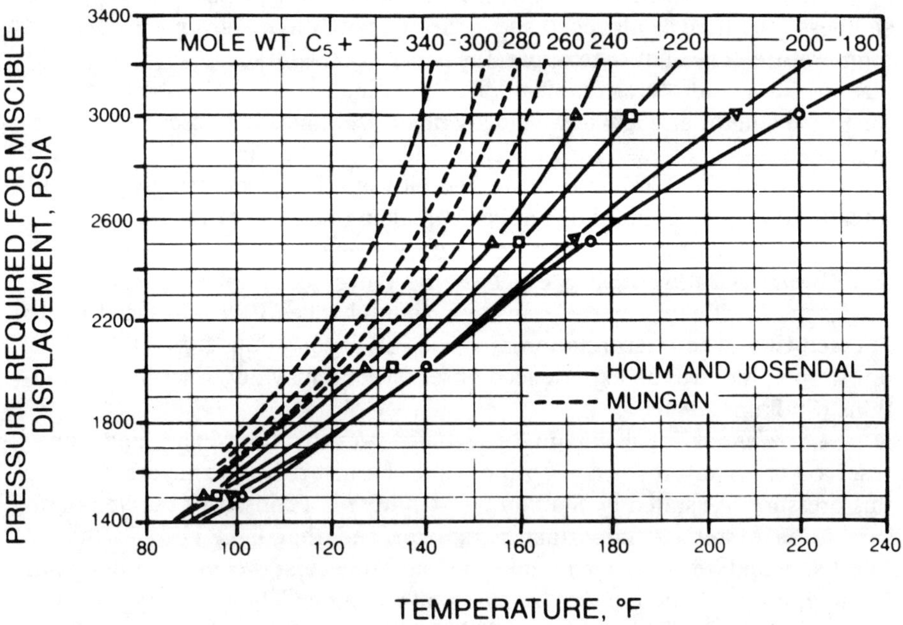


**Fig. 10-8** Pressure required for miscible displacement in $CO_2$ flooding (From Holm and Josendal, 1974, and Mungan, 1981)

are the correlating parameters. Given the complexity of the phenomena and the fact that miscibility pressure is affected by various contaminants in the $CO_2$, it is difficult to predict the miscibility pressure for all reservoirs accurately. The correlations should only be used for screening and preliminary calculations.

## 10-4 $CO_2$ IMMISCIBLE FLOODING

Immiscible $CO_2$-oil displacement is best suited to medium and heavy oils since the oil viscosity reduction is greater and more significant. The $CO_2$ flooding process involves alternating injections of $CO_2$ and water until a certain amount of $CO_2$ has been injected, then water is injected continuously. The water-alternating-gas (WAG) process is characterized by an improved mobility ratio and additional recovery over that of waterflooding without $CO_2$. In addition, the swelling effect of crude oil with $CO_2$ increases the oil formation volume factor so that residual oil behind the waterflood is smaller in volume at surface conditions. Also, oil swelling within the pore spaces displaces water out of the pores, resulting in a decrease in the wetting phase saturation (drainage process). For water-wet porous media, relative permeabilities of the drainage oil are higher than imbibition values, thus favoring additional oil recovery (Mungan, 1983).

## 10-5 DESIGN CONSIDERATIONS

### General

The field of reservoir engineering of miscible flood design and performance prediction is very large and complex. As Stalkup stated (1984) in his detailed SPE monograph, an accurate reservoir description and the extent of miscible sweepout throughout the total swept volume should be taken into account when designing a miscible flood project. The miscible sweepout is affected by many factors such as pore volumes of solvent and drive fluid injected, pressure distribution, size of the solvent slug, type of drive fluid, mobility of the solvent, drive fluid and reservoir fluid, and the displacement efficiencies achieved in both miscible and immiscible swept areas. Laboratory tests to determine the miscibility conditions have been performed. Physical and numerical reservoir models to predict the behavior and flow of fluids in the reservoir have been developed (Stalkup, 1984; Mungan, 1983; Christian et al., 1981; Bilhartz et al., 1978) and continue to be improved.

Before any calculations are made one must consider whether displacement can be miscible or immiscible and whether flow is vertical or horizontal. If the crude oil gravity is medium or light and the reservoir is deep or of medium depth, then miscible displacement should be considered. When the reservoir is of medium or shallow depth with medium or heavy gravity oil the miscibility pressure, if it exists, cannot be reached without surpassing the formation parting pressure. In this case displacement has to be immiscible but will include the beneficial effects of viscosity reduction and oil swelling.

The direction of displacement depends on reservoir geometry and characteristics. The *horizontal displacement* developed in nondip reservoirs with not so thick pay zone is controlled by the mobility ratio of $CO_2$ to oil. To avoid or diminish $CO_2$ fingering, the process should probably use alternating $CO_2$-water injection. *Vertical displacement* is a characteristic of pinnacle reef or salt dome flank reservoirs and is controlled by the effect of gravity. To have a gravity-stable process, upward vertical displacement is achieved using water as a chasing fluid. Downward displacement is accomplished by using gas.

## Flood Design and Performance Predictions

$CO_2$ flood design and performance predictions differ from reservoir to reservoir and for different operation strategies. However, there are features of $CO_2$ flooding that are more or less common to any field application. Examples of some of these features are presented in the paragraphs that follow.

*Reservoir pressurization* is an initial phase of the $CO_2$ miscible flood in which the reservoir pressure, reduced by primary depletion or some other recovery mechanism, has to be increased so that miscible displacement can occur. The reservoir pressure is usually increased by injecting water.

**Example 10-1.** Water is to be injected at an available rate of 12,580 bbl/day (2000 $m^3$/day) for reservoir pressurization. Although the original reservoir pressure of 2143 psia was above the MMP of 2114 psia, primary depletion had reduced the reservoir pressure to 1143 psia. The decision was made to return the reservoir to its original pressure before the start of $CO_2$ injection.

Calculate the total amount of injected water, $W$, and the time necessary for reservoir pressurization. The following data are known:

**RESERVOIR DATA:**

| | |
|---|---|
| Oil formation volume factor | $B_{oi}$ = 1.53 at initial pressure |
| | $B_o$ = 1.33 at actual pressure |
| Gas formation volume factor | $B_{oi}$ = 0.010 at initial pressure |
| | $B_g$ = 0.014 at actual pressure |
| Solution ratio | $R_{si}$ = 778 scf/bbl (137 $m^3/m^3$) at initial pressure |
| | $R_s$ = 522 scf/bbl (92 $m^3/m^3$) at actual pressure |

**PRODUCTION DATA:**

| | |
|---|---|
| Cumulative oil produced | $N_p$ = 2.516 × $10^6$ bbl (0.4 × $10^6$m$^3$) |
| Actual recovery factor | $E_R$ = 15% of OOIP |
| Cumulative water produced | $W_p$ = 14 × $10^4$ bbl (2.3 × $10^4$m$^3$) |
| Oil production rate (actual) | $q_o$ = 1352 STB/day (215 m$^3$/day) |
| Gas oil ratio | GOR (average) = 200 STB/bbl (m$^3$/m$^3$) |
| Water production rate (actual) | $q_w$ = 126 bbl/day (20 m$^3$/day) |

SOLUTION

To pressurize the reservoir it is necessary to inject a total amount of water, $W$, that corresponds to the cumulative volume of fluids already produced $F_p$, plus the amount of fluids that will be produced during the pressurization phase, $F_{pp}$.

$$W = F_p + F_{pp} \times \text{pressurization time} \tag{10-1}$$

1. Calculate $F_p$. After 2.516 × $10^6$ bbl of oil has been produced the reservoir pore volume left free, $V_g$, is occupied by gas at actual reservoir pressure. Pressurization of these gases causes them to redissolve and occupy the volume $V_{gs}$. Also considering the amount of water produced, $W_p$, we have

$$V_g = NB_{oi} - (N - N_p)B_o \tag{10-2}$$

where $N$ is the original oil in place

$$V_{gs} = \frac{V_g}{B_g} \times \frac{1}{(R_i - R_s)/(B_{oi} - B_o)} \tag{10-3}$$

and

$$F_p = V_g - V_{gs} + W_p \tag{10-4}$$

$$V_g = \frac{2.516 \times 10^6 \text{ bbl}}{0.15} \times 1.53 - (16.773 - 2.516)$$

$$\times 10^6 \text{ bbl} \times 1.33 = 6.7 \times 10^6 \text{ bbl}$$

$$V_{gs} = \frac{6.7 \times 10^6 \text{ bbl}}{0.014} \times \frac{5.614 \text{ scf/bbl}}{(778 - 522) \text{ scf/bbl}/(1.53 - 1.33)} = 2.1 \times 10^6 \text{ bbl}$$

$$F_p = 4.74 \times 10^6 \text{ bbl}$$

2. Calculate $F_{pp}$. The reservoir volume of fluids produced daily, $q_f$, is given by

$$q_f = q_o \overline{B}_o + q_o(\text{GOR} - \overline{R}_s)\overline{B}_g + q_w \tag{10-5}$$

where $\overline{B}_o$, $\overline{B}_g$, and $\overline{R}_s$ are the average formation volume factors and the average solution ratio, respectively.

$\overline{B}_o = 1.43$, $\overline{B}_g = 0.011$, and $\overline{R}_s$ = 650 scf/bbl (114 m$^3$/m$^3$)

between actual and pressurization pressure.

$$q_f = 1352 \times 1.43 + 1352\left(200 - \frac{650}{5.614}\right)0.011 + 126$$

$$= 3312 \text{ bbl/day}$$

From the total daily amount of 12,580 bbl of water injected 3312 bbl are considered necessary to replace the fluids produced during pressurization.

**3.** Calculate the time necessary for pressurization.

$$t_{\text{pressurization}} = \frac{F_p}{q_i - q_f} \tag{10-6}$$

where $q_i$ = 12,580 bbl/day, the available rate of water injection

$$t_{\text{pressurization}} = 511 \text{ days or } 1.4 \text{ years}$$

**4.** Calculate the total amount of water injected:

$$W = 4.74 \times 10^6 \text{ bbl} + 3312 \text{ bbl/day} \times 511 \text{ days}$$

$$= 6.433 \times 10^6 \text{ bbl}$$

Since the reservoir is produced under the dissolved gas drive mechanism, the water-injected downstructure (below the initial water-oil contact) will advance and displace the oil upward. The strategy of $CO_2$ miscible displacement will have to be considered as a secondary operation in zones unswept or poorly swept by water and as a tertiary operation in the waterswept zones.

*$CO_2$ requirements* are determined in many ways depending on the reservoir geometry and displacement direction and on the miscibility conditions and injection strategy.

To illustrate one of the ways to estimate the volume of $CO_2$ needed for injection, the hypothetical simple example given by Mungan (1983) is given next.

**Example 10-2.** Estimate the total amount of $CO_2$ needed for injection into a reservoir whose geometry (pinnacle reef) requires vertical, downward, and gravity-stabilized displacement and in which the $CO_2$ is driven by nitrogen. The following data and information are known:

| | |
|---|---|
| Pilot area | $A$ = 40 acres |
| Average thickness for oil zone | $h$ = 300 ft |
| Average porosity | $\phi = 0.09$ |
| Areal sweep efficiency | $E_a = 1.0$ |
| Vertical conformance | $E_v = 0.8$ |
| Residual oil saturation in $CO_2$-swept zone | $S_{or}$ = 0.05 pore volume |
| Injection rate | $q_i$ = 4000 bbl/day |

It is assumed that at the time of $CO_2$ breakthrough the $CO_2$ displacing bank appears as a mixing zone. The $CO_2$ concentration decreases by diffusion from 90 percent in the middle of the bank to 10 percent at both the $CO_2$/oil downward and $CO_2$/nitrogen upward ends.

The length of the diffusion zone X is given (Perkins and Johnston, 1963) by

$$X = 3.625\sqrt{D_e t} \tag{10-7}$$

where

$t$ = the time that the $CO_2$ front has been moving through the reservoir, seconds

$D_e$ = the effective diffusion coefficient, cm²/s

It is assumed that the diffusion coefficients for $CO_2$/oil and $N_2/CO_2$ are on the same order as for hydrocarbon solvent floods and so

$$D_{c\text{-}o} \simeq 3.5 \times 10^{-5} \text{ cm}^2/\text{s}$$

$$D_{n\text{-}c} \simeq 65 \times 10^{-5} \text{ cm}^2/\text{s}$$

SOLUTION

The volume of $CO_2$ required is

$$V_{CO_2} = V_d + V_s \tag{10-8}$$

where $V_d$ is the amount corresponding to the mixing (diffusion) zone and $V_s$ is the amount that is soluble in the oil and water left behind the displacement front at floodout.

1. Calculate the amount of $CO_2$ in the diffusion zone, $V_d$. The time required for the displacement front to reach the production wells is given (Mungan, 1983) by

$$t = (6.7 \times 10^8)\frac{Ah\phi E_a E_v(1 - S_{or})}{q_i} \text{ s} \tag{10-9}$$

where the number $6.7 \times 10^8$ is the conversion constant for the units used

$$t = 6.7 \times 10^8 \frac{40 \times 300 \times 0.09 \times 1 \times 0.8(1 - 0.05)}{4000}$$

$$= 137.49 \times 10^6 \text{ s}$$

The length of the diffusion (mixing) zone is written twice, once for $CO_2$/oil ($c$-$o$) and once for $N_2/CO_2$ ($n$-$c$):

$$X_{90\text{-}10} = 3.625\left(\sqrt{3.5 \times 10^{-5}} + \sqrt{65 \times 10^{-5}}\right)\sqrt{137.49 \times 10^6}$$

$$= 1337 \text{ cm} \simeq 43.9 \text{ ft}$$

The $CO_2$ volume in the diffusion zone

$$V_d = \frac{A\phi X(7758)}{2} \text{ bbl} \tag{10-10}$$

where the factor $\frac{1}{2}$ is used to integrate the concentration between the 10 and 90 percent values (Mungan, 1983)

$$V_d = \frac{40 \times 0.09 \times 43.9(7758)}{2} = 613{,}037 \text{ bbl } (97{,}473 \text{ m}^3)$$

or 7.37% pore volume

2. Calculate the amount of $CO_2$ dissolved behind the front, $V_s$. The amount of $CO_2$ dissolved in the oil and water left behind the displacement front is calculated knowing the oil and water saturations in the swept and unswept zones, the extent of the zones and the solubilities of $CO_2$ in oil and water under the conditions of reservoir pressure and temperature, respectively.

   Usually, when the residual oil saturation is low (efficient miscible displacement) an amount of 5 to 10 percent pore volume of $CO_2$ is required to saturate the reservoir fluids. Assuming for our example that $V_s = 7.5\%$ pore volume of $CO_2$, the total volume of $CO_2$ required is

$$V_{CO_2} \simeq 15\% \text{ of pore volume}$$

or

$$1.25 \times 10^6 \text{ bbls } (\simeq 0.2 \times 10^6 \text{m}^3) \text{ of } CO_2$$

**$CO_2$ injection pressure.** When $CO_2$ is injected, care must be taken to ensure that the injection pressures are always below formation parting pressure. The surface $CO_2$ injection pressure value is calculated to assure the required miscibility pressure in the reservoir. The pressure exerted by the weight of the column of $CO_2$ gas at high pressure must be taken into consideration.

**Example 10-3.** Calculate the $CO_2$ static wellhead pressure, $P_{ts}$, when the static bottom hole pressure is the miscibility pressure of 2114 psia. The following additional information is available:

| | |
|---|---|
| Bottom hole temperature | $T_R = 170°$ F (76° C) |
| Surface temperature | $T_s = 70°$ F (21° C) |
| $CO_2$ specific gravity | SG = 1.529 (air = 1) |
| $CO_2$ deviation factor | $Z = 0.56$ is assumed to be practically constant between reservoir pressure and temperature range |
| Reservoir depth | $D = 4264$ ft (1300 m) |

SOLUTION

The pressure exerted by the weight of a $CO_2$-gas column under static conditions can be calculated by the average pressure and temperature method (Beggs, 1984). In conventional field units,

$$P_{ws} = P_{ts} \exp\left[\frac{0.01875 \text{ SG}(D)}{TZ}\right] \tag{10-11}$$

where $P_{ws}$ is the static bottom hole pressure, psia, $P_{ts}$ is the static tubing pressure, psia, and $T$ is the average temperature in the tubing, °$R$

$$2114 = P_{ts} \exp\left[\frac{0.01875(1.529)(4264)}{580(0.56)}\right]$$

and the $CO_2$ static wellhead pressure is

$$P_{ts} = \frac{2114}{1.45697} = 1451 \text{ psia}$$

So the pressure exerted by the weight of a $CO_2$-gas column under static conditions of $p$ and $T$ is 663 psia.

**Example 10-4.** Calculate the tubing $CO_2$ injection pressure $P_{ti}$ when the injection bottom hole pressure $P_{wi}$ is 2300 psia. The $CO_2$ injection rate averages $q$ = 1 MM scf/day per injection well. Other data are

- Well tubing inside diameter, $d = 2.441$ in.
- The measured depth MD = TVD (true vertical depth)
- The tubing roughness $n = 5 \times 10^{-4}$ in.
- The $CO_2$ viscosity at 120° F and 2000 psia = 0.05 cp (from Stalkup, 1984). Other data are those given in Example 10-3.

SOLUTION The tubing $CO_2$ injection pressure can be calculated using the same average pressure and temperature method (Beggs, 1984). The equation is

$$p_{wf}^2 = p_{tf}^2 \exp(S) + \frac{25(\text{SG})q^2\,TZf(\text{MD})[\exp(S) - 1]}{Sd^5} \qquad (10\text{-}12)$$

where

$p_{wf}$ = bottom flowing pressure (in this case, $P_{wi}$)
$p_{tf}$ = tubing flowing pressure (in this case, $P_{ti}$)
$S = 0.0375(\text{SG})(\text{TDV})/TZ$
$f$ = friction factor is given by

$$\frac{1}{\sqrt{f}} = 1.14 - 2\log\left(\frac{n}{d} + \frac{21.25}{N_e^{0.9}}\right) \qquad (10\text{-}13)$$

where the Reynolds number $N_e$ is obtained from

$$N_e = \frac{\rho v g}{\mu} = \frac{20{,}011(\text{SG})q}{\mu d} = \frac{20{,}011(1.529) \times 1}{0.05(2.441)} = 250{,}691$$

and

$$\frac{1}{\sqrt{f}} = 1.14 - 2\log\left(\frac{5 \times 10^{-4}}{2.441} + \frac{21.75}{(250{,}691)^{0.9}}\right) = 8.516$$

$$f = 0.01379$$

$$S = \frac{0.0375(1.529)(4264)}{580(0.56)} = 0.7527$$

$$\exp(S) = \exp(0.7527) = 2.1227$$

Now replacing in Eq. 10-12:

$$(2300)^2 = p_{ti}^2 \times 2.1227 + \frac{25(1.529)1^2(580)0.56(0.01379)4264(2.1227 - 1)}{0.7527(2.441)^5}$$

$$P_{ti} = 1577 \text{ psia}$$

The compressor horsepower required is that which is calculated to compress 1 MM scf/day of a 1.529-gravity gas ($CO_2$) from a given pressure and temperature to 1577 psia, plus the pressure drop in the well flow line and surface choke.

## 10-6 $CO_2$ DEMAND, SOURCES, AND TRANSPORTATION

### $CO_2$ Demand

Carbon dioxide miscible displacement is one of the most promising EOR methods. Estimates of potential $CO_2$ demand in each of the four major U.S. basins studied in the DOE report (Anada et al., 1982) show the projected $CO_2$ requirement (based on 300 days injection per year) is as follows:

| | | |
|---|---|---|
| Permian Basin and Texas Gulf Coast | = | 10,011 MM scf/day |
| Williston Basin | = | 236 MM scf/day |
| Appalachian Basin | = | 83 MM scf/day |
| Los Angeles Basin | = | 376 MM scf/day |

The DOE report also analyzed the injected $CO_2$ required per additional barrel of oil recovered. For example, in a secondary recovery process using $CO_2$, 13.5 M scf of $CO_2$ is needed to recover one additional barrel of light oil and 28.4 M scf of $CO_2$ is needed to recover one additional barrel of heavy oil. In a tertiary type of recovery using $CO_2$ (after waterflooding) 16.4 M scf of $CO_2$ is needed to recover one additional barrel of light oil. One can anticipate these figures being lowered to 6.8, 13.9, and 7.8 M scf/bbl, respectively, since over time a percentage of injected $CO_2$ is provided from a recycling process.

### $CO_2$ Sources

A reliable source of supply for $CO_2$ is very important because the gas must be available on a continuous basis in large volumes for long periods of time, between 5 to 10 years or more. The $CO_2$ gas used must have a purity of 90 percent or more. If other gases such as methane or nitrogen are present with the $CO_2$, a higher injection pressure is needed to render the gas miscible with the oil.

The presence of $H_2S$ is undesirable because it is hazardous and detrimental to the environment. The presence of water vapor with the gas leads to corrosion so the $CO_2$ has to be dried.

The best $CO_2$ sources are naturally occurring high-pressure gas reservoirs with high-purity $CO_2$, mostly found while exploring for oil and gas. In the United States the oil producing basins of Wyoming, Utah, Colorado, and New Mexico have the largest $CO_2$ reserves. The economics of a $CO_2$ miscible project are improved if $CO_2$ wells are located in the same geologic basins as those that produce oil, since the $CO_2$ transportation and injection pressure can then be partially supplied by the $CO_2$ reservoir pressure.

However, a DOE report (Anada et al., 1982) shows that currently known naturally occurring high-purity $CO_2$ sources can only provide less than 15 percent of the total $CO_2$ demand. Other natural $CO_2$ sources are natural gas fields containing $CO_2$ which is removed in gas processing plants (Delaware and Val Verde basins of Southwest Texas). These, along with any sources which incidentally produced $CO_2$ as a by-product, are more than sufficient to satisfy the $CO_2$ demand. The sources which produce $CO_2$ as an offgas are coal-, oil-, and gas-fired power plants (which generate large quantities of flue gas containing $CO_2$), cement and ammonia plants, refineries, and ethanol and ethylene oxide plants, among others.

The most convenient $CO_2$ source must be considered for each particular process application, and favorable circumstances must be exploited in time to reduce the technology cost for extraction of $CO_2$ and the cost of compression, transportation, and injection.

### Transportation of $CO_2$

The method of transportation of $CO_2$ from its source to the oil field depends on whether the $CO_2$ is liquid or gas. For small injection rates of 1 to 5 MM scf/day and short injection periods, $CO_2$ is liquefied at its source and transported to the project sites by refrigerated trucks, tank trucks, tank cars on rail, or in storage tanks located on barges. Transporting the $CO_2$ liquid at 0° F and 300 psi using existing insulated steel containers is the least costly method of transportation (Anada et al., 1982). The $CO_2$ necessary for large long-term projects is transported most economically through a pipeline as vapor at pressures between 1400 to 2000 psi (which are above the critical pressure) so that two-phase flow does not occur.

## 10-7 FIELD PROJECTS

Since the early 1970s when higher oil prices began to generate widespread interest in $CO_2$ injection, numerous articles have been written about field projects using $CO_2$ in both miscible and immiscible processes. State-of-the-art reviews of $CO_2$ recovery methods have been successively presented by Holm (1976), Stalkup (1984), Brock and Bryan (1989), and Pautz et al. (1990), among others. Complete reference lists attached to these presentations give a well-oriented image regarding the effort focused on laboratory tests and field applications designed to develop and improve the technologies of using $CO_2$ to increase oil recovery.

The miscible and immiscible carbon dioxide projects started since 1980 are shown in Tables 10-1 and 10-2.

Most of the carbon dioxide miscible projects were started in the Permian Basin of West Texas and New Mexico. Most immiscible projects are in Louisiana and Texas. As can be observed, more than 50 percent of the total $CO_2$ miscible projects and practically all the immiscible $CO_2$ projects started since 1980 are operated by major oil companies in oil reservoirs with a large variety of characteristics (Pautz et al., 1990).

**TABLE 10-1.** Carbon Dioxide Miscible Project Starts, 1980–1986

| Year started | (1) Majors #/Total | (2) Area (acres) | (3) Avg. Depth | (4) Range Depth | (5) Avg. °API | (6) Range °API | (7) #SS/ Proj | (8) Avg. Porosity | (9) Range Porosity |
|---|---|---|---|---|---|---|---|---|---|
| 1980 | 14/26 | 1581 | 5679 | 100–10,400 | 35.7 | 14–45 | 11/9 | 17.0 | 9–37 |
| 1981 | 15/28 | 2430 | 6030 | 1,300–11,530 | 36.4 | 14–44 | 13/11 | 16.9 | 6–37 |
| 1982 | 4/10 | 991 | 7220 | 2,300–13,000 | 34.2 | 14–49 | 5/1 | 17.1 | 8.5–27 |
| 1983 | 4/6 | 6169 | 6724 | 4,900– 8,500 | 38.8 | 33–43 | 1/3 | 14.0 | 8–30 |
| 1984 | 7/8 | 3936 | 6515 | 5,050–13,275 | 33.9 | 28–45 | 1/6 | 13.8 | 6.4–30 |
| 1985 | 8/11 | 5094 | 5981 | 1,270–10,750 | 36.8 | 20–41 | 4/5 | 16.5 | 7.7–29 |
| 1986 | 6/8 | 3410 | 6438 | 800–12,000 | 35.5 | 28–46 | 1/2 | 13.0 | 10–15 |

(1) Number of projects started by major oil companies/total EOR projects started.
(2) Average reported area in acres.
(3) Average depth to top of producing formation in feet.
(4) Shallowest project–deepest project.
(5) Average API gravity.
(6) Range of API gravities.
(7) Number of projects reported in sandstone/number in limestone.
(8) Average reported porosity in %.
(9) Range of porosities in %.
From Pautz et al. (1990)

## Operational Problems

Corrosion, asphaltene deposition, and handling of the produced $CO_2$ are some of the operational problems that are encountered in field applications during $CO_2$ flooding (Mungan, 1983). *Steel corrosion* is a result of the corrosive environment created by the carbonic acid formed when $CO_2$ is in the presence of water. This can be reduced by using dual water and $CO_2$ lines, dehydrating the $CO_2$ before compression and transportation, and using corrosion inhibitor programs. *Asphaltene deposition* can be a serious problem when the crude oil is highly asphaltic and the reservoir permeability is low. Asphaltene precipitation in the producing wells can be successfully cleaned out with soak treatments of about 1000 gal of xylene. In *handling the produced* $CO_2$ reinjection of the gas after separation appears to be the best method. The produced $CO_2$ can also be vented to the atmosphere. Because the density of $CO_2$ is greater than the density of air, careful measures must be taken to prevent $CO_2$ from collecting at lower levels (valleys, ditches, etc.) where it may be harmful to humans and animals, even deadly if the $CO_2$ contains any $H_2S$ at all.

To illustrate how $CO_2$ is used in field applications two case histories with miscible and immiscible oil displacement are presented.

**TABLE 10-2.** Immiscible Carbon Dioxide Project Starts, 1980–1986

| Year started | (1) Majors #/Total | (2) Area (acres) | (3) Avg. Depth | (4) Range Depth | (5) Avg. °API | (6) Range °API | (7) #SS/ Proj | (8) Avg. Porosity | (9) Range Porosity |
|---|---|---|---|---|---|---|---|---|---|
| 1980 | — | — | — | — | — | — | — | — | — |
| 1981 | 1/1 | — | 12,000 | — | 47.0 | — | — | 5.0 | — |
| 1982 | 4/4 | 766 | 5,184 | 3,785– 9,000 | 23.7 | 23–25 | 1/0 | 27.0 | 25–30 |
| 1983 | 15/16 | 540 | 4,948 | 2,600–10,000 | 25.6 | 14–39 | 2/0 | 25.8 | 13–31 |
| 1984 | 10/11 | 231 | 6,745 | 1,300–10,200 | 32.9 | 14–47 | 5/0 | 21.1 | 8–31 |
| 1985 | 11/11 | 2106 | 8,350 | 1,400–13,125 | 33.7 | 26–42 | 6/1 | 22.4 | 4.5–32 |
| 1986 | 20/21 | 264 | 9,728 | 5,200 14,000 | 37.0 | 31–45 | 20/0 | 22.0 | — |

(1) Number of projects started by major oil companies/total EOR projects started.
(2) Average reported area in acres.
(3) Average depth to top of producing formation in feet.
(4) Shallowest project–deepest project.
(5) Average API gravity.
(6) Range of API gravities.
(7) Number of projects reported in sandstone/number in limestone.
(8) Average reported porosity in %.
(9) Range of porosities in %.
From Pautz et al. (1990)

## Miscible $CO_2$ Flood Kelly-Snyder Field, SACROC Unit, Texas, United States

The Kelly-Snyder Field of Scurry County, Texas, with its 50,000 acre ($202.3 \times 10^6$ m$^2$) SACROC unit is the world's largest $CO_2$ miscible flood. Since the start of the project in 1972, many technical papers have been written and presented regarding the reservoir description, the process design and implementation, and the performance evaluation (Smith, 1971; DiCharry et al., 1973; Brummett et al., 1976; Kane, 1979; Stalkup, 1984; Langston, 1988).

**Characteristics.** The oil reservoir, at an average depth of 6700 ft in Canion Reef formation, is an anticline of a series of shelf limestones stratified into layers of porous and dense zones and ranging in thickness from about 10 ft on the flanks to about 800 ft on the crest, with an average gross thickness of 268 ft.

The reservoir rock and fluid properties are summarized in Table 10-3.

**TABLE 10-3.** Reservoir Rock and Fluid Properties, SACROC Unit

| | |
|---|---|
| Initial reservoir pressure @ −4300 ft | 3122 psig |
| Bubble point pressure | 1850 psia |
| Reservoir temperature @ −4300 ft | 130° F |
| Average porosity | 9.41% |
| Average permeability | 19.4 md |
| Average initial water saturation | 21.9% |
| Oil viscosity | 0.35 cp |
| Oil gravity | 41 °API |
| Oil volume factor | 1.5 RB/STB |
| Solution GOR | 1000 scf/STB |
| Water-oil mobility ratio | 0.3 |
| $CO_2$-oil mobility ratio | 8.0 |

From Kane (1979)

**Production history.** The reservoir, discovered in 1948, had 2.113 billion bbl original oil in place and was developed to production through over 1600 wells by 1951. The primary recovery mechanism was fluid expansion and solution gas drive. The rapid reservoir pressure drop to 50 percent of its initial pressure and the corresponding low recovery of 4.5 percent of OOIP are explained by the reservoir oil's unsaturated condition and by an inactive aquifer. To prevent excessive loss of reserves, a water injection pressure maintenance program began in September 1954, with large volumes of water injected through 53 wells initially, up to a total of 144 wells located along the crest into the thickest portion of the reef (Figure 10-9).

At the end of 1971, 771 million bbl [$122.6 \times 10^6$ $m^3$] of water had been injected. The oil recovered was 24.5 percent of OOIP, the reservoir was repressurized to 2350 psi in most areas, and the producing GOR was stabilized at near-solution ratio (Kane, 1979). It is interesting to note that the effect of waterflood on the oil production rate was observed late, starting in 1966, after 12 years of water injection. This late response to waterflooding is explained by the crestal injection made in the thickest portion of the reef and by the low

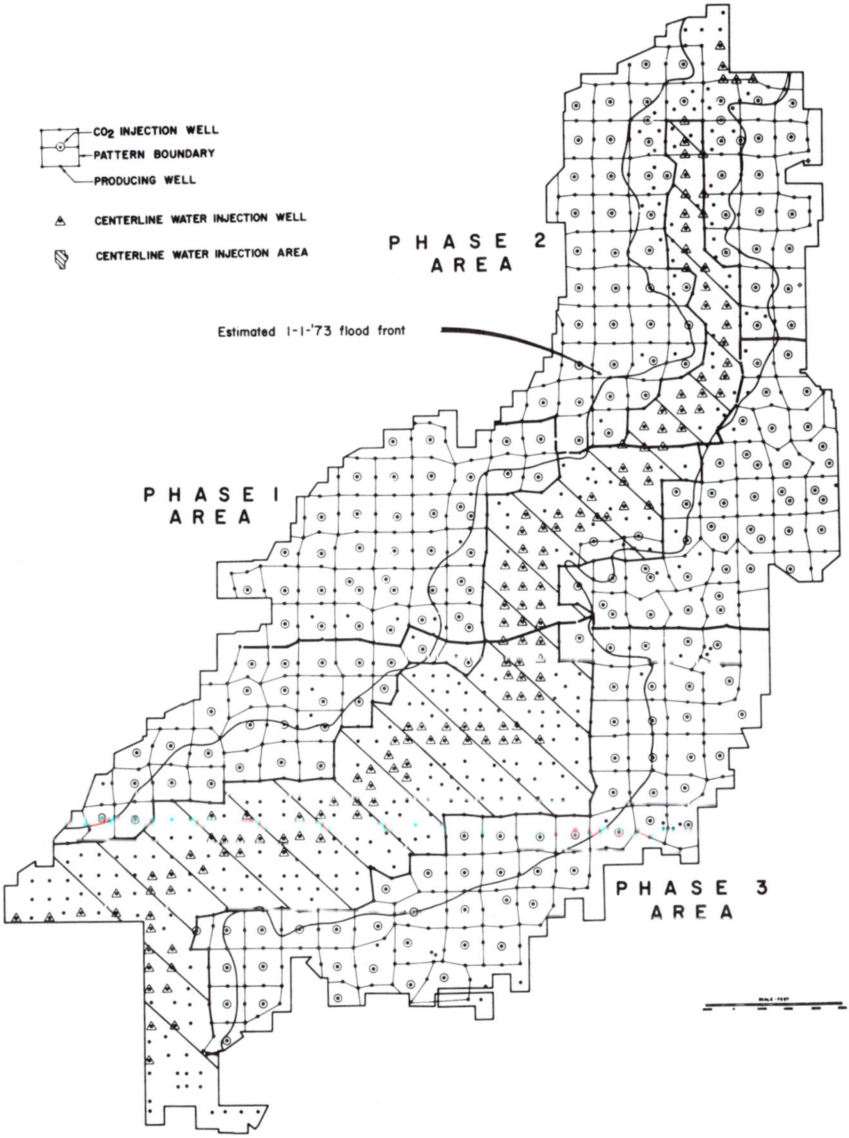


**Fig. 10-9** Map of SACROC unit area (From Kane, 1979)

water injection rate. The oil production rate increased threefold in the following four years from an average of 45,000 bbl/day to 140,000 bbl/day as the amount of water injected was increased threefold in the same period, from 100,000 bbl/day to 320,000 bbl/day by the end of 1971 (Figure 10-10).

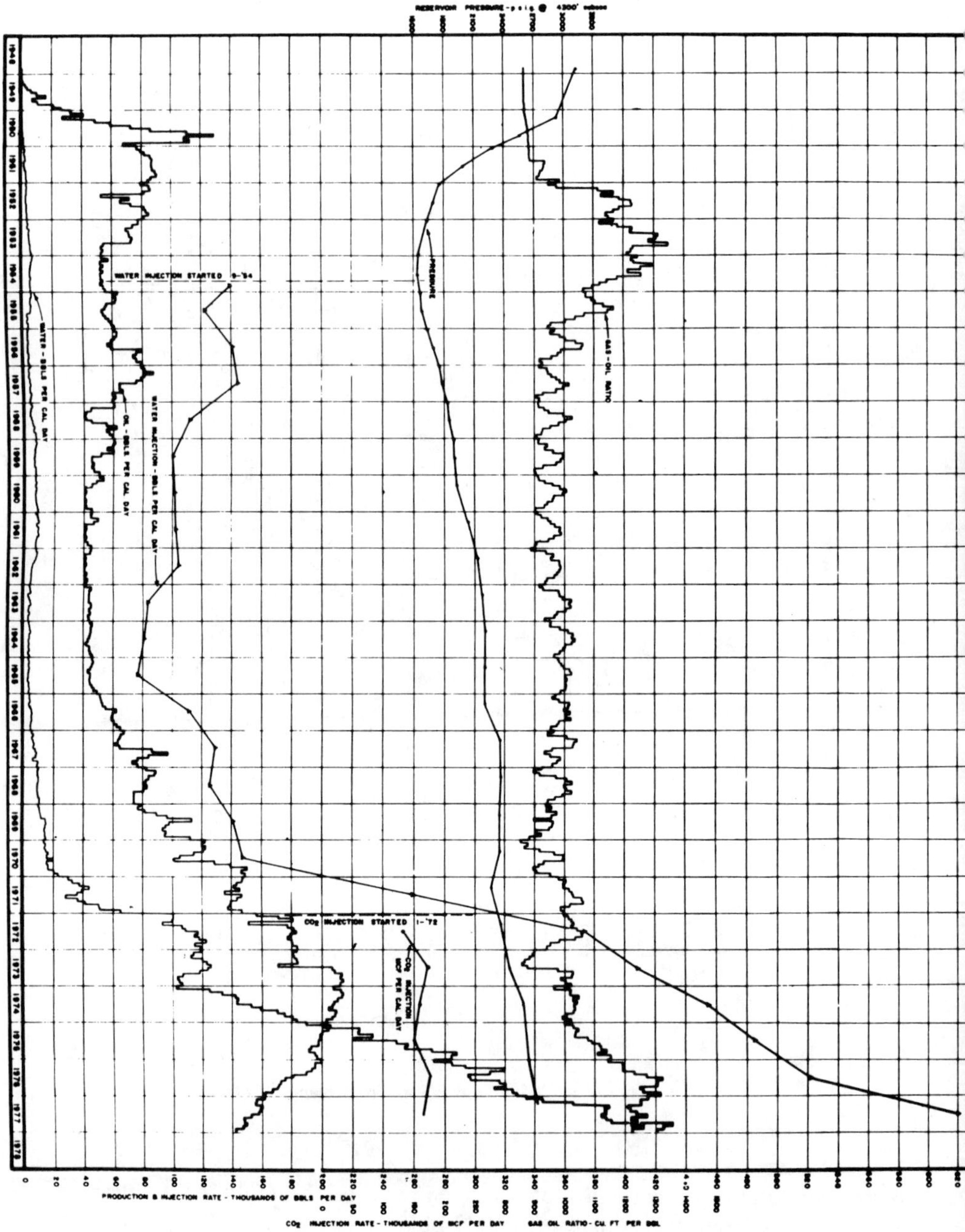


Fig. 10-10 Kelly-Snyder Field performance summary (From Kane, 1979)

### $CO_2$-WAG Project

Since the center-line injection system was considered incapable of supporting production on the flanks of the reef (Langston et al., 1988), the feasibility of modifying or replacing the center-line system was investigated, and an inverted 9-spot injection pattern flooding with $CO_2$ in a WAG as a secondary process was considered to be the most efficient alternative (Hull, 1970). Since the delivery rate of 200 MMcf/day [$5.66 \times 10^6$ $m^3/d$] of $CO_2$ was about one-third the total pattern area injection requirement, the $CO_2$ pattern injection scheme was implemented by developing the field area in three "phases" of similar hydrocarbon pore volume (Figure 10-9), which were to be flooded sequentially. To assure that the minimum miscibility pressures of about 2300 psia were attained, pre-$CO_2$ water slugs of 6 percent HCPV were injected in low-pressure patterns, followed by $CO_2$-WAG cycles. The WAG-cycle slug volumes were 6 percent HCPV $CO_2$ followed by 2.8 to 3.6 percent HCPV water for an average WAG ratio of 0.55 : 1.

The $CO_2$ obtained as by-product at several gas plants was delivered to the site through 220 miles of pipeline in supercritical condition provided by four compressor and two booster stations. Dual water and $CO_2$ injection lines were installed to the $CO_2$-WAG injection wells. The water was injected at 2000 psi by four high-pressure water injection stations using centrifugal pumps. The $CO_2$ compressor discharge pressure ranges from 2200 to 2400 psig. After separation, the produced gas was processed at three gasoline plants using hot potassium carbonate and amine $CO_2$ removal systems with a total $CO_2$ removal capacity of 56 MMcf/day ($1.59 \times 10^6$ $m^3/d$). The $CO_2$-WAG injection process started in January 1972 in the phase 1 area, and the cumulatives and rates through December 1977 are summarized in Table 10-4 (Kane, 1979).

**TABLE 10-4.** Injection Status in Nine-Spot Pattern Areas CO2-WAG Project, SACROC Unit

| CO2-WAG injection began | Phase 1<br>I, 1972 | Phase 2<br>III, 1974 | Phase 3<br>XI, 1976 | Total |
|---|---|---|---|---|
| Cumulatives, through December 1977 | | | | |
| CO2, BSCF | 184 | 142 | 18 | 344 |
| CO2, % HCPV | 14.3 | 10.0 | 1.4 | av 8.6 |
| Water, million bbl | 290 | 200 | 193 | 683 |
| Water, % HCPV | 49.9 | 30.3 | 31.2 | av 37.1 |
| Rates, as of December 1977 | | | | |
| CO2 MM scf/day | 20 | 93 | 34 | 147 |
| Water, million bbl | 189 | 150 | 141 | 480 |

From Kane (1979)

*Early performances* are shown as total field production and injection history in Figure 10-10 and as oil production for the pattern area of phases 1, 2, and 3 in Figure 10-11. The reservoir response to pattern area $CO_2$-WAG injection was a continuous increase of reservoir pressure as a result of the sharp increase in the injected water rate, separately and within the $CO_2$-WAG cycles. The oil production rate reached a peak of 212,000 bbl/day in 1973–1974, declining afterward as a result of the increased GOR and rising watercuts. The cumulative oil production at the end of 1977 was 937 million STB or 44.34 percent of OOIP compared with 24.5 percent of OOIP obtained only eight years before (1971). Whether to attribute the reservoir's or the individual phases' response to the waterflood effect or to the effect of $CO_2$ injection is difficult to determine. The project was implemented as a secondary project, and the bulk of the early response must be attributed to the waterflood component of the $CO_2$-WAG process (Langston et al., 1988). The possibility of clearly attributing oil production response to $CO_2$ injection is realized when $CO_2$ miscible is a tertiary process. This situation was found in phase 3, where a long period of pattern waterflooding existed prior to $CO_2$ injection.

### Four and Seventeen Pattern Areas

$CO_2$ performances were analyzed by Langston et al. (1988) who found this project to have good potential for serving as an example of $CO_2$ processing

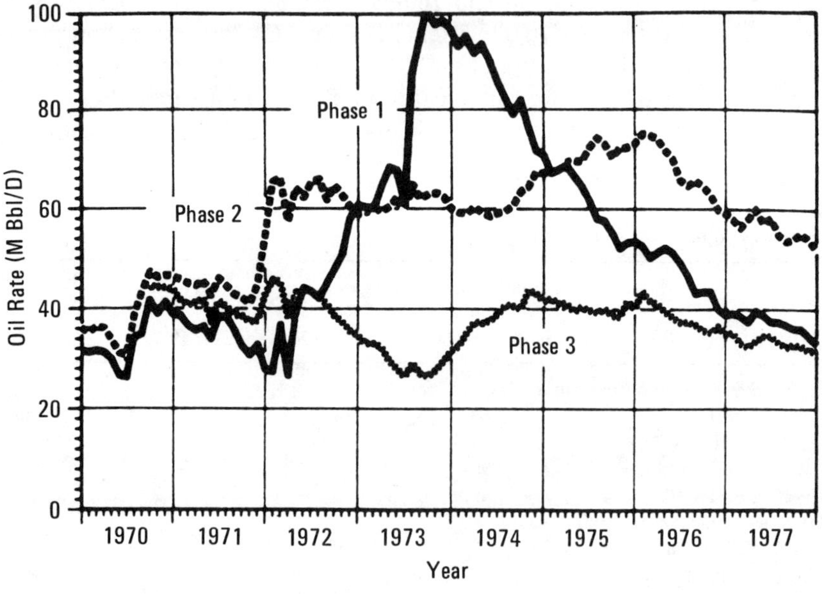


**Fig. 10-11** Oil production for pattern area of phases 1, 2 and 3 (From Langston et al., 1988)

performance at SACROC. The 4PA area and the 17PA area are located on the south flank of phase 3. 4PA covers approximately 603.1 acres ($2.441 \times 10^6$ m$^2$) with 24 wells, and 17PA covers 2,700 acres ($10.93 \times 10^6$ m$^2$) with over 100 wells (Figure 10-12). The corresponding 4PA original oil in place was 26,764 M bbl ($4.255 \times 10^6$ m$^3$). The 17PA OOIP was 79,100 M bbl ($12.58 \times 10^6$ m$^3$).

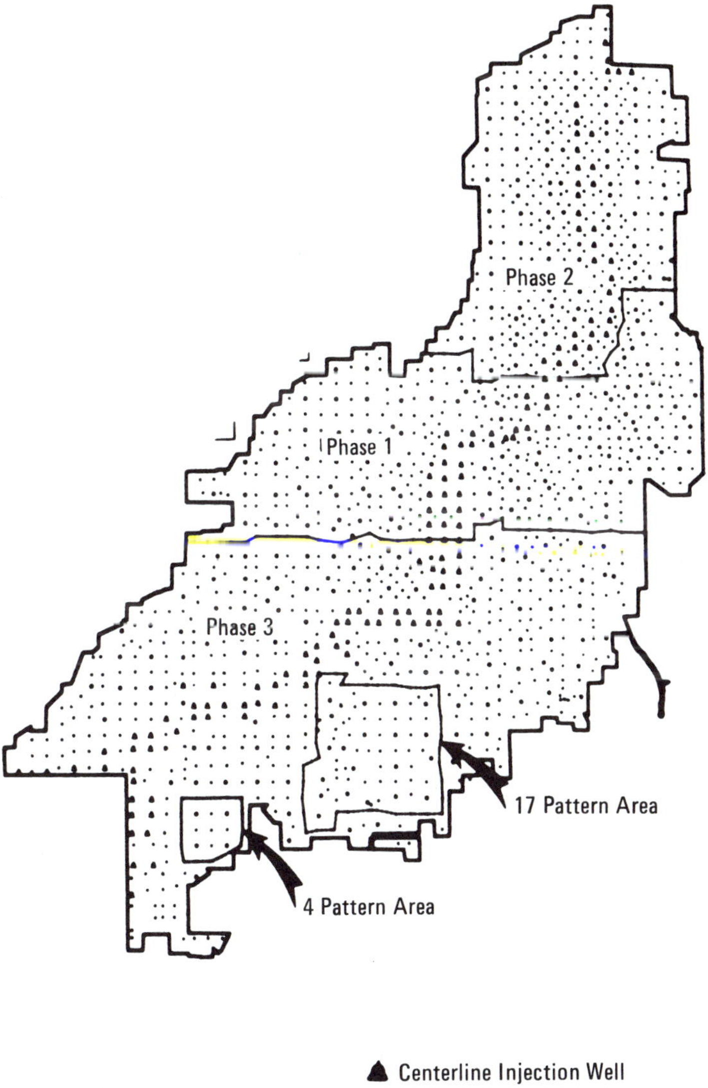


**Fig. 10-12** Map of SACROC unit (From Langston et al., 1988)

Before the start of the $CO_2$-WAG process in May–June 1981, the 4PA area was placed on a pattern waterflood in April 1973 and the 17PA area was influenced by center-line secondary recovery injection. Cumulative $CO_2$ injection as of December 1987 represented 29.2 percent of the 4PA hydrocarbon pore volume and 17.5 percent of the 17PA HCPV. The performances of both pattern areas are shown in Figures 10-13 and 10-14. The "factored production" refers to production allocated to a pattern from a pattern side well or from a well located on a pattern corner (Langston et al., 1988).

The decline curve analysis from which a production decline rate was determined starting with the pre-$CO_2$ injection period and extrapolated into the future shows the quantitative response to $CO_2$ injection as follows:

The incremental oil production in the 4PA as result of the $CO_2$ miscible process is estimated to be 10.2 percent of OOIP and 7.5 percent of OOIP in the 17PA. The ratio of cumulative $CO_2$ injection to cumulative incremental oil production, which is the cumulative utilization of the injected $CO_2$, is about 9600 $ft^3/bbl$ (1718 $m^3/m^3$).

The ultimate recovery fieldwide can be estimated as being 19 percent of OOIP in primary as studies indicated, plus about 22 percent in secondary as result of waterflood, plus at least 9 percent in tertiary with the $CO_2$ miscible process, for a total of 50 percent of OOIP.

The performance of the large-scale $CO_2$-WAG project conducted at the SACROC Unit–Kelly-Snyder Field showed that effective $CO_2$ miscible processes can be developed and implemented to recover additional residual crude oil using large volumes of $CO_2$ transported long distances, distributed to injectors, and injected successfully.

## Immiscible $CO_2$ Project, Tar Zone, Wilmington Field, California, United States

This project is an example of the use of $CO_2$ for immiscible flooding in heavy oils. The presentation and the data that follow are summarized from the article review written by Spivak et al. (1990).

**Characteristics.** The oil reservoir—the Tar Zone—is the uppermost producing zone in the Fault Block V of the Wilmington anticline structure and includes subzones S, T, and F, which consist of a series of loosely consolidated sands with an average gross thickness of 300 ft. The reservoir rock and fluid characteristics are given in Table 10-5.

**Production history.** The original oil in place in the project area was $69.465 \times 10^6$ STB ($11.045 \times 10^6$ $m^3$). The total oil produced, from 1961 under waterflood, was 23.27 percent of OOIP up to 95 percent watercut. In March 1982, an immiscible $CO_2$ displacement was started as a tertiary recovery mechanism.

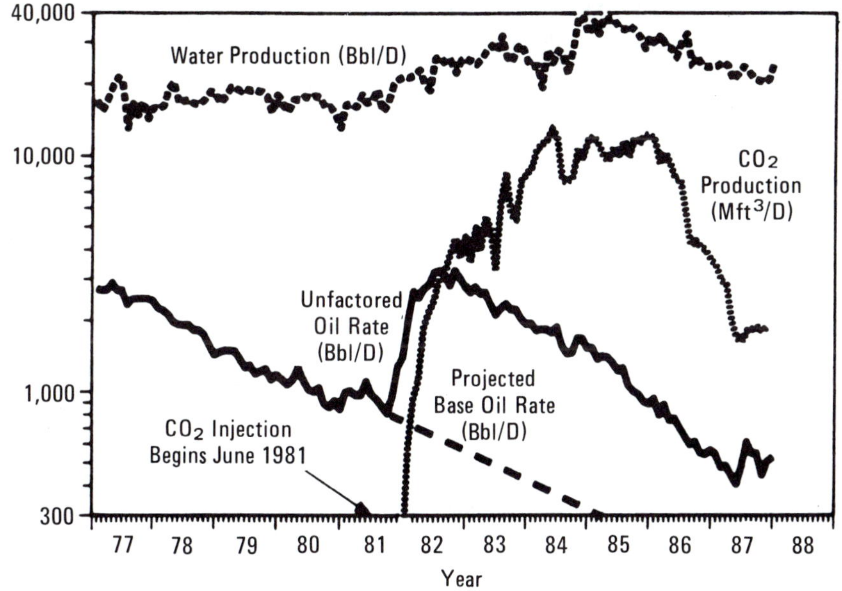


**Fig. 10-13** Four pattern area performance summary (From Langston et al., 1988)

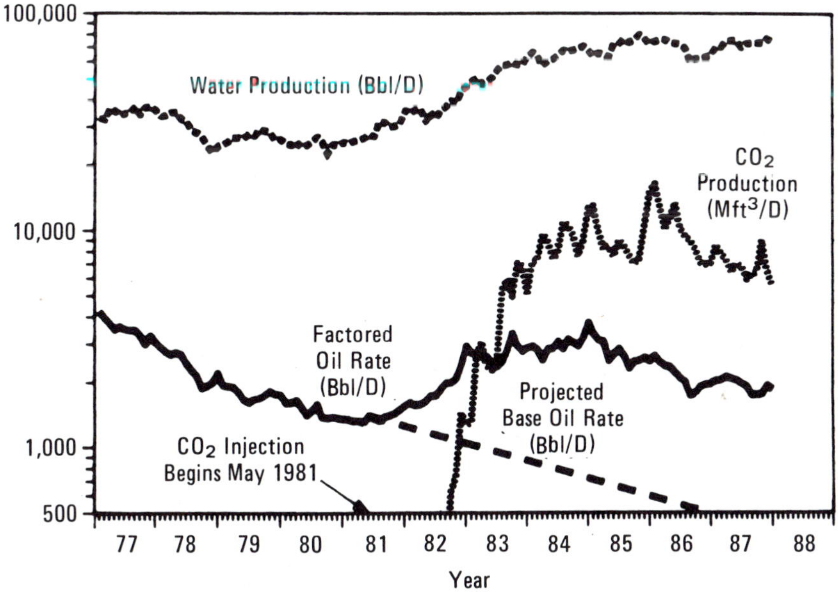


**Fig. 10-14** Seventeen pattern area performance summary (From Langston et al., 1988)

**TABLE 10-5.** Reservoir Rock and Fluid Properties Tar Zone, Wilmington Field, California

| | |
|---|---|
| Project surface area | 330 acres |
| Depth | 2300 ft |
| Initial pressure | 1000 psig |
| Reservoir temperature | 120° F |
| Reservoir dip | 9.5° to the south* |
| Average net oil pay | 141 ft |
| Porosity | 27% |
| Permeability | 100 md (S) |
| Permeability | 500 md (T) |
| Permeability | 400 md (F) |
| Initial oil saturation | 75% |
| Oil gravity | 14 °API |
| Oil viscosity (res. cond.) | 180 to 410 cp |
| Initial solution GOR | 87 scf/STB |
| Oil volume factor | 1.05 |

* In the southern portion of the project area.
From Spivak et al. (1990)

*The $CO_2$ injection* was begun downstructure and half central as a WAG process (Figure 10-15).

Water and $CO_2$ were injected on a two-week cycle schedule. The gas-to-water ratio was reduced after an early breakthrough to the order of 1.0 on a reservoir-volume basis, to control the adverse mobility ratio. After the alternate injection of 12 percent HCPV $CO_2$, water was injected continuously to displace the less viscous oil (with $CO_2$ in solution). The improved relative permeability conditions and the oil swelling in addition to the viscosity reduction were the beneficial effects of the $CO_2$-WAG process.

The injected gas was an 85 percent $CO_2$ and 15 percent $N_2$ mixture provided as stack gas from hydrogen generation units in the area. The total amount of 8.22 billion scf ($230 \times 10^6$ m$^3$) of gas $CO_2$ was purchased and injected as of July 1986 when gas injection was terminated. This amount represents only 30 percent of the theoretical maximum gas requirements (40 percent of HCPV). Water injection rates averaged about 20,000 STB/day (3180 m$^3$/d) since 1984.

*The theoretical calculation* of the incremental oil recovery achieved by the immiscible $CO_2$ process was made using the Buckley–Leverett-type two-phase immiscible analysis. The technique was to compare the oil recovery through continued waterflooding with the oil recovery obtained using the modified flow properties expected from immiscible $CO_2$ flood. The water-oil mobility ratio is improved when the oil viscosity at maximum oil saturation is reduced from about 406 cp to 54 cp. The results showed that cumulative recovery at watercut 95 percent (WOR = 20) was 21.9 percent of OOIP for the continued waterflood compared with 31.7 percent of OOIP for the immiscible $CO_2$ process or $6.66 \times 10^6$ STB [$1.06 \times 10^6$ m$^3$] additional oil over a 20-year period. By inject-

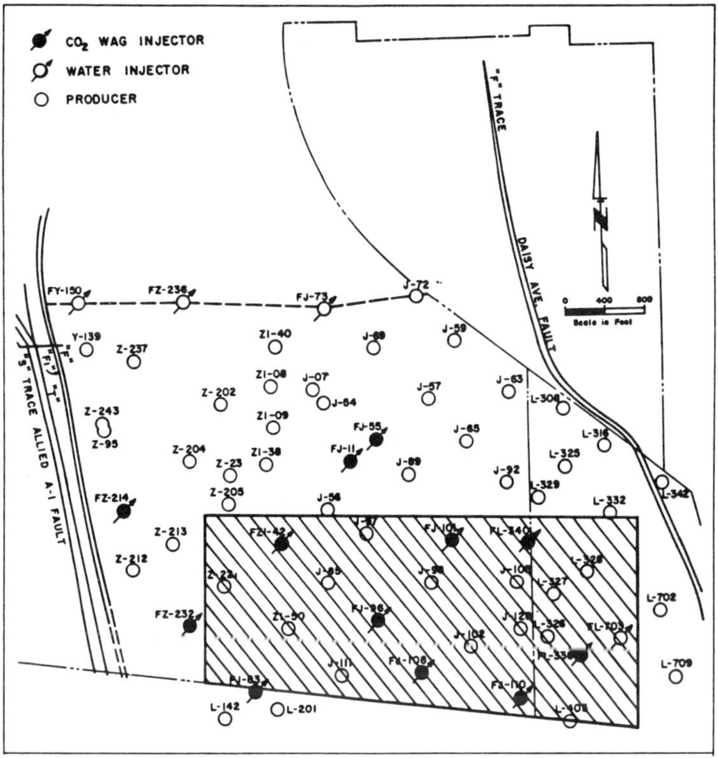


Fig. 10-15 Subzone $F_1$ scaled-down area (From Spivak et al., 1990)

ing less $CO_2$ than required to saturate all the oil fully, the oil viscosity would be proportionally higher and the recovery lower. Figure 10-16 shows the oil rate and the gas production rate for the period from March 1982.

**Performance evaluation.** As can be observed from Figure 10-16, gas breakthrough occurred after about one year of $CO_2$ injection and the gas production increased significantly.

Because of downstructure injection the gravity override effect that causes gas to migrate updip was probably greater in Subzone S, a massive clean sand with high permeability. To reduce this effect the gas-to-water ratio of 1.23 M scf/STB (222 $m^3/m^3$) was changed to less than 1.0 on a reservoir-volume basis and WAG cycles were reduced to about 2 weeks. As can be observed, gas production was brought under control in late 1983. The oil rate performance shows dependence on gas injection as the driving agent, as the oil rate increased sharply at the beginning of 1985 when injection was concentrated in the scaled-down area (Figure 10-15) and then dropped soon after curtailment of $CO_2$ purchases in the spring of 1986. Incremental tertiary oil recovery as of the end of August 1987 was 488,000 STB (77,590 $m^3$) and is most probably the oil

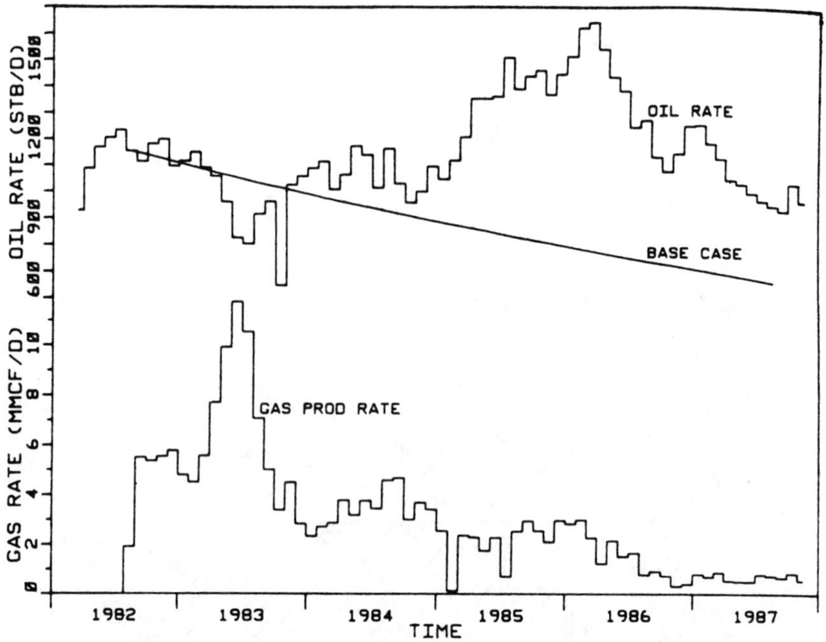


**Fig. 10-16** Oil and gas production rates (From Spivak et al., 1990)

displaced by gas from zones unswept or poorly swept by water. The effect of immiscible $CO_2$ injection on viscosity reduction, swelling, and relative permeability alteration is realized gradually over many years (Spivak, 1990).

## QUESTIONS AND PROBLEMS

**10-1** How can the multiple uses of $CO_2$ in different industries and processes be explained?

**10-2** Enumerate the factors that make $CO_2$ an EOR agent.

**10-3** Suppose that $CO_2$ causes an eightfold reduction in oil viscosity in reservoir conditions. What is the effect on a water-oil mobility ratio of 7.2?

**10-4** Explain quantitatively, assuming appropriate values, how oil swelling as a result of $CO_2$ dissolved in crude oil increases the oil recovery.

**10-5** The production performance of an oil reservoir indicates that the oil recovery of 15.2 percent of OOIP was the result of solution gas drive as the primary recovery mechanism. To improve recovery, $CO_2$ is injected into the reservoir and production is resumed until the same residual oil saturation is reached. Find the incremental oil produced and the secondary oil recovery. (Ignore the presence of free gas saturation in the reservoir.) The following information is given:

| | |
|---|---|
| Porous volume | $PV = 15.7 \times 10^6$ bbl ($2.5 \times 10^6$ m$^3$) |
| Interstitial water saturation | $S_{wi} = 0.20$ |
| Oil formation volume factor | $B_{oi} = 1.14$ |

Oil formation volume factor at $S_{or}$ $B_o = 1.08$
Oil formation volume factor after $CO_2$ injection $B_o\text{-}CO_2 = 1.26$

**10-6** To make the preliminary calculations, estimate the miscibility pressure of an oil reservoir characterized by having the molecular weight of the pentanes and heavier fractions of 170 and a temperature of 240° F.

**10-7** The most convenient way to pressurize a reservoir is to inject water before the start of $CO_2$ injection for miscible displacement. Calculate the total amount of injected water and time needed to repressurize a reservoir to its initial pressure knowing the following information:

| | |
|---|---|
| Original oil in place | $N = 12 \times 10^6$ bbl ($1.9 \times 10^6$ m$^3$) |
| Actual recovery factor | $E_R = 35\%$ of OOIP |
| Cumulative water produced | $W_p = 2 \times 10^6$ bbl ($318 \times 10^3$ m$^3$) |
| Reservoir pressure | |
| Initial | $P_i = 2600$ psia (17.96 MPa) |
| Actual | $P = 1300$ psia (8.978 MPa) |
| Solution ratio | |
| Initial | $R_{Si} = 500$ scf/bbl (88 m$^3$/m$^3$) |
| Actual | $R_S = 300$ scf/bbl (52.82 m$^3$/m$^3$) |
| Oil formation volume factor | |
| Initial | $B_{oi} = 1.46$ |
| Actual | $B_o = 1.25$ |
| Gas formation volume factor | |
| Initial | $B_{gi} = 0.009$ |
| Actual | $B_g = 0.012$ |
| Gas-oil ratio (average) | GOR = 180 STB/bbl (m$^3$/m$^3$) |
| Actual oil production rate | $q_o = 1850$ STB/day (294 m$^3$/day) |
| Actual water production rate | $q_w = 600$ STB/day (95.4 m$^3$/day) |
| Available water injection rate | $q_i = 10{,}000$ bbl/day (1590 m$^3$/day) |

**10-8** A pilot $CO_2$ miscible project must be designed. The project will be conducted in an isolated, inverted five-spot pattern 10 acres in extent, with a 150 ft thickness and 16 percent porosity. The $CO_2$ gravity-stabilized displacement slug will form a 31-ft diffusion zone. Find the total amount of $CO_2$ needed for injection assuming 6 percent pore volume of $CO_2$ will saturate the reservoir fluids.

**10-9** Calculate the pressure exerted by the weight of a 9500-ft $CO_2$ gas column in a well with 2700 psia (18.64 MPa) tubing pressure and 140° F average temperature.

## REFERENCES


AALUND, L. R., "EOR Projects Decline But $CO_2$ Pushes Up Production (Production/Enhanced Oil Recovery Report)," *Oil and Gas Journal* (April 18, 1988), pp. 33–73.

ANADA, H., J. SEARS, et al., *Feasibility and Economics of By-product $CO_2$ Supply for Enhanced Oil Recovery*, *Final Report*, Vol. 1, Technical Report, DOE Contract No. DE-AT21-78MC08333-3 (Bartlesville, OK: U.S. Department of Energy, January 1982), pp. 96–98.

BEGGS, H. D., *Gas Production Operations* (Tulsa, OK: Oil & Gas Consultants, 1984), pp. 102–03.

BILHARTZ, H. L., et al., "A Method for Projecting Full-Scale Performance of $CO_2$ Flooding in the Willard Unit," Society of Petroleum Engineers, Paper SPE 7051 presented at the 1978 SPE Symposium on Improved Methods for Oil Recovery, Tulsa, Oklahoma, April 16–19, 1978.

BROCK, W. R., and L. A. BRYAN, "Summary Results of $CO_2$ EOR Field Tests, 1972–1987," SPE 18977, Paper presented at the SPE Joint Rocky Mountain Regional/Low Permeability Reservoirs Symposium and Exhibition, Denver, Colorado, March 6–8, 1989.

BRUMMETT, W. M., A. S. EMANUEL, and T. D. RONQUILLE, "Reservoir Description by Simulation at SACROC—A Case History," *Journal of Petroleum Technology* (October 1976), pp. 1241–55.

CHRISTIAN, L. D., et al., "Planning a Tertiary Oil Recovery Project for Jay-Little Escambia Creek Fields Unit," *Journal of Petroleum Technology* (August 1981), pp. 1535–44.

DICHARRY, R. M., T. L. PERRYMAN, and J. D. RONQUILLE, "Evaluation and Design of a $CO_2$ Miscible Flood Project—SACROC Unit, Kelly-Snyder Field," *Journal of Petroleum Technology* (November 1973), pp. 1309–1918.

HENRY, R. L., and R. S. METCALFE, "Multiple Phase Generation During $CO_2$ Flooding," SPE Paper 2812 presented at the 1980 SPE/DOE Enhanced Oil Recovery Symposium, Tulsa, Oklahoma, April 20–23, 1980.

HOLM, L. W., "Status of $CO_2$ and Hydrocarbon Miscible Oil Recovery Methods," *Journal of Petroleum Technology* (January 1976).

HOLM, L. W., and V. A. JOSENDAL, "Mechanism of Oil Displacement by Carbon Dioxide, *Journal of Petroleum Technology* (December 1974), p. 1427.

HULL, P., "SACROC: An Engineering Conservation Triumph," *Oil and Gas Journal* (August 17, 1970), pp. 57–62.

KANE, A. V., "Performance Review of a Large-Scale $CO_2$-WAG Project, SACROC-Unit, Kelly-Snyder Field, *Journal of Petroleum Technology* (February 1979), pp. 217–31.

KENNEDY, J. T., and G. THODAS, "The Transport Properties of Carbon Dioxide," *American Journal of Chemical Engineers*, Vol. 7 (December 1961), p. 625.

LANGSTON, M. V., S. F. HOADLEY, and D. N. YOUNG, "Definitive $CO_2$ Flooding Response in the SACROC Unit," Paper SPE/DOE 17321 presented at the SPE/DOE Enhanced Oil Recovery Symposium, Tulsa, Oklahoma, April 17–20, 1988.

MUNGAN, N., Carbon Dioxide Flooding—Fundamentals," *Journal of Canadian Petroleum Technology* (January–March 1981), pp. 87–92.

MUNGAN, N., "Improved Oil Recovery," Chapter IV in *Carbon Dioxide Flooding* (Oklahoma City, OK: Interstate Oil Compact Commission, March 1983), pp. 113–60.

PAUTZ, J. F., et al., NIPER-471 *Review of EOR Project Trends and Thermal EOR Technology—Topical Report*, performed under cooperative agreement No. FC22-83FE60149-DOE (Bartlesville, OK: IIT Research Institute, NIPER/DOE, March 1990).

PERKINS, T. K., JR., and O. C. JOHNSTON, "A Review of Diffusion and Dispersion in Porous Media," *SPE Journal* (March 1963), p. 70.

SAGE, B. H., and W. N. LACEY, *Some Properties of the Lighter Hydrocarbons, Hydrogen Sulfide and Carbon Dioxide*, Monograph on API Research Project 37 (Dallas, TX: American Petroleum Institute, 1955).

SIMON, R., and D. J. GRAUE, "Generalized Correlations for Predicting Solubility, Swelling and Viscosity Behavior of $CO_2$-Crude Oil System," *Journal of Petroleum Technology* (January 1965), pp. 102–06.

SMITH, R. L., "Sacroc Initiates Landmark $CO_2$ Injection Project," *Petroleum Engineering* (December 1971), pp. 43–47.

SPIVAK, A., W. H. GARRISON, and J. P. NGUYEN, "Review of an Immiscible $CO_2$ Project, Tar Zone, Fault Block V, Wilmington Field, California," *SPE Reservoir Engineering* (May 1990), pp. 155–62.

STALKUP, F. I., JR., *Miscible Displacement*, SPE Monograph Series (SPE, Richardson, TX, 1984), pp. 137–56.

VAN POOLLEN, H. K., and associates, *Fundamentals of Enhanced Oil Recovery* (Tulsa, OK: PennWell, 1980), p. 133.

YELLING, W. R., and R. S. METCALFE, "Determination and Prediction of $CO_2$ Minimum Miscibility Pressures, *Journal of Petroleum Technology* (January 1980), pp. 160–68.

Chapter 11

# Oil Mining, Microbial EOR, and Electrothermal Processes

## 11-1 GENERAL

The steady decline in domestic crude oil production causes an increased reliance on oil imports. This is a major contributing factor in the national trade imbalance. Although domestic production is declining, the potential domestic petroleum resource is large. The national average for possible ultimate recovery with actual recovery methods is 34 percent of the 492 billion bbl ($78.22 \times 10^9$ $m^3$) original oil in place. An amount representing 28 percent of OOIP has been already produced. What remains as recoverable oil is only 6 percent of OOIP or 29.5 billion bbl ($4.69 \times 10^9$ $m^3$). The remaining 325 billion bbl ($51.67 \times 10^9$ $m^3$) of oil, considered residual oil or oil that cannot be recovered economically is being trapped or bypassed by current use of conventional technology. Enhanced oil recovery methods can produce a portion of this residual oil. Research and development efforts are in place to increase the fundamental understanding of the residual oil resource and the EOR technologies that can be used to produce it. The methods to increase the effectiveness of oil recovery include

- using foaming agents to improve reservoir conformance.
- waterflood after steam drive and after in situ combustion.
- downhole steam generation and injection of $CO_2$ with steam.
- alkaline flooding with synthetic surfactant and polymer.
- using oxygen-enriched air for in situ combustion.
- infill drilling and horizontal drilling with EOR methods.

Other innovative techniques such as mining, microbial flooding, and electrothermal processes have been proposed in the past few years.

## 11-2 OIL MINING METHODS

### Historical

Oil mining methods combine mining and petroleum technologies to produce oil. This combination method can be achieved by approaching the rock reservoir through underground drill sites or by excavating the reservoir rock and then processing it on the surface by chemical or thermal means to recover the oil. The second procedure is used to produce tar sands and oil shale and is not the subject of this chapter.

The first procedure called "underground oil recovery" or "mine assisted oil recovery" has its origin in ancient Egypt and Persia in areas with oil surface spots where shallow holes were dug into the ground to collect oil. In Europe, at Pechelbronn in France, oil has been produced by underground recovery since 1785, and at Wietze in Germany in the 1920s wells were drilled on a fan pattern from underground excavated mine openings directly beneath the reservoir. In Romania, at Sarata Monteoru, a vertical shaft was sunk to 700 ft in the early 1930s, mine galleries were excavated in the production formation, and wells were drilled horizontally into the reservoir to collect the oil. In the Soviet Union, at Yarega, heavy oil is produced through mine drifts and underground wells using steam injection. In Canada, the Srania mine-assisted gravity drainage produces oil from a 430-ft depth. It should be pointed out that at Athabasca, Alberta, the shallow tar sand deposits are currently produced by the world's largest surface mining process.

### Oil Mining in the United States

In the United States the first patent issued in 1865 and many others granted before 1930 were used as references in the *U.S. Bureau of Mines Bulletin 351* (Rice, 1932). This was a state-of-the-art description of mine-assisted oil production and also a review of the techniques used in Germany and France.

Interest in underground oil recovery increased during the 1970s and numerous detailed studies of the subject have been prepared and some published (Energy Development Consultants, 1978; Golder Associates, 1978; Albayrak and Protopapas, 1984).

Recently, as a result of joint investigations made by the National Institute for Petroleum and Energy Research (NIPER), by K&A Energy Consultants, and by Terra Quest Recovery specialists, in a research program supported by the U.S. Department of Energy, a detailed report has been released entitled "Improved Oil Mining: A Feasibility Report." After a review of oil mining technology in the United States (24 projects are listed) and elsewhere in the world, the report analyzes geological aspects, mining and drilling technology, reservoir and production engineering, and the applicability of EOR methods in improved oil mining operations. The report recommends that a research field project be performed at Caddo Pine Island field to confirm the technical and economic feasibility of the improved oil mining concept.

*The oil mining concept* is to sink or drill shafts from which oil mines are developed in the producing formation or outside, preferably located immediately below the producing formation. Room is provided for subsurface drilling and petroleum production operations as well as a life-supporting atmosphere and safe working conditions (Tham et al., 1988). Directional (horizontal or slant) wells are drilled into the reservoir. Oil is produced by pressure depletion and gravity drainage and the recovery efficiency is improved by applying enhanced oil recovery methods. Figure 11-1 illustrates the shaft, the mine below the producing formation, and the directional wells arrangement.

There are several forms of the basic configuration for an underground drill site depending on the reservoir characteristics and economic conditions (Streeter et al., 1989). One of these forms in which long horizontal or low-angle wells are drilled into the reservoir along the base of the production zone is shown in Figure 11-2.

**Candidate reservoirs.** Referring to reservoir parameters only, the criteria for initial selection of reservoirs for improved oil mining (IOM) applications are as follows:

- Reservoir depth between 500 and 300 ft
- Reservoir temperature less than 140° F
- Reservoir thickness of 40 ft minimum
- API gravity higher than 18–20 °API

Reservoirs that are potential candidates should be pressure depleted and must be kept below low-pressure limits. Abandoned shallow reservoirs with high remaining oil saturation and a low content of gas in solution can be good candidate reservoirs for application of underground oil recovery. The method can be a possible alternative for developing reservoirs where surface conditions or environmental restrictions produce constraints.

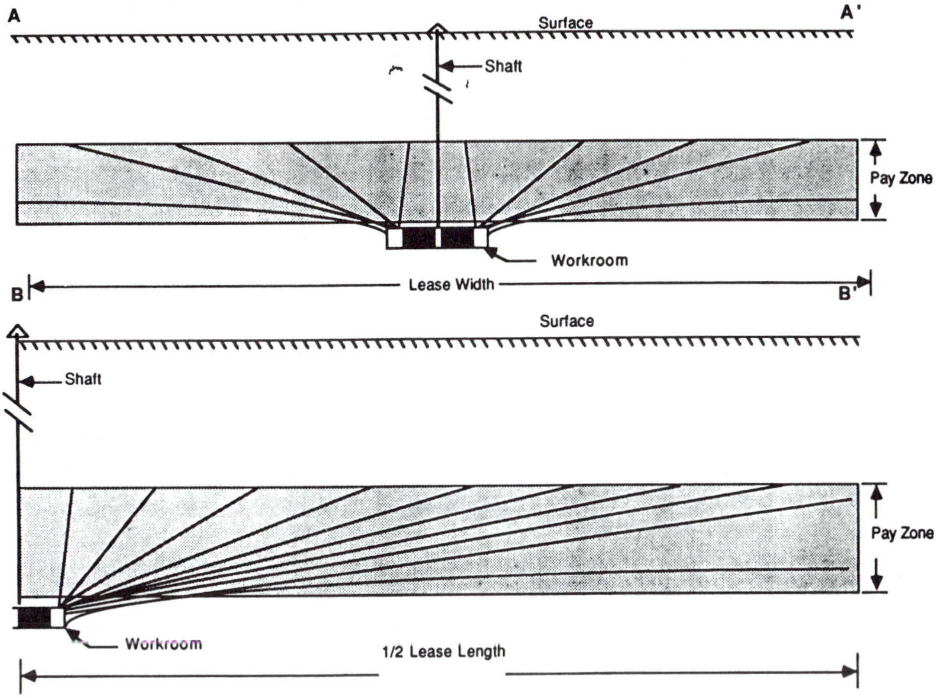


**Fig. 11-1** Mine and directional wells arrangement in improved oil mining (From Tham et al., U.S. Department of Energy, 1988)

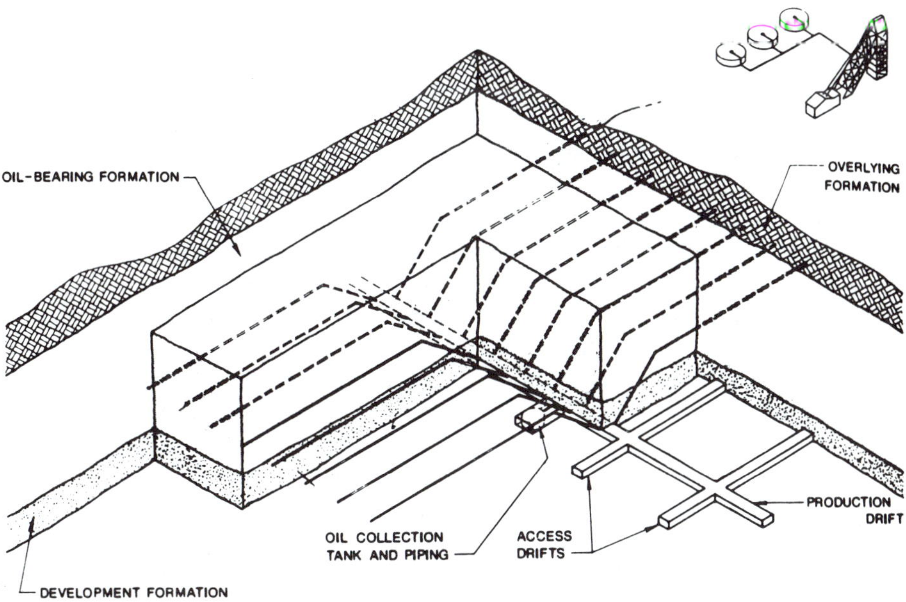


**Fig. 11-2** Long horizontal wells (From Streeter et al., 1989)

After a number of reservoirs satisfying these criteria are selected, another screening step must be considered. This next screening step includes criteria such as reservoir size; the ratio of remaining oil saturation to initial oil saturation, $S_{or}/S_{oi}$, which has to be a minimum of 0.7; average concentration of remaining oil; and $S_{or} \times \phi$, which has to be higher than 0.07. In addition, qualitative information about the formation consolidation or about the presence of horizontal barriers within the formation determines the selection of one reservoir over another for a feasibility study for improved oil mining technology (Tham et al., 1988).

*Applicability of EOR methods* in oil mining operations is considered to have real potential for improving oil recovery by increasing the effectiveness of the gravity drainage process. For example, injection of carbon dioxide at the top of the producing zone under gravity-stable conditions provides additional pressure energy to the gravitational drainage and increases the producing rate of the horizontal or inclined wells drilled along the base of the production zone. Additional energy in the form of heat can also be provided to the reservoir by cyclic steam stimulation operations which reduce oil viscosity and increase the fluid flow. The chemical EOR processes, especially surfactant injection applied within improved oil mining, has a heightened effect since the close well spacing increases the sweep area of a surfactant slug designed for high displacement efficiency.

**Safety precautions.** One of the most important aspects to be considered in underground oil recovery operations is assuring safe conditions for oil mine excavations and oil production operations. The main concern is the presence of the contaminating gases in the oil mine atmosphere. These gases evolve from the oil, and even at low-solution ratios oil releases gases which can create explosive conditions and have negative physiological effects on human beings. The ventilation system must provide a quality fresh air supply of suitable volume to assure optimum working conditions. The shaft and mine galleries must be sealed off from any other uncontrolled inflow of gases or water. All existing vertically drilled holes from surface abandoned or producing oil wells must be sealed off above and below the mining zone. If formations with prolific aquifers are adjacent to the oil formation, water having a higher pressure could enter an oil mine through fractures, joints, ruptured lines, or abandoned wells. This problem and others dealing with gases, shafts constructed through the oil and gas reservoir, gallery excavated, and directional wells drilled have been successfully solved with methods and equipment currently available.

*The future for mine-assisted oil recovery* is considered bright and is sustained by the experience gained, by projects already underway, and by existing private and DOE research programs.

The most successful U.S. project has been the North Tisdale Field in Wyoming. This oil mining project produced over 500,000 bbl of oil using a vertical shaft with a drilling room, an incline built into the reservoir, and long, 3000-foot horizontal wells (Streeter et al., 1989).

The Research Report of NIPER (Tham et al., 1988) based on analyses of geological data, on a close examination of production data, and on economic and technical considerations selected the Annona Chalk section of Caddo Pine Island Field for a feasibility study for improved oil mining recovery and recommended that a demonstration field project be performed.

The reservoir lies at a depth of 1600 ft from the surface, is 160 to 185 ft thick, has 15 to 28 percent porosity, and has low permeability values (between 0.1 to 1.5 md). The oil of 34 to 40 °API gravity flows through the existing fractures system at initially high rates that decline rapidly as the fractures surrounding the matrix are drained and depleted. The flow rate of the oil from the matrix into the fractures system is low, and the oil well production of 1 bbl/day or less has been sustained over a long period. The recovery efficiency is low, between 6 and 8 percent of the OOIP, and attempts made to increase the oil recovery have failed to recover additional oil economically. The high existing oil saturation corresponding to 90 to 95 percent of OOIP, the low current reservoir pressure estimated between 200 to 300 psig, and the low gas-oil ratio (less than 30 $ft^3/ft^3$) are also favorable characteristics for applying IOM. Indeed, by using improved oil mining methods the reservoir has what is needed to yield more oil:

- Greater communication between matrix and the wellbore, obtained by high well density and greater wellbore length exposed to the matrix
- Reduced backpressure against the matrix to improve the flow from matrix to fractures to wellbore

Improved oil mining or underground oil recovery is a method that has been developed to the point where it has to be considered a type of EOR, even if there are no precise published data regarding the amount of additional oil, percentage of OOIP, recovered as a result of this method. Demonstration field projects could show that reservoirs, equipment, and technologies are available to prove that EOR potential could be increased in an economically viable way.

## 11-3 MICROBIAL EOR

### General

The concept of using microorganisms to increase oil production as a single-well stimulation treatment is more than 40 years old. The treatment involves injecting potential bacterial species into a stripper well (which produces less than 10 bbl/day) along with appropriate nutrients. After a shut-in period, the biometabolites generated during fermentation, primarily organic acids and solvents, surfactants, and gases, could release more oil. However, early field trials have not been conclusive. Even when oil producing rate increases were claimed to be a result of the bacterial treatments, the lack of adequately controlled studies led to disputes regarding the role of biometabolites in any improvement in oil production (Sheehy, 1990).

In considering the feasibility of using microbial biometabolites in enhanced oil recovery the concept of the mobility of bacteria and their penetration deep within the reservoir in connection with waterflood reservoirs has been pursued, but few laboratory studies have been carried out to date.

### New Developments and Field Tests

Starting in the early 1980s, research programs, laboratory studies, and field trial results presented at different symposia and meetings reviewed microbial EOR and indicated the beneficial effect of bacterial metabolites in both well stimulation and oil displacement (Yen, 1986; King and Stevens, 1987; Bryant et al., 1989; Knapp et al., 1989; Sheehy, 1990). Surfactant producing aerobe *Bacillus subtilis* and acid and solvent producing anaerobe *Clostridium acetobutylicum* were found to displace residual heavy crude after waterflooding (Yen, 1986).

Described next are two field trials which determined and documented the effectiveness of a microbial system introduced in an oil reservoir.

**Mink Unit of the Delaware Childers Field, Oklahoma, United States.** This site was chosen for a field pilot to determine if injection of a microbial system could increase oil production in a mature waterflood. The progress of the ongoing microbial-enhanced waterflood field experiment, initiated October 1, 1986, is described in the NIPER-356 topical report prepared for the U.S. Department of Energy (Bryant et al., 1989).

*The reservoir* in the Mink Unit is a Bartlesville sandstone covering a 160-acre area at 600-ft average depth, with a 30-ft net pay thickness, 20 percent porosity, and 60-md permeability as average parameters.

*The oil content* has 34 °API gravity and 7-cp viscosity at a 77°F temperature. Discovered in 1906, the reservoir was produced by primary and secondary recovery methods, a waterflood being in continuous operation until the present time. The average oil saturation at the start of the project was 36.2 percent.

*The pilot area* of 17.8 acres within the Mink Unit consists of four adjacent inverted five-spot patterns drilled on 5-acre spacing with four injectors and eight production wells.

*Microbial formulation* was established using several microorganisms grown with the Mink reservoir fluids and tested on reservoir core samples and simulated porous media. The tests showed that the microbial system efficiently displaced the residual oil remaining after waterflooding. To check the injection pressure and conditions of mobility, a single-well injection test was performed in an off-pattern well. After the injection of 26 gal of inoculum and a shut-in period of 12 days, backflush samples showed that the microbes were still growing. No plugging was observed when injection was resumed.

**Validity control.** To make the assessment of the process scientifically valid baseline studies were executed. These studies showed consistency in different parameter values (total dissolved solids, pH, viscosities) during the period before the process was started. Chemical tracer studies with fluorescein showed communication between all producing wells and the four injectors. A waterflood history match was obtained using simulation models, and the effectiveness of the microbial system was determined against the injection baseline model. All possible efforts were made to ensure that no changes in operating conditions or workovers were made during the pilot test.

*Injection* of 26 gal of microbial formulation per injector was started in March 1987 along with 500 gal of nutrient per well (molasses 4 percent concentration) during and after the microbial injection, followed by a 2-week shut-in period. Water injection was then resumed along with 2 gal of pure molasses per well per day.

*The first results* shown in Figure 11-3 indicate a significant increase in oil production since 1982, representing an increase in oil production rate with about 25 percent in the pilot area and 13.5 percent in the whole Mink Unit (Bryant et al., 1989).

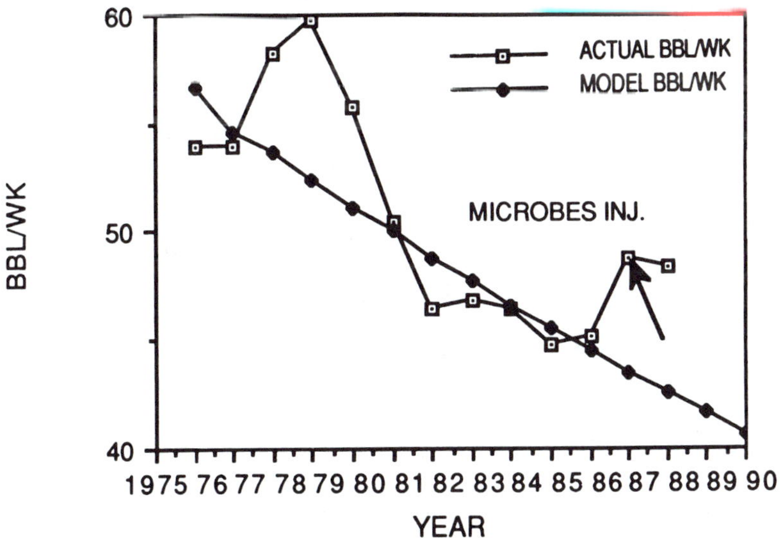


**Fig. 11-3** History match of Mink Unit oil production (From Bryant et al., 1989)

*The water-oil ratios* in all producing wells have dropped below the averages in the baseline period, with some wells showing a significant decrease. The oil viscosity in all wells has not changed significantly, and the surface tension is not lowered enough from the baseline data.

The oil mobilization appears to be caused by the growth of microorganisms in the highly permeable water channels. Their presence blocks the pores, diverts the water to the poorly swept zones and increases the volumetric sweep efficiency.

**Alton Field microbial trial, Australia.** This field pilot is described in the SPE/DOE 20254 technical paper (Sheehy, 1990) and shows not only the feasibility of stimulating oil production using microbial injection but also the capability of the process to operate at higher reservoir temperatures.

*The reservoir* in the Alton Field is a Boxvale sandstone covering 1840 acres at 6100-ft average depth with thin sands having average porosity of 17.2 percent and permeability of 260 md.

*The original oil in place* estimated at approximately 10 MM STB ($1.59 \times 10^6$ $m^3$) is a medium light oil on a paraffin base at 169° F (76° C) reservoir temperature. The Alton reservoir was discovered in 1966 and was produced by fluid expansion and a weak water drive. The initial production rate of 1000 bbl/day declined, nearing its economic limit of 15 bbl/day obtained at present from five wells on a beam pump.

*Microbial stimulation* has been considered a more viable alternative than expensive waterflood or other EOR methods, since high-risk factors are involved due to the stratified nature of the sands. The high residual oil saturation estimated at 50 percent of the porous volume, the possibility of improving the sweep efficiency of the thin sands, and the desire to generate surfactants in situ sustained the decision to stimulate a producing well with microorganisms.

*Detailed laboratory studies* and coreflood experiments have been conducted to select and isolate desirable microbial species, to enrich the biometabolites produced in reservoir simulated conditions, and to check the compatibility of introduced microorganisms with the resident reservoir microorganisms.

To assure the *validity of the results* all evaluations of the effectiveness of the microbial system were determined by comparison with the natural baseline established by a trial shut-in period (point A, Figure 11-4) to log post shut-in production (Sheehy, 1990). Since a workover program was conducted on the Alton #3 well, an injection baseline was also established by injection of production water, followed by well shut-in (point B, Figure 11-4).

*Injection* of 86 bbl of microbial solution followed by 35 bbl of produced water to displace the system is designated as point C on Figure 11-4. After a 3-week shut-in period, the Altona #3 well was put back into production.

The *first results* are illustrated in Figures 11-5 and 11-6. As can be clearly observed with reference to the control lines, oil production increased approx-

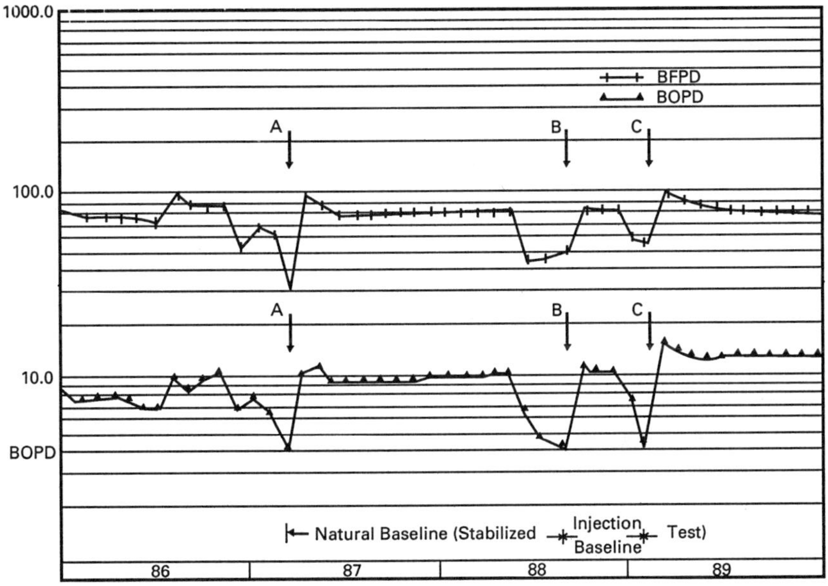


**Fig. 11-4** Graph 1: Production history of Alton no. 3, 1986 to 1990 (Adapted from Sheehy, 1990)

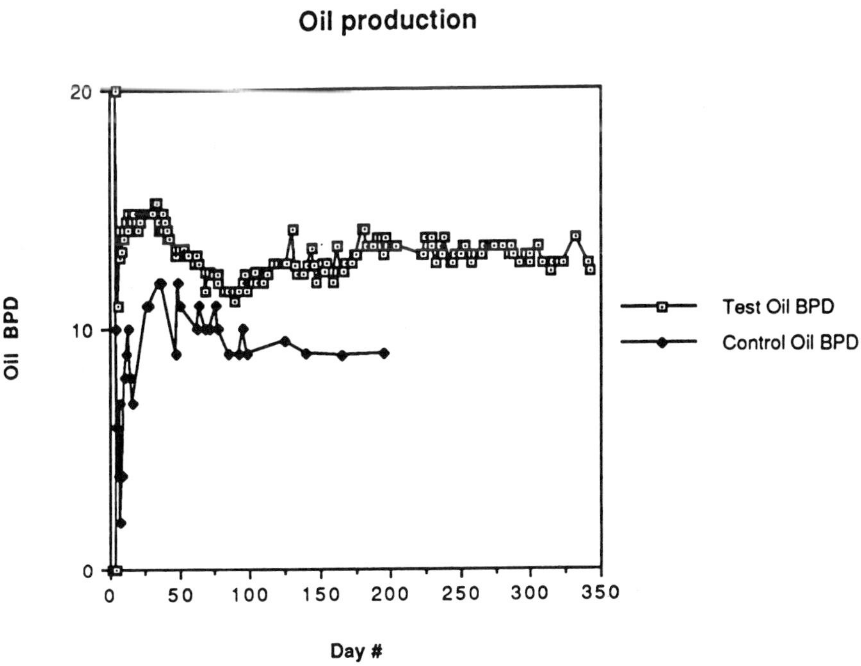


**Fig. 11-5** Microbial EOR-daily production (From Sheehy, 1990)

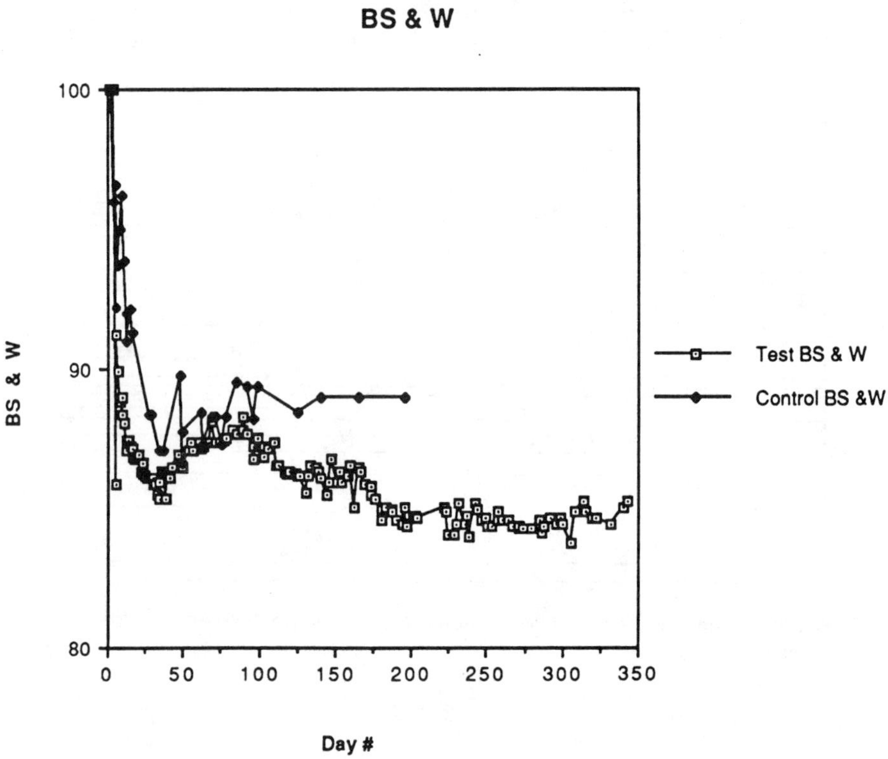


**Fig. 11-6** Microbial EOR-water cut (From Sheehy, 1990)

imately 40 percent and the watercut was reduced. The result of biological activity in the reservoir was also sustained by the increase in annulus pressure due to $CO_2$ and methane accumulation, the increase in the ion level of the production water, the 1000-fold increase in microbial numbers, and a significant decrease (10 to 25 percent) in the water-oil interfacial tension.

**Comments.** Although there is no given explanation as to how the microbial system injected into the reservoir induced the production of biometabolites, the conclusive results of this field trial are that oil production was beneficially stimulated. However, numerous problems need to be resolved and more research and tests performed in controlled field projects to establish different stimulation technologies and to prove quantitatively that microbial biometabolites can indeed enhance oil recovery.

## 11-4 ELECTROTHERMAL PROCESSES

Electrothermal processes utilize electricity or electromagnetic energy to stimulate heavy oil reservoirs and tar sands. Several processes have been proposed, and an overview of the existing state of specific technologies is presented by Chute and Vermeylen (1988) and by Pautz and coworkers (1990).

The processes involve heating of the formation to a temperature that will lower the oil viscosity to the point where it can flow or be displaced by steam. This temperature increase is achieved with electromagnetic energy produced by using adjacent wells as electrodes. Reservoirs can also be heated by radio frequency (RF). In this case monopole or dipole antenna structures are introduced downhole to apply RF energy to the reservoir.

Few of these proposed procedures have been tested successfully in the field, and more innovation is needed in the area of electrode design and siting before the process can become commercially viable (Pautz et al., 1990).

## QUESTIONS

**11-1** Why has interest in underground oil recovery increased only recently (during the last two decades) even though mining for oil is a very old method?

**11-2** Describe the oil mining concept.

**11-3** What are the characteristics and considerations on which to base the selection of an oil reservoir as a good candidate for underground oil recovery?

**11-4** Enumerate some of the most important safety precautions in underground recovery operations.

**11-5** Considering the reservoir characteristics, history, and present conditions, explain why applying IOM at the Annona Chalk section of Caddo Pine Island Field would produce more oil from the reservoir.

**11-6** What are the biometabolites generated by bacterial activity and how can oil production be improved by them?

**11-7** The microbial stimulation process needs a two- to three-week well shut-in period for fermentation. How can the results obtained be conclusively attributed to microbial stimulation when it is known that a well produces more oil after a period of shut-in time, even without any stimulation process applied?

## REFERENCES

ALBAYRAK, F. A., and T. E. PROTOPAPAS, *Detailed Technical and Economic Feasibility of Light Oil Mining in the United States*, DOE/BC/10704-8 (Washington, D.C.: U.S. Department of Energy, August 1984).

BRYANT, R. S., et al., NIPER-356 *Microbial-Enhanced Waterflood Field Experiment, Topical Report*, prepared for U.S. DOE under Coop. Agreement No. FC22-83FE60149, NIPER (Bartlesville, OK: U.S. Department of Energy, January 1989).

BRYANT, R. S. et al., "Optimization of Microbial Formulations for Oil Recovery: Mechanism of Oil Mobilization, Transport of Microbes and Metabolites, and Effects of Additives," Society of Petroleum Engineers: Paper SPE 19686 presented at the Annual Technical Conference and Exhibition, SPE 89, San Antonio, Texas, October 8–11, 1989.

CHUTE, F. S., and F. E. VERMEYLEN, "Present and Potential Applications of Electromagnetic Heating in the In-Situ Recovery of Oil," Vol. 4, no. 1, *AOSTRA Journal of Research* (1988), pp. 19–33.

ENERGY DEVELOPMENT CONSULTANTS, *Mining for Petroleum Feasibility Study,* U.S. Bureau of Mines, OFR 56-79, DOE/PB 297133 (July 1978).

GOLDER ASSOCIATES, *Oil Mining—A Technical and Economic Feasibility Study of Oil Production by Mining Methods*, U.S. Bureau of Mines, OFR 55-79, PB 297134 (October 1978).

KING, J. W., *MEOR Technical Status and Assessment of Needs—1986*, performed for U.S. DOE under Contract No. AC 19-85/BC10852-2, Hardin Simmons University, Abilene, Texas (Bartlesville, OK: U.S. Department of Energy, 1987).

KING, J. W., and D. A. STEVENS, *Proceedings of the First International MEOR Workshop, April 1–3, 1986*, Prepared for U.S. DOE under Contract No. AC 19-85/BC10852-1, Hardin-Simmons University, Abilene, Texas (Bartlesville, OK: U.S. Department of Energy, January 1987).

KNAPP, R. M., et al., *Microbial Field Pilot Study, Final Report for the Period Dec. 15, 1986–March 31, 1988*, Prepared for U.S. DOE under Contract No. AS 19-86/BC14084-6, University of Oklahoma, Norman (Bartlesville, OK: U.S. Department of Energy, January 1989).

PAUTZ, J. F., P. SARATHI, and R. THOMAS, *Review of EOR Project Trends and Thermal EOR Technology, Topical Report*, prepared for DOE under Coop. Agreement No. FC 22-83FE60149, by NIPER-461 (Bartlesville, OK: U.S. Department of Energy, March 1990).

RICE, G., "Mining Petroleum by Underground Methods: A Study of Methods Used in France and Germany and Possible Application to Depleted Oil Fields Under American Conditions," *Bureau of Mines Bulletin 351* (Washington, D.C.: U.S. Department of Commerce, June 1932).

SHEEHY, A. J., "Field Studies of Microbial EOR," Paper SPE/DOE 20254, presented at the SPE/DOE Seventh Symposium on Enhanced Oil Recovery, Tulsa, Oklahoma, April 22–25, 1990.

STREETER et al., "Recovery of Oil from Underground Drillsites," SPE 19344, 1989 SPE Eastern Regional Conference and Exhibition, Morgantown, West Virginia, October 24–27, 1989.

THAM, M. K., H. B. CARROLL, JR., et al., *Improved Oil Mining: A Feasibility Study—Final Report*, NIPER-328, performed under Coop. Agreem. No. FC22-83FE60149 (Bartlesville, OK: U.S. Department of Energy, September 1988).

YEN, T. F., *Bacteria Transport Through Porous Media, Annual Report*, prepared for U.S. DOE under Contract No. AS19-81/BC10508, University of Southern California, Los Angeles, California, March and September 1986.

# Chapter 12

# EOR Could Offset Oil Production Decline

## 12-1 ENERGY CONSUMPTION

The distribution of U.S. energy consumption is shown in Figure 12-1. The pyramid represents the percentage of total U.S. consumption of each energy source provided during 1988 ("Energy Focus," 1989). Petroleum products represent 42.5 percent of total U.S. energy consumption and about half are imported.

The energy forecast through the year 2000 indicates that the United States will remain highly dependent on crude oil since its consumption by the primary energy sources that use petroleum liquids will remain constant if not increase.

Growth in coal consumption is projected, but the main sources of non-hydrocarbon energies such as nuclear, hydroelectric, solar, and wind are not considered to have a significant impact in the near future (National Petroleum Council, 1984).

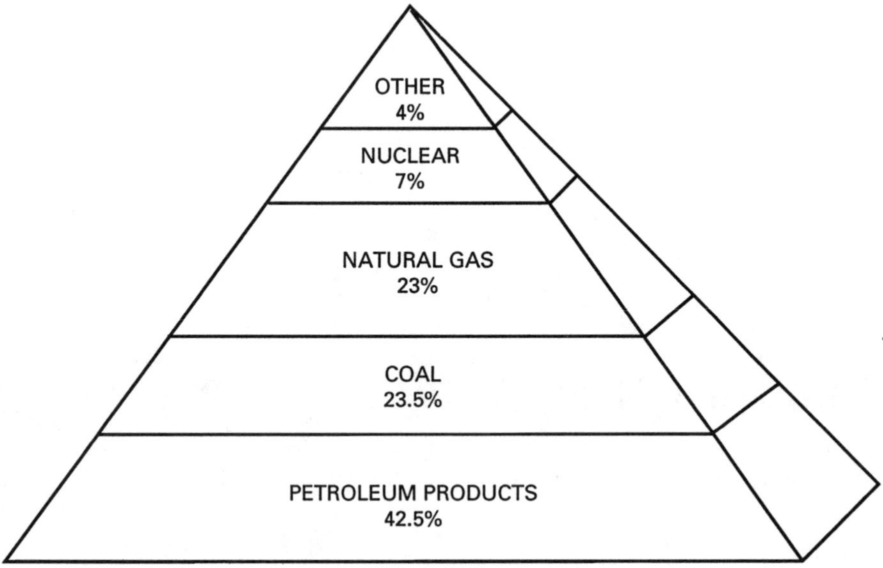


**Fig. 12-1** The dynamics of U.S. energy consumption (Adapted from *Journal of Petroleum Technology,* July 1989)

## 12-2 ENERGY SUPPLY

Petroleum products are supplied by domestic crude oil production and by foreign sources.

### Domestic Crude Oil Production

**Existing oil reservoirs.** Domestic crude oil production is obtained from existing oil reservoirs which, as hydrodynamic units, are characterized by four productive stages (Figure 12-2):

| | |
|---|---|
| Stage I. | Increased production rate as the wells pattern is drilled |
| Stage II. | Peak production rate |
| Stage III. | Sharp decline as reservoir pressure declines |
| Stage IV. | Slow decline of the long and final production phase |

Even if this scenario is improved by injecting water or gas to supply the reservoirs with energy, only an average of one-third of the original oil in place is recovered. A substantial amount of oil, nearly 325 billion barrels (66 percent of OOIP) remains trapped in the swept areas of existing oil reservoirs or is bypassed by the injected agents.

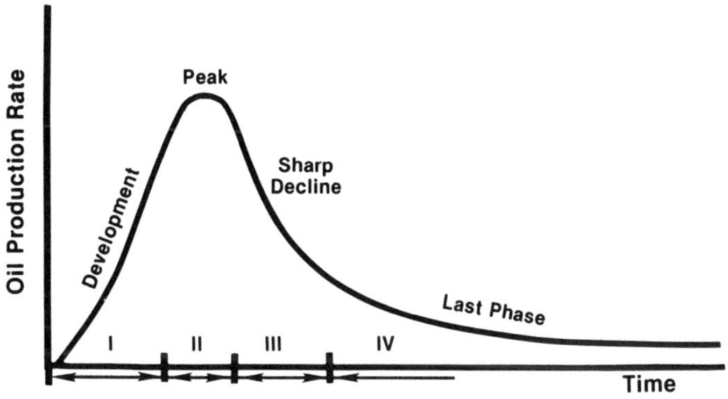


**Fig. 12-2** Oil reservoir productive stages

**New oil reservoirs.** The domestic crude oil production can be maintained at a constant level only if new oil reservoirs are discovered by exploration and recoverable reserves are added annually in the same percentage as they are consumed.

To find, explore, and develop new oil reservoirs for production is much more difficult now than in the past years. The trend of reserve additions per foot of exploration drilling is declining (Figure 12-3).

Today exploration drilling results are less and less attractive (one good well out of seven to nine dry holes), and harsher environments raise the expenses of drilling (National Petroleum Council, 1984). The discovery of any huge new oil field is unlikely.

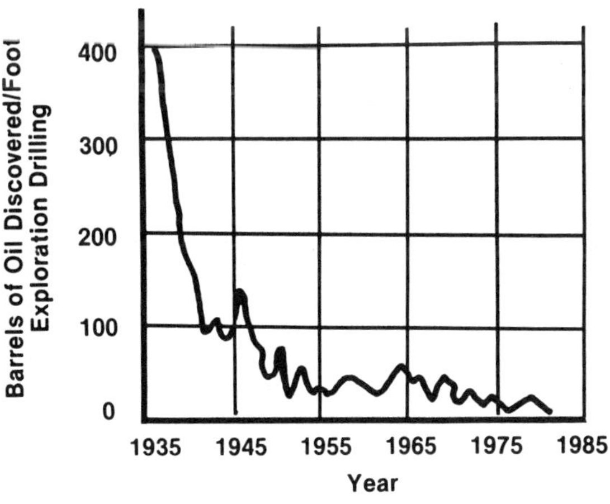


**Fig. 12-3** Barrels of oil discovered per foot of exploratory well in United States (From Ivanhoe, 1983)

**Consequences.** Domestic crude oil production cannot be maintained, and the major U.S. oil producing states show production declines from peak years (Figure 12-4). Alaska is the only remaining domestic location with oil reservoirs mostly in stages I and II that still have not shown production declines ("Energy Focus," 1987).

### Foreign Sources

**Future trends.** The balance of crude oil needed to satisfy the level of consumption must be met by imports. Since there are still no ready alternatives to petroleum as an energy source and since domestic crude oil production is declining, there will be an even greater reliance on foreign supply.

*Lower crude oil prices* also lead to higher U.S. dependency on imports, and countries offering the greatest export potential cannot necessarily be considered reliable sources, politically or economically (King, 1987).

As is stated in the federal EOR research report (U.S. DOE, 1985) increased imports are detrimental since they

- will reduce domestic employment, productivity, and economic activity;
- will reduce Federal and State tax bases;
- will exacerbate balance of payment deficits;
- will increase vulnerability to import disruptions and pressure from exporting countries; and
- the US national economic and military security demands a significant level of domestic petroleum production.

One example of activity highly dependent on oil prices is the enhanced oil recovery project starts in the last decade (Figure 12-5). As can be observed, the numbers of project starts correspond to the fall of the oil price (Pautz et al., 1990) and are also expected to follow the rise in oil prices beginning with 1990.

## 12-3 EOR: THE ANSWER FOR OFFSETTING OIL PRODUCTION DECLINE

Enhanced oil recovery can be applied to the already known oil reservoirs to recover more oil that is trapped or unreachable by conventional technology.

The expert in-depth evaluation by the National Petroleum Council (1984) has estimated that EOR could add

- From 7.4 to 34 billion barrels ($1.176 \times 10^9$ to $5.4 \times 10^9$ $m^3$) to U.S. reserves—representing 25 to 120 percent of currently proven reserves and
- From 600,000 to 2.8 million barrels ($25.4 \times 10^3$ to $445.2 \times 10^3$ $m^3$) per day

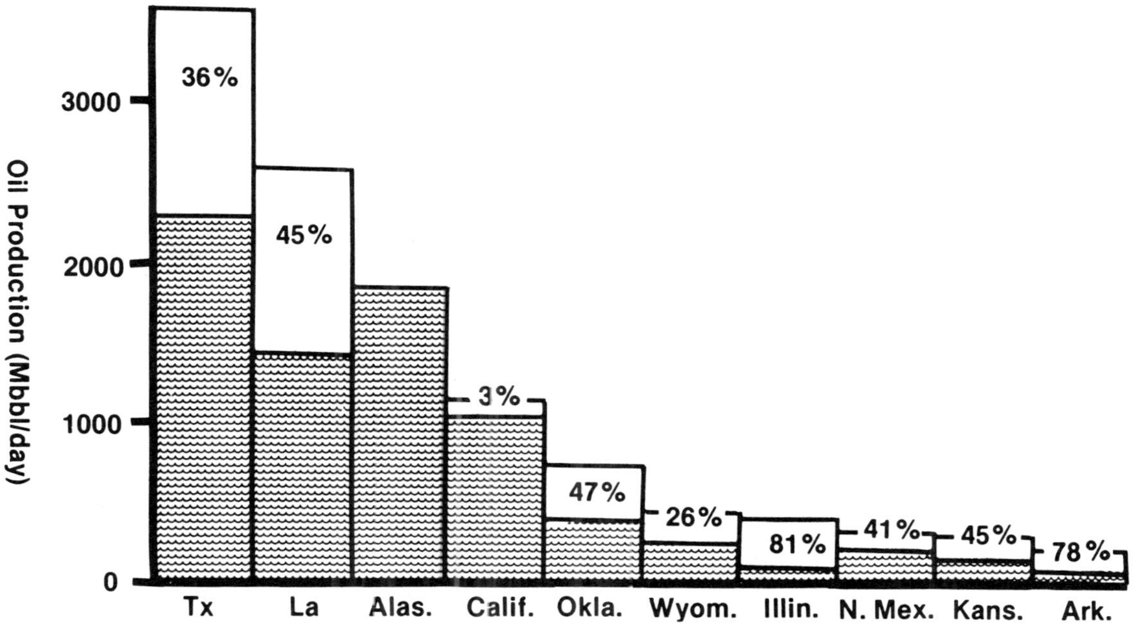


**Fig. 12-4** Percentage production decline since peak years (Adapted from *Journal of Petroleum Technology*, April 1987)

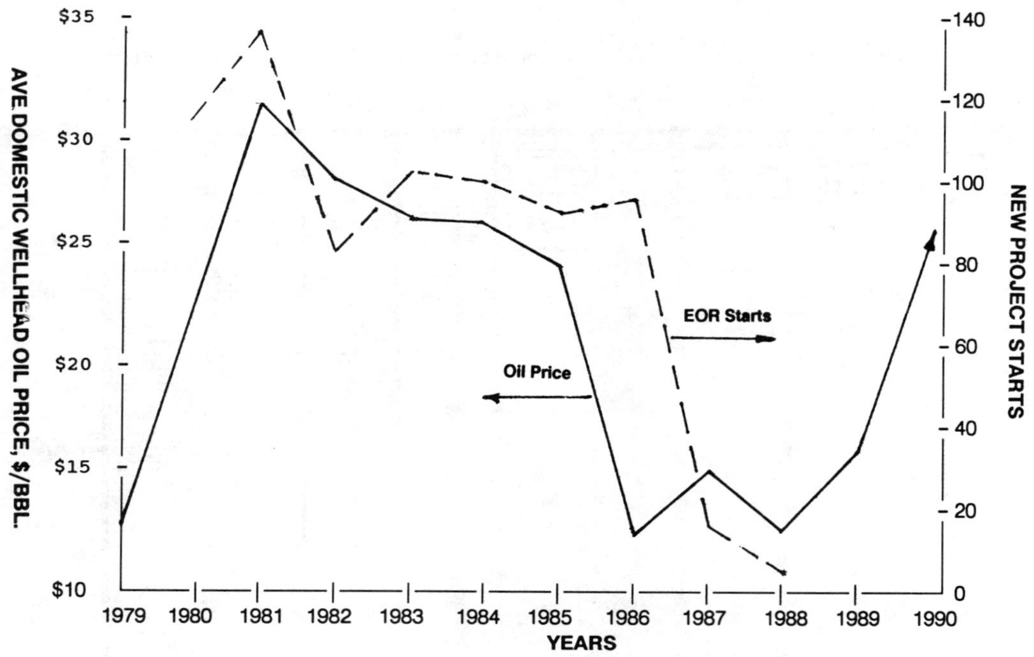


**Fig. 12-5** Oil prices compared to EOR project starts (Adapted from Pautz et al., U.S. Department of Energy, 1990)

in production (U.S. DOE, 1985), depending on future oil prices and reflecting different degrees of technological advancement.

However, the overall potential for EOR production is highly uncertain since efforts for more research and development activity are disturbed by oil price fluctuations in the low range. To attenuate this oil price stress, the amount of capital and incentives available to the petroleum industry must be increased. We must continue to analyze, understand, and improve the recovery technology of domestic crude oil's potential as the nation's main source of energy. Only then can we be better prepared for the energy needs of the twenty-first century.

## REFERENCES

"Energy Focus," *Journal of Petroleum Technology* (April 1987).

"Energy Focus," *Journal of Petroleum Technology* (July 1989), p. 688.

IVANHOE, L. B., "Free World Oil Discovery Indexes—1945–1981," *Oil and Gas Journal* (November 21, 1983), pp. 88–90.

KING, J. W., *MEOR Technical Status and Assessment of Needs—1986*, prepared for US/DOE under Contract No. AC19-85BC10852-2, Hardin-Simmons University, Abilene, Texas (Bartlesville, OK: US-DOE, March 1987).

NATIONAL PETROLEUM COUNCIL, *Enhanced Oil Recovery* (Washington, D.C.: NPC, June 1984), pp. 89–91.

PAUTZ, F. S., et al., *Review of EOR Project Trends and Thermal EOR Technology—Topical Report*, prepared for DOE by NIPER-461 under Coop. Agreement No. FC 22-83FE60149 by NIPER (Bartlesville, OK: U.S. DOE, March 1990).

U.S. DEPARTMENT OF ENERGY, *Federal EOR Research: Increased Understanding of the 300-Billion-Barrel U.S. Residual Oil Resource and the Technologies to Produce It*, DOE/BC-85/6/SP (Bartlesville, OK: U.S. DOE, distribution category UC-92a, October 1985).

# Index